# History of Analytic Philosophy

Series Editor
Michael Beaney
King's College London
Humboldt University Berlin
Berlin, Germany

The main aim of this series is to create a venue for work on the history of analytic philosophy, and to consolidate the area as a major branch of philosophy. The 'history of analytic philosophy' is to be understood broadly, as covering the period from the last three decades of the nineteenth century to the end of the twentieth century, beginning with the work of Frege, Russell, Moore and Wittgenstein (who are generally regarded as its main founders) and the influences upon them, and going right up to the recent history of the analytic tradition. In allowing the 'history' to extend to the present, the aim is to encourage engagement with contemporary debates in philosophy, for example, in showing how the concerns of early analytic philosophy relate to current concerns. In focusing on analytic philosophy, the aim is not to exclude comparisons with other earlier or contemporary traditions, or consideration of figures or themes that some might regard as marginal to the analytic tradition but which also throw light on analytic philosophy. Indeed, a further aim of the series is to deepen our understanding of the broader context in which analytic philosophy developed, by looking, for example, at the roots of analytic philosophy in neo-Kantianism or British idealism, or the connections between analytic philosophy and phenomenology, or discussing the work of philosophers who were important in the development of analytic philosophy but who are now often forgotten.



More information about this series at
http://www.palgrave.com/gp/series/14867

Landon D. C. Elkind • Gregory Landini
Editors

# The Philosophy of Logical Atomism

## A Centenary Reappraisal

palgrave
macmillan

*Editors*
Landon D. C. Elkind
University of Iowa
Iowa City, IA, USA

Gregory Landini
University of Iowa
Iowa City, IA, USA

History of Analytic Philosophy
ISBN 978-3-030-40487-1 ISBN 978-3-319-94364-0 (eBook)
https://doi.org/10.1007/978-3-319-94364-0

Library of Congress Control Number: 2018948844



This Palgrave Macmillan imprint is published by the registered company Springer Nature Switzerland AG
The registered company address is: Gewerbestrasse 11, 6330 Cham, Switzerland

# Series Editor's Foreword

During the first half of the twentieth century analytic philosophy gradually established itself as the dominant tradition in the English-speaking world, and over the last few decades, it has taken firm root in many other parts of the world. There has been increasing debate over just what 'analytic philosophy' means, as the movement has ramified into the complex tradition that we know today, but the influence of the concerns, ideas and methods of early analytic philosophy on contemporary thought is indisputable. All this has led to greater self-consciousness among analytic philosophers about the nature and origins of their tradition, and scholarly interest in its historical development and philosophical foundations has blossomed in recent years, with the result that history of analytic philosophy is now recognized as a major field of philosophy in its own right.

The main aim of the series in which the present book appears, the first series of its kind, is to create a venue for work on the history of analytic philosophy, consolidating the area as a major field of philosophy and promoting further research and debate. The 'history of analytic philosophy' is understood broadly, as covering the period from the last three decades of the nineteenth century to the start of the twenty-first century, beginning with the work of Frege, Russell, Moore and Wittgenstein, who are generally regarded as its main founders, and the influences upon them, and going right up to the most recent developments. In allowing the 'history' to extend to the present, the aim is to encourage engagement with contemporary debates in philosophy, for example, in showing how the concerns of early analytic philosophy relate to current concerns. In focusing on analytic philosophy, the aim is not to exclude comparisons with other—

earlier or contemporary—traditions, or consideration of figures or themes that some might regard as marginal to the analytic tradition but which also throw light on analytic philosophy. Indeed, a further aim of the series is to deepen our understanding of the broader context in which analytic philosophy developed, by looking, for example, at the roots of analytic philosophy in neo-Kantianism or British idealism, or the connections between analytic philosophy and phenomenology, or discussing the work of philosophers who were important in the development of analytic philosophy but who are now often forgotten.

The present volume, edited by Landon D. C. Elkind and Gregory Landini, celebrates the centenary of a series of lectures titled 'The Philosophy of Logical Atomism' that Bertrand Russell (1872–1970) gave in London in the first three months of 1918. These lectures are often seen as offering the most characteristic and mature statement of Russell's 'analytic philosophy'. But Russell's philosophy was always in development, and these lectures, in particular, must be viewed in the broader context not only of Russell's own thinking but also of its connections with the work of others, and especially, at this time, with the work of Ludwig Wittgenstein (1889–1951). Wittgenstein had gone to study with Russell at Cambridge in 1912, and had written some 'Notes on Logic' for Russell in October 1913 before leaving for Norway. On the outbreak of the First World War Wittgenstein joined the Austrian army, and Russell did not see him again until after the war. Wittgenstein's *Tractatus Logico-Philosophicus*, one of the canonical texts of analytic philosophy, which undoubtedly offers the main statement of his own early thinking, was not itself published until 1921, although it had more or less been completed in the summer of 1918. Russell claimed in his lectures on the philosophy of logical atomism to be explaining ideas he had learnt from Wittgenstein, but exactly what he had learnt, how he had interpreted them and the precise connections, similarities and differences between their respective ideas are controversial issues. These issues provide one of the central themes of the present volume.

In his lectures, Russell does indeed discuss many of the ideas and problems that are now recognized as characteristic of analytic philosophy. 'Logical atomism' may only have represented one particular phase in the history of analytic philosophy, but debate continues about the nature of acquaintance, the distinction between particulars and universals, truthmakers and the treatment of 'negative facts', foundationalism in metaphysics, conceptions of judgement and proposition, the theory of descriptions, and the various forms of analysis that can be found in the work of Russell, Wittgenstein and other analytic philosophers. All these topics are discussed in the present volume, and a great deal of light is shed not only on the history of the ideas and debates but also on the philosophical issues themselves. A century later, it is clear that the questions opened up and explored in Russell's pioneering lectures on the philosophy of logical atomism are at the very centre of analytic philosophy today.

King's College London
Humboldt University Berlin
Berlin, Germany
May 2018

Michael Beaney

# Introduction: *The Philosophy of Logical Atomism*—A Centenary Reappraisal

In spring 1918 Bertrand Russell delivered eight Tuesday evening lectures, from January 22 to March 12, in a room in Dr. William's Library, now known as University Hall, in London's Gordon Square on Grafton Street. Russell seems to have been paid £10 for the lectures (*CPBR 8*: 157–158). Converting that to 2018 terms, this comes to $771.21 for the lot, or $96.40 for each hour-long lecture. Russell published these eight lectures, now known as the logical atomism lectures, in 1918–1919, in four issues of *The Monist*, under the title *The Philosophy of Logical Atomism*.

The logical atomism lectures have been the subject of many scholarly commentaries and debates in the twentieth century and they played a central role in defining analytic philosophy itself. This volume is a critical reappraisal of the logical atomism lectures and the ideas contained in them as we approach their centenary.

In reading the essays of this volume, it should be understood that Russell's logical atomism, considered as a philosophical position, is not exhausted by the logical atomism lectures. His logical atomism both predates and postdates the 1918 lectures. So, in reappraising his logical atomism lectures, one must appreciate the various positions Russell held, and that he may mean, by the phrase "logical atomism". Accordingly, we shall introduce the collection of papers with an attention to the relation of the logical atomism lectures to Russell's logical atomism more broadly.

Logical atomism originates in Russell's 1911 paper "Analytic Realism", and its ties to Whitehead and Russell's 1910 *Principia Mathematica* are evident. Russell's logical atomism in this "*Principia* era" does not embrace characteristic theses of the 1918 logical lectures, such as the existence of general facts and of negative facts. Similarly, Russell's logical atomism in his mature work in the 1940s does not posit general facts or negative facts. Understanding logical atomism, as a view, cannot then consist solely in engaging with the 1918 lectures. One might even distinguish *early*, *middle*, and *late* versions of Russell's logical atomism.

One distinctive role of the 1918 logical atomism lectures in Russell's development of logical atomism is that they offer a striking record of, one the one hand, Russell's complicated relationship with the ideas of his former student, Ludwig Wittgenstein, and, on the other hand, original ideas on the nature of philosophy. Naturally, the relationship between Russell's logical atomism and Wittgenstein's pre-*Tractatus* views is a deeply engaging and ongoing scholarly topic, and our reappraisals focus a good amount of attention here.

Another distinctive role of the logical atomism lectures is that they record a peak between Russell's early, pre-Wittgenstein views and his later, post-Wittgenstein views. A core thesis of Russell's *Principia* era philosophy, and of his early logical atomism, is that logic in some rather strong sense, a logic at least as strong as the mathematical logic of *Principia* itself, is the essence of a new research program in scientific philosophy. Russell calls just this thesis "logical atomism" in 1914.

This raises a central question: What (if anything) is the canonical form of Russell's logical atomism? Do the core theses of the early, *Principia* era logical atomism pervade every variant of logical atomism? Is logical atomism a philosophical method only? It might be thought that one must embrace some or other combination of views found in the logical atomism lectures, such as the following: logical atoms, sense-data, universals, logical particulars, general facts, negative facts, logicism, the thesis that logical necessity is the only necessity, the elimination of all conceptions of necessity, the thesis that *truth* is a correspondence relation, and the existence of a logically perfect language whose syntactic forms align with the logical forms of their truth-maker(s). We think it helpful to view the reappraisals in this volume as being dedicated, in various ways, to answering these questions. The reappraisals all look toward the logical atomism lectures, with varying degrees of favorability, in answering these questions. They assess the merits and demerits of Russell's theses in the lectures.

The resulting multi-faceted discussion of logical atomism, in various understandings of that phrase, all holding the 1918 lectures in view, adds to our scholarly understanding of Russell's logical atomism and its history, and of the viability of logical atomism as a philosophical program. We group the reappraisals into five broad categories:

1. History of Russell's Logical Atomism (Elkind, Landini, Maclean)
2. Influences on Russell's Logical Atomism (Garavaso, Stern, Wahl)
3. Metaphysics: Fundamentality and Negative Facts (Klement, Linsky, Perović)
4. Language: The Theory of Judgment and Descriptivism (Korhonen, Orilia)
5. Epistemology: Acquaintance and Analysis (Fisher and McCarty, Fumerton, Levine)

We next briefly summarize the contents of the chapters in each grouping.

## History of Russell's Logical Atomism

The first three pieces directly engage the question, "What, if anything, is the canonical form of Russell's logical atomism?" Perhaps the most traditional answer, due largely to the influential work of D. F. Pears, is that Russell was concerned with providing an empiricist foundation for human knowledge. This foundation, Pears maintains, was based on the relations of acquaintance that one has to universals and to particulars of sense-experience. Pears views logical atomism as a reductive empiricism constrained by what is possible to be acquainted with: it is concerned with reducing the more or less confused statements we make to clear and definite statements about complexes composed of entities with which we are acquainted. As Elkind says, logical atomism is on this interpretation a search for acquaintance-complexes. Elkind criticizes this answer, arguing that interpreting logical atomism as a search for acquaintance-complexes poorly fits both the history of logical atomism and the text of the logical atomism lectures. He focuses on the logical nature and mathematical inspiration for logical atomism and what these tell us about the essence of logical atomism. In his view, the essence of logical atomism is an expressive and powerful logic: logical atomism on his reading is a search for logical forms.

A closely related question is "When was Russell a logical atomist?" How one answers this question is tied to one's account of what, if anything, is the canonical form of Russell's logical atomism. For someone who finds Pears' answer compelling, Russell abandoned logical atomism when Russell abandoned the notion of acquaintance upon which logical atomism as a search for acquaintance-complexes rests. Maclean argues that Russell was a logical atomist in an important sense in his later work, even after abandoning acquaintance relations. Maclean makes a crucial distinction between logical atomism as a *method* and logical atomism as an *ontology* including a commitment to metaphysical particulars. Maclean argues at length that, in *Human Knowledge: Its Scope and Its Limits*, Russell rejected the existence of particulars. Maclean further argues that Russell in his later works retained logical atomism as a method while abandoning logical atomism as an ontology with a commitment to metaphysical particulars.

Landini treats as canonical Russell's earlier conception of logical atomism as a new scientific method in philosophy that makes *Principia*'s mathematical logic its essence. The logical atomist's method is to apply *Principia's* study of structure to undermine the metaphysicist's indispensability arguments for abstract particulars and their purportedly specialized kinds of necessity. *Principia* had shown the way by eliminating abstract particulars (numbers, classes, geometric spatial objects, propositions) from every branch of mathematics and from logic itself. Scientific method in philosophy endeavors to eliminate the metaphysician's abstract particulars from every branch of science. The early logical atomist philosophy, as Russell says in 1914, is the science of what is (logically) *possible*. Landini argues that the pristine form of logical atomism is not found in the 1918 lectures at all, but in Russell's 1911 "Analytic Realism" and it finds its canonical articulation in his 1914 *Our Knowledge of the External World as a Field for Scientific Method in Philosophy*. The logical atomism lectures reflect Russell in flux, with his acquaintance epistemology at an impasse. Russell could not imagine how the multiple-relation theory and the acquaintance epistemology so eloquently set out in *The Problems of Philosophy* could find the truth-bearers for general and molecular belief required for the epistemology of mathematical logic. With what is one acquainted in understanding *all*, *some*, *and*, *or*, *not*? Introducing abstract particulars as objects of acquaintance to ground the epistemology of logic directly conflicts with scientific method in philosophy. Both *Principia* and its epistemology must be privileged for it to be the essence of a scientific philosophy. Ultimately, to get over his impasse Russell turned to neutral monism and its alliance with behaviorism. Landini suggests there is hope for the original acquaintance epistemology of *Problems* and its multiple-relation theory. He goes on to argue that in spite of its hostility to modern non-logical necessities (especially metaphysical necessity *de re*), Russell's logical atomism remains the last and best hope for a genuinely scientific philosophy.

## Influences on Russell's Logical Atomism

Wittgenstein's influence looms large in Russell's lectures: Russell explicitly acknowledged his impact and much scholarly attention has been given to Wittgenstein's influence on the logical atomism lectures. In contrast, Frege is not mentioned by Russell and little scholarly attention has been given to Frege's influence on Russell's 1918 lectures. To fill this gap, Garavaso discusses points of agreement and disagreement between Frege and Russell as regards doctrines in the logical atomism lectures, especially on the role of the Compositionality Principle, the importance of logical symbolism to both thinkers, and the increasingly critical role of symbolism to Russell.

Stern discusses, with a focus on certain stages of Wittgenstein and Russell's relationship, the rapid developments in the respective philosophies of Russell and Wittgenstein from the time of their first contact through the 1918 lectures. Stern argues that since the views in the lectures Russell credits Wittgenstein with inspiring come largely from the 1913 *Notes on Logic*, Russell engages there almost solely with ideas from their early interaction before Wittgenstein left for Norway. Stern also discusses the composition history of the *Prototractatus* (i.e., its relationship to manuscript 104), and especially McGuinness's suggestion that the first 71 pages of that manuscript are closer to a Russellian logical atomism than to the *Tractatus*. Stern suggests that historians use the phrase "logical atomisms", which is helpful in distinguishing versions of logical atomism Russell and Wittgenstein may have developed.

Focusing on the impact of Wittgenstein's 1913 *Notes on Logic*, Wahl discusses Russell's development between 1913 and 1918. He argues that a great many shifts in Russell's thought were due to Wittgenstein's influence well before the 1918 logical atomism lectures. These shifts include the thesis that facts cannot be named, that logic consists in tautologies, the abandonment of free variables, and the application of logical analysis to physics. On the other hand, Wahl maintains that Russell's method of analysis, fully developed through applications in the philosophy of mathematics, remained unaltered by Wittgenstein's influence.

## Metaphysics: Fundamentality and Negative Facts

In coming to understand the how logical atomism bears on contemporary metaphysics, one must address the questions surrounding Russell's notion of ontology. One such question is, "What is ontology?" There is a Quine-inspired view of metaphysics according to which ontology studies what there is. Klement argues that Russell and Quine share a quantificational account of the logical form of existence claims, and reject all terms besides individual variables, but that Russell departs from Quine in holding that ontology is not the study of what there is, but of what is fundamental. Klement goes on to argue that Russell understood quantification in the modern substitutional sense, which Quine found incapable of yielding answers to questions of what there is.

Another question in contemporary metaphysics is, "What, if anything, are the truth-makers for true negative statements?" The ontology of Russell's logical atomism lectures is directly connected with this contemporary discussion of truth-makers. Russell notes in his 1918 lectures that he is inclined toward accepting negative facts. He quips that discussing the issues in his class at Harvard in 1914 "nearly produced a riot". Thanks to Linsky's discovery of a small notebook of Harry T. Costello's and to some corroborating letters, we now know that the "riot" producing lecture likely occurred on 11 April 1914. In that lecture, however, the discussion is limited to features of Wittgenstein's 1913 *Notes on Logic* concerning facts which point "positively" toward a fact and those that point "negatively" away from it. This is a far cry from accepting negative facts as truth-makers. Linsky offers readers a chance to see the notes and decide for themselves how Russell's views on the issue evolved to produce Russell's inclination to accept negative facts as truth-makers in the logical atomism lectures. Raphael Demos seems to have been a leader of the rioters in 1914. Linsky gives an explanation of his view and argues on some historical evidence that Demos attended the logical atomism lectures in 1918, thus prompting Russell to make his quip about a "riot".

Russell's inclination toward negative facts in the 1918 lectures yields no positive characterization of them in the work. He argues that the quest for truth-makers seems to trap us into accepting them. Perović argues from the textual evidence from Russell's post 1921 work that Russell's commitment to negative facts was brief and always highly tentative. Some remarks in the first of Russell's logical atomism lectures even seem in tension with his stated inclination toward accepting negative facts. Perović critically discusses the viability of various positive characterizations of Russell's negative facts. In the lectures, Russell himself admits that if there are negative facts, their negativeness is ultimate and unamenable to a general definition. Perović agrees and concludes that metaphysicians can't be justified in adopting a positive attitude about the existence of negative facts.

## Language: The Theory of Judgment and Descriptivism

Russell's multiple-relation theory of judgment and its connection to his *Theory of Knowledge* book manuscript remain controversial topics for historians. When did he finally abandon the multiple-relation theory? Russell modified his multiple-relation theory several times: once in connection with Stout's 1910 criticism, once in *The Problems of Philosophy*, and yet again in his 1913 *Theory of Knowledge* manuscript. Russell eventually abandoned *Theory of Knowledge* but there remain hints in his 1914 *Our Knowledge of the External World* that suggest that he thought some form of the multiple-relation theory was still viable. Russell's discussion of judgment in logical atomism Lecture IV has attracted attention partly because he says that one cannot "make a map-in-space" of a belief-fact (*PLA*: 198). (See Perović (2016) for a discussion of Russell's diagrams of understanding complexes.) Korhonen's chapter offers evidence that that Russell's discussion of judgment in the 1918 logical atomism lectures is actually a new version of the multiple-relation theory that embraces some version of the idea of Wittgenstein's 1913 *Notes on Logic* concerning *pointing positively* toward a fact and *pointing negatively* away from a fact.

The theory of descriptions is one of Russell's paradigmatic examples of logical analysis, and it is given as a paradigm example in these lectures. In Russell's philosophy, the theory of descriptions is bound up with the thesis, known as *descriptivism*, that most sentences that use ordinary proper names and indexicals are expressible, and even really express, sentences in which definite descriptions, but not the ordinary proper names or indexicals, occur. Descriptivism is a minority position. Orilia, building on his 2010 book *Singular Reference*, argues that descriptivism can be modified so as to account for indexicals and in a way that frees descriptivism of the content-subjectivism that makes it vulnerable to referentialist objections.

## Epistemology: Acquaintance and Analysis

Fisher and McCarty unify various discussions of analysis in Russell and Wittgenstein under what they call *schemes of analysis*: these schemes operate using replacement rules to apply contextual definitions. They argue that Russell and Wittgenstein both held that such schemes of analysis converge in completely analyzed sentences. Fisher and McCarty show that there is no one proof procedure that can determine, in general, whether a given scheme of analysis converges on completely analyzed sentences. Accordingly, they conclude that there are schemes of analysis that cannot converge and, perhaps more importantly, procedures of analysis that do converge on all relevant inputs but cannot be proven to do so.

An *acquaintance theory of meaning*, on which the meaning of a genuine term is the referent of that term and the referent of any genuine term used as a term by a subject is known to that subject by acquaintance, is a captivating thesis from the logical atomism lectures. Fumerton draws a novel analogy between Russell's theory of meaning and the argument for foundationalism in meta-epistemology by applying the distinction between direct thought and indirect thought. Fumerton argues that Russell was led to this view of meaning by the simple but compelling consideration that, if we ever think of anything at all, then some thoughts of things must be direct rather than indirect.

Another important theme of the logical atomism lectures is Occam's Razor. Levine traces Russell's notion of Occam's Razor in the logical atomism lectures to Russell's earlier notion of logical analysis in *The Principles of Mathematics*, which Levine argues was inspired by Peano's conception of analysis. Levine argues that Russell leverages his theory of acquaintance to bring precision to the vague data with which analysis must begin. Levine goes on to argue that Russell understands generality in a way that allows quantifying over entities with which we cannot be acquainted. Thus, Russell takes Occam's Razor to enjoin us to find constructions so as to avoid our having to make precarious inferences to objects with which we cannot have acquaintance. He argues, appealing to writings of Ramsey's, that Wittgenstein's notion of generality in 1921 is much more restrictive, resulting in a quite different notion of Occam's Razor.

This concludes our brief description of the essays in this volume. The occasion for these essays was a seminar conference we organized. On 23 February 2016, Elkind, then a doctoral student in Philosophy writing his dissertation on Russell's logical atomism, noticed the upcoming centenary of Russell's 1918 lectures. He proposed co-sponsoring a conference to Landini. The University of Iowa's Obermann Center for Advanced Studies welcomed the opportunity to fund the event. The seminar was held from 12 to 16 June 2017 in Room E254 of the Adler Journalism Building. The Obermann Center has funded many important projects, including a 2015 seminar organized by David G. Stern devoted to G. E. Moore's notes on

Wittgenstein's 1930–1933 Cambridge Lectures. We owe a debt of gratitude to the Obermann Center for their generosity, and we hope that a new State Legislature will in the future see the wisdom of continuing to fund the Obermann Center's projects.

We are grateful to Teresa Mangum, director of the Obermann Center for Advanced Studies, and Erin Hackathorn, director of operations. Without their expertise none of this would have been possible. We also thank the Bertrand Russell Society—Iowa Chapter, for co-sponsoring the seminar, and for designing and purchasing the seminar's banner.

Tragically, Erik Banks, author of the outstanding book *The Realistic Empiricism of Mach, James and Russell* (Cambridge 2014), passed away just a few months after delivering his paper at the seminar. As editors of this volume, it is difficult to express our heartfelt regret that he was not able to refine and amend his excellent paper as was his clear intent: we have only this acknowledgment and appreciation of his work, and not his paper, to include in our volume.

We close with some remarks on the spirit and significance of Russell's lectures. At the turn of the twentieth century, Russell and Moore had broken with Bradley's Idealism and were exploring a new philosophical frontier grounded in the realist metaphysics of external relations and a new acquaintance epistemology suited to that metaphysics. The powerful new logic of Whitehead and Russell's *Principia Mathematica* was a novel extension of revolutionary nineteenth and early twentieth-century discoveries in mathematics and logic. In Russell's mind, some combination of these breakthroughs in philosophy is combinable into a new philosophy capable of producing final solutions, or dissolutions, of many philosophical problems. His conviction that a powerful new logic is adequate to dissolving many insuperable philosophical quandaries, more than anything else, is what underlies the spirit of logical atomism. It is why Russell, after many more shifts in his philosophical views, still claimed in his *My Philosophical Development* to be a logical atomist even, at the age of 87, just ten years before his death. It is in this logical atomist spirit that we worked on a critical reappraisal of *The Philosophy of Logical Atomism*.

University of Iowa — Landon D. C. Elkind
Iowa City, IA — Gregory Landini

## References

Perović, Katarina (2016). "Mapping The Understanding Complex in Russell's *Theory of Knowledge*". *Russell: The Journal of Bertrand Russell Studies* 36 (2): 101–127.

Orilia, Francesco (2010). *Singular Reference: A Descriptivist Perspective*. Dordrecht: Springer.

Wittgenstein, Ludwig (1913). "Notes on Logic." In L. Wittgenstein, *Notebooks 1914–16*. Ed. G. H. von Wright and G. E. M. Anscombe. Oxford: Basil Blackwell: 93–106.

# Contents

# Notes on Contributors

**Landon D. C. Elkind** (University of Iowa) is a treasurer of the Bertrand Russell Society and of the Society for the Study of the History of Analytical Philosophy. He is the founder of the Bertrand Russell Society—Iowa Chapter and President of the Iowa Lyceum. He has published in *Russell* and *Logos & Episteme*. His dissertation, *The Search for Logical Forms: In Defense of Logical Atomism*, focuses on what logical atomism is, arguing for the thesis that logical atomism is a search for logical forms, as opposed to a search for acquaintance-complexes, and that one's logic characterizes one's logical atomism.

**David Fisher** (Indiana University—Bloomington) is a doctoral candidate at Indiana University. He is completing his dissertation, titled *Loosen Up! Logical Tools for Metaontology*. He has research interests in metaphysics, logic, and the history of analytic philosophy, and seeks to combine them as much as possible.

**Richard Fumerton** (University of Iowa, F. Wendell Miller Professor) has authored multiple books, including *Metaphysical and Epistemological Problems of Perception* (University of Nebraska Press 1985); *Reason and Morality: A Defense of the Egocentric Perspective* (Cornell 1990); *Metaepistemology and Skepticism* (Rowman & Littlefield 1995); *Realism and the Correspondence Theory of Truth* (Rowman & Littlefield 2002); *Epistemology* (Blackwell 2006); *Mill* (with Wendy Donner (Wiley-Blackwell 2009)), and *Knowledge, Thought and the Case for Dualism* (Cambridge 2013). He has also published many scholarly articles, including "Russelling Causal Theories of Reference" in *Rereading Russell: Essays in Bertrand Russell's Metaphysics and Epistemology.*

**Pieranna Garavaso** (University of Minnesota—Morris) is the author of *Filosofia della matematica. Numeri e Strutture* (Guerini 1998) and the editor of *Arithmetic and Ontology*. She has co-authored *Filosofia delle Donne* (Laterza 2007) and *Frege on Thinking and Its Epistemic Significance*. She has edited a special issue of *Paradigmi* devoted to Contemporary Perspectives on Frege. She has published numerous scholarly articles, including "Psychological Continuity and Trauma" and "The Argument from Agreement and Mathematical Realism" in *The Journal of Philosophical Research*, "On Frege's Alleged Indispensability Argument" and "Hilary Putnam's Consistency Objection against Wittgenstein's Conventionalism in Mathematics" in *Philosophia Mathematica*, and "Frege on the Analysis of Thought" in *History and Philosophy of Logic*.

**Kevin C. Klement** (University of Massachusetts—Amherst) is editor-in-chief of the *Journal for the History of Analytic Philosophy* and a director of the Bertrand Russell Society. He has a number of scholarly publications on Russell, including "The Constituents of Logical Propositions" in *Acquaintance, Knowledge, and Logic: New Essays on Bertrand Russell's* The Problems of Philosophy, "PM's Circumflex, Syntax and Philosophy of Types" in *The Palgrave Centenary Companion to* Principia Mathematica, "The Paradoxes and Russell's Theory of Incomplete Symbols" in *Philosophical Studies*, "Neo-logicism and Russell's Logicism" in *Russell: The Journal of Russell Studies*, "Early Russell on Types and Plurals" in *Journal for the History of Analytic Philosophy*, and "The Functions of Russell's No-Class Theory" in *The Review of Symbolic Logic*. He wrote the *Stanford Encyclopedia of Philosophy* entry on Russell's Logical Atomism.

**Gülberk Koç Maclean** (Mount Royal University) has published one of the leading scholarly studies on the metaphysics of Russell's later thought, *Bertrand Russell's Bundle Theory of Particulars* (Bloomsbury 2014). She also received the Bertrand Russell Society Book Award for this work. Her scholarly work on Russell includes "Ramsey's Influence on Russell's Construction of Points" in *Russell: The Journal of Bertrand Russell Studies*. She is also a director of the Bertrand Russell Society.

**Anssi Korhonen** (University of Helsinki) has published *Logic as Universal Science: Russell's Early Logicism and Its Philosophical Context* (Palgrave Macmillan 2013) and "Frege, the Normativity of Logic, and the Kantian Tradition" in *New Essays on Frege*. He has also published a number of scholarly articles, including "Logic as a Science and Logic as a Theory: Some Remarks on Frege, Russell and the Logocentric Predicament" in *Logica Universalis*, "Russell's Early Metaphysics of Propositions" in *Prolegomena: Journal of Philosophy*, and "Russell and Poincaré on Logicism and Mathematical Logic" in *Henri Poincaré—Science and Philosophy*.

**Gregory Landini** (University of Iowa) is the author of books on the founding figures of analytic philosophy, including the groundbreaking *Russell's Hidden Substitutional Theory* (Oxford 1998), and *Wittgenstein's Apprenticeship with Russell* (Cambridge 2007); *Russell* (Routledge 2011); each book earned a Bertrand Russell Society Book Award. He has written many scholarly articles on Russell, including "Whitehead's (Badly) Emended *Principia*" and "Typos of *Principia Mathematica*" in *History and Philosophy of Logic*, "Russellian Facts About the Slingshot" in *Axiomathes*, and "Zermelo and Russell's Paradox: Is There a Universal Set?" and "Logicism and the Problem of Infinity: The Number of Numbers" in *Philosophia Mathematica*. He is a director of the Bertrand Russell Society.

**James Levine** (Trinity College—Dublin) has published extensively in the history of analytical philosophy. His recent scholarly publications include "The Place of Vagueness in Russell's Philosophical Development" in *Early Analytic Philosophy: New Perspectives on the Tradition*, "Russell, Pragmatism, and the Priority of Use over Meaning" in *Pragmatism and the European Traditions*, "Prior, Berkeley, and the Barcan Formula" in *Synthese*, "*Principia Mathematica*, the Multiple-Relation Theory of Judgment and Molecular Facts" in *The Palgrave Centenary Companion to Principia Mathematica*, and "Logic and Solipsism" in P. Sullivan & M. Potter eds., *Wittgenstein's Tractatus: History and Interpretation* (Oxford 2013).

**Bernard Linsky** (University of Alberta) is co-editor of three collections of critical essays on Russell's work: *"On Denoting": 1905–2005* (with Guido Imaguire, Philosophia 2005); *The Palgrave Centenary Companion to* Principia Mathematica (with Nicholas Griffin, Palgrave Macmillan 2013), and *Acquaintance, Knowledge, and Logic: New Essays on Bertrand*

*Russell's* The Problems of Philosophy (with Donovan Wishon, CSLI 2015). He has written a book devoted to Russell's and Whitehead's: Russell's Metaphysical Logic (CSLI 1999); and he has edited *The Evolution of* Principia Mathematica: *Bertrand Russell's Manuscripts and Notes for the Second Edition* (Cambridge 2011), which won the Bertrand Russell Society Book Award. He has published many scholarly articles on Russell, including "Logical Constructions" and "The Notation in *Principia Mathematica*" in the *Stanford Encyclopedia of Philosophy.*

**David Charles McCarty** (Indiana University—Bloomington) is a professor of philosophy specializing in intuitionistic mathematics and logic. He publishes regularly on topics in logic and philosophy, including Wittgenstein's *Tractatus.*

**Francesco Orilia** (University of Macerata) has published six monographs, including *Singular Reference: A Descriptivist Perspective* and *Predication, Analysis and Reference* (Springer 2010). He has also edited four volumes and published numerous scholarly papers, including "Property Theory and the Revision Theory of Definitions" in *Journal of Symbolic Logic*, "Moderate Presentism" in *Philosophical Studies*, "Positions, Ordering Relations and O-Roles" in *Dialectica*, and "Properties" in *The Stanford Encyclopedia of Philosophy.*

**Katarina Perović** (University of Iowa) has published numerous scholarly works, including "Bradley's Regress" in *The Stanford Encyclopedia of Philosophy*, "Mapping the Understanding Complex in Russell's *Theory of Knowledge*" in *Russell*, "A Neo-Armstrongian Defense of States of Affairs: A Reply to Vallicella" in *Metaphysica*, "Bare Particulars Laid Bare" in *Acta Analytica*, "The Importance of Russell's Regress Argument for Universals" in *Acquaintance, Knowledge, and Logic: New Essays on Bertrand Russell's* The Problems of Philosophy, and "The Import of the Original Bradley Regress(es)" in *Axiomathes.* She is a director of the Bertrand Russell Society and a feature contributor to its biannual *Bulletin.*

**David G. Stern** (University of Iowa, Collegiate Fellow in the College of Liberal Arts and Sciences) is the author of *Wittgenstein on Mind and Language* (Oxford 1995) and *Wittgenstein's Philosophical Investigations: An Introduction* (Cambridge 2004). He has also co-edited *The Cambridge Companion to Wittgenstein* (first edition, 1996; second edition 2018); *Wittgenstein Reads Weininger: A Reassessment* (with Be'la Szabados,

Cambridge 2004), *Wittgenstein: Lectures, Cambridge 1930–1933, From the Notes of G. E Moore* (with B. Rodgers & G. Citron, Cambridge 2016) and is the editor of *Wittgenstein in the 1930s: Between the* Tractatus *and the* Investigations (Cambridge 2018). Together with Landon Elkind and Phillip Ricks, he has recently been working on the University of Iowa *Tractatus* Map, http://tractatus.lib.uiowa.edu/.

**Russell Wahl** (Idaho State University) is the editor of the forthcoming *Bloomsbury Companion to Bertrand Russell.* His recent scholarly work includes "The Axiom of Reducibility" in the 2011 *Russell* journal and "Sense-Data and the Inference to Material Objects: The Epistemological Project in *Problems* and Its Fate in Russell's Later Work" in the 2015 collection of critical essays, *Acquaintance, Knowledge, and Logic: New Essays on Bertrand Russell's* The Problems of Philosophy (CLSI 2015). Wahl has written extensively on Russell's principle of acquaintance, his theory of judgment, and his sense-datum theory. He is director of Philosophy at Idaho State University and a director of the Bertrand Russell Society.

PART I

# History of Russell's Logical Atomism

CHAPTER 1

# On Russell's Logical Atomism

*Landon D. C. Elkind*

## 1 Introduction

Among historians of philosophy, there have been significant developments in our understanding of Russell's logical atomism. But setting aside disagreements over the details, I think it may be said that the interpretation of logical atomism, which I describe below, remains the dominant view of logical atomism among professional philosophers. I here argue against the dominant interpretation.

By way of introduction, I start with common ground between the dominant interpretation and my own. Logical atomists do claim that "there are many separate things":

> The logic which I shall advocate is atomistic, as opposed to the monistic logic of the people who more or less follow Hegel. When I say that my logic is atomistic, I mean that I share the common-sense belief that there are many separate things; I do not regard the apparent multiplicity of the world as consisting merely in phases and unreal divisions of a single indivisible Reality. (*PLA*: 160)

L. D. C. Elkind (✉)
University of Iowa, Iowa City, IA, USA
e-mail: landon-elkind@uiowa.edu



L. D. C. Elkind, G. Landini (eds.), *The Philosophy of Logical Atomism*, History of Analytic Philosophy,
https://doi.org/10.1007/978-3-319-94364-0_1

Similarly, any atoms of logical atomism are, as the name suggests, logical and not physical:

> The reason that I call my doctrine *logical* atomism is because the atoms that I wish to arrive at as the sort of last residue in analysis are logical atoms and not physical atoms. Some of them will be what I call "particulars"—such things as little patches of colour or sounds, momentary things—and some of them will be predicates or relations and so on. The point is that the atom I wish to arrive at is the atom of logical analysis, not the atom of physical analysis. (*PLA*: 161)

These two quotes from the beginning of Russell's 1918 logical atomism lectures are jointly the focus of multiple influential scholarly commentaries on Russell's lectures:

> When they are true, atomic sentences stand for the simplest complex constituents of reality, which are facts consisting of objects named by logically proper names bearing the properties, and standing in the relations, designated by predicates. (Soames 2014: 577, see also 574)

> Russell's lectures on *The Philosophy of Logical Atomism* begin with an outline sketch of what he means by 'logical atomism': Its basic premise is that the world contains many different things, and that to find out what these things are we need to practise what he calls 'analysis'. The idea is that almost all familiar things are in one or another way complex, but by analysis we can find out what simpler components these complex things are put together from, and if we continue this process we should end by reaching the ultimately simple things from which all else is composed. These are the 'atoms', and they are called 'logical atoms' because they are the last residue of 'logical analysis', which is said to be something quite different from physical analysis. (Bostock 2012: 252)

> The 'logical' in the label signals that the atoms are arrived at as the 'last residue of analysis' where the analysis is logical rather than physical … Logical atomism is the view that in theory, if not in practice, analysis takes us down to the ultimate simples out of which the world is built. (Grayling 1996: 50–51)

> Logical atomism … is a theory about the fundamental structure of reality and so it belongs to the main tradition of western metaphysics. Its central claim is that everything that we ever experience can be analyzed into logical atoms. [Pears then quotes Russell: "the atom I wish to arrive at is the atom of logical analysis".] (Pears 1985: 1)

> The basic thesis of logical atomism … was that the world consists of simple particulars which have only simple qualities and stand only in simple relations to one another. (Ayer 1972: 103–104)

I see here a pattern of taking the above two quotes from Russell's lectures as statements of the essence, that is, the critical theses, of logical atomism. We should add to the above two quotes Russell's views on a logically perfect language, and on acquaintance and meaning, which have led some to call logical atomism a "theory of meaning":

> In a logically perfect language the words in a proposition would correspond one by one with the components of the corresponding fact, with the exception of such words as "or", "not", "if", "then", which have a different function. In a logically perfect language, there will be one word and no more for every simple object, and everything that is not simple will be expressed by a combination of words, by a combination derived, of course, from the words for the simple things that enter in, one word for each simple component. (*PLA*: 176)

Soames, Urmson, and Pears state the dominant interpretation's view of such statements:

> It is a central aim of logical atomism to replace unanalyzed terms, predicates, and sentences-cum-propositions—which may stand in conceptual relations to one another—with logically proper names, simple unanalyzable predicates, and fully analyzed propositions. When this aim is achieved, the conceptual properties of, and relations holding among, unanalyzed expressions and sentences are traced to genuinely logical properties of, and relations holding among, fully analyzed propositions of the agent's logically perfect language. (Soames 2014: 586)

> The shortest account of logical atomism that can be given is that the world has the structure of Russell's mathematical logic … The structure of the world would thus resemble the structure of *Principia Mathematica*. That is the simple argument of the plot. (Urmson 1956: 6–7)

> [Logical atomism does] start from the assumption that there is a general correspondence between language and reality, which ensures that the complete analysis of words will match the complete analysis of things … The idea is that, when we analyze the words in our vocabulary, we soon reach a point at which we find that we cannot analyze them any further, and so we conclude that we have reached the bottom line where unanalyzable words correspond to unanalyzable things. (Pears 1985: 4–5)

The dominant interpretation of logical atomism thus emerges from an elegant synthesis of Russell's views. That is, logical atomism on the dominant reading merges (1) an ontological pluralism on which there are multiple logically independent entities, (2) a theory of meaning on which the referents of predicates and names are logically simple and mean their objects, and (3) an acquaintance epistemology that guarantees both the reference of predicates and names, and the truth of fully analyzed sentences in a philosophically ideal language. To these doctrines is added (4) an analytic program that picks apart the meaning of logically complex ordinary words and traces their meaning to logically simple words. It is this interpretative synthesis that I describe as the reading of logical atomism as "a search for complexes composed of entities with which we have acquaintance", or just "a search for acquaintance-complexes" for short.

This is the dominant reading of logical atomism. Some historians of philosophy reject this interpretation in its particulars, and some reject it in its essentials. Still, the widespread impression of logical atomism seems to be that it is essentially tied to Russell's views on acquaintance and meaning, and to an ontology of logical atoms, or perhaps necessarily existing simples.

I argue the dominant interpretation with its correlated and widespread impression of logical atomism as bound up with Russell's acquaintance epistemology is a misleading characterization of logical atomism. It is misleading because what is really crucial to logical atomism is logic. And not just any logic will do: a logical atomist needs a logic that is quite powerful, one at least that has expressive capacity sufficient to logically analyze and synthesize philosophical notions at least as complex as those of higher mathematics. A logical atomist also holds a certain view about the critical assistance of such a logic in philosophizing. A failure to make logic the essence of logical atomism produces a grossly mistaken history of logical atomism. That is, the textual data of the 1918 lectures and historical data we have about logical atomism conflict with the dominant interpretation's account of logical atomism. By *modus tollens*, the dominant interpretation is wrong.

Scholars have recently made advances in our understanding of logical atomism. This is partly because scholars have critically considered what counts as a logical atomist text, and have given increasing significance to texts beyond the 1918 lectures in their readings of logical atomism.[1] Some of them have previously criticized the dominant interpretation, as I do here.[2] All this work is reflected in the current *Stanford Encyclopedia of Philosophy* article on logical atomism, which carefully details the historical origins and textual sources of logical atomism (Klement 2016: §2).[3]

My discussion of logical atomism here builds on and extends this work. But that work does not go far enough. I want a clean break from the dominant interpretation. The dominant interpretation's reading of logical atomism is interesting in itself. But it is just wrong. And I do not argue here that it is wrong on internal grounds. I mean that it does not correspond to what we find in the historical record and in the logical atomist texts. The dominant interpretation is a bad history of logical atomism.

I proceed as follows. In the next section, I present the dominant interpretation of logical atomism, which I call "the search for acquaintance-complexes reading". In the following two sections, I criticize interpreting logical atomism as a search for acquaintance-complexes on two grounds. First I argue that it badly fits the history of logical atomism, especially its genesis from nineteenth-century mathematics. Then I argue that it conflicts with many of Russell's remarks in the logical atomism lectures themselves, including Russell's own descriptions of logical atomism.

## 2 The Search for Acquaintance-Complexes Reading

In this section, I present the dominant interpretation of Russell's logical atomism. If we set aside the various minor discrepancies between authors, then we can characterize the dominant interpretation of logical atomism as a search for acquaintance complexes. I mean by an *acquaintance-complex* a complex constituted by simples that are objects of acquaintance. I first describe Russell's notions of *complex*, *logical atom*, and *acquaintance* from March 1911 to March 1918. I will keep this brief as many of the details will be familiar to readers. Then I will present the argument for the dominant interpretation of logical atomism.

### *2.1 Russell's Notion of a Complex*

In the logical atomism lectures, Russell's notion of a *complex* is explicated within an ontology of facts.[4] For Russell, a *fact* is taken as a primitive notion to be described but not defined:

> When I speak of a fact—I do not propose to attempt an exact definition, but an explanation, so that you will know what I am talking about—I mean the kind of thing that makes a proposition true or false … We express a fact, for example, when we say that a certain thing has a certain property, or that it has a certain relation to another thing … (*PLA*: 163–164)[5]

Russell's facts, being taken as primitive, are not for him amendable to definition using other terms. But the sort of entity Russell means is as comprehendible as any philosophical notion. And Russell describes facts somewhat further in the 1918 lectures (and in his 1914 book—see notes). He states that they have the following four features: facts (a) are objective, (b) are the sort of entity that makes a statement true or false,[6] (c) have constituents, and (d) have various logical forms:

1. It is important to observe that facts belong to the objective world. They are not created by our thoughts or beliefs except in special cases. (*PLA*: 164)[7]
2. When we speak falsely it is an objective fact that makes what we say false, and it is an objective fact that makes what we say true when we speak truly. (*PLA*: 164)[8]
3. The things in the world have various properties and stand in various relations to each other. That they have these properties and relations are *facts*, and the things and their qualities and relations are quite clearly in some sense or other components of the facts that have those qualities or relations. (*PLA*: 171)[9]
4. There are a great many different kinds of facts … (*PLA*: 164)
   There are, of course, a good many forms that facts may have, a strictly infinite number, and I do not wish you to suppose that I pretend to deal with all of them. (*PLA*: 191)[10]

Russell's examples of facts in 1918 include *it is raining*, *Socrates is dead*, *gravitation varies inversely as the square of the distance*, *two and two are four*, *this is white*, and *this is to the left of that* (*PLA*: 163, 176).[11]

Now facts are not all alike in Russell's view: facts are rather of different sorts and are distinct in virtue of either their constituents[12] or their form (*OKEW*: 53, footnote 1). For example, the logical atomism lectures discuss *atomic facts*, of which there arguably two species: *positive facts* and *negative facts*. Now as Russell tells us in his 1918 lectures, he entertained negative facts in 1914, and by 1918 he is inclined, with hesitation, to accept them (*PLA*: 187, 189–190).[13] Russell heavily qualifies even this tentative endorsement of negative facts: "It is a difficult question. I really only ask that you should not dogmatize. I do not say positively that there are, but there may be" (*PLA*: 187).[14] As another example, there are arguably also facts that are not atomic, like *general facts* and *existence facts*, and also *belief facts* and even *molecular facts*.

We need not delve into which of these Russell posited in the lectures and which he rejected. What we have so far as an explication of Russell's notion of a complex, as accounted for through his notion of a fact, is enough for our purposes here.

### 2.2 Russell's Notion of a Logical Atom

In Lecture I, Russell offers examples of logical atoms:

> Some of them [the logical atoms] will be what I call 'particulars'—such things as little patches of colour or sounds, momentary things—and some of them will be predicates or relations and so on. (*PLA*: 161)

So we have two varieties of logical atoms: particulars and relations.[15] He defines *particular* as "terms of relations in atomic facts" (*PLA*: 177).[16] He immediately comments on this definition to clarify that this definition is "purely logical" and that logicians do not care whether there are any particulars at all.[17] He adds they are logically independent:

> Particulars have this peculiarity, among the sort of objects you have to take account of in an inventory of the world, that each of them stands entirely alone and is completely self-subsistent. (*PLA*: 201)[18]

The notion of logical independence is critical to Russell's notion of a logical atom: it is essential to being a logical atom that it is logically independent of all other entities. As he says in 1911, complexes presuppose logical atoms, but not conversely (*AR*: 134).[19] Logical atoms also exist in logical independence of each other.[20] Russell further suggests that the only way in which one entity depends on another is that one is a part of another.[21] These remarks about logical independence equally apply to particulars and to relations.

From 1911 to 1918, Russell accounts for relations through an ontology of universals (*AR*: 133). He discusses universals extensively in the second and third logical atomism lectures. In 1911, he calls known universals *concepts*. It is not quite clear that he keeps this locution in the 1918 lectures, though he sometimes describes as "concepts" what are likely universals, including "the concept of humanity" (*PLA*: 231). Russell also says in a 1960 interview with Woodrow Wyatt that logical atomism is concerned with breaking apart "ideas out of which a thing is built up".[22]

In this same period, Russell accounts for logical particulars—in the sense of logically independent particulars, not in the sense of logical entities—through an ontology of sense-data. In 1911, he describes particulars that are known as *sense-data*.[23] In his 1914 "The Relation of Sense-Data to Physics", Russell indicates that sense-data are logically independent in the required sense.[24] This mindset seems to carry over into the 1918 lectures, as he describes particulars as "little patches of colour or sounds" above.

We need not adjudicate these issues over Russell's ontology here. The vital point is that a logical atom is logically independent of all other entities while all complexes are composed of them, and that simples are to be designated by the terms of a logically perspicuous language. Russell eschews giving them essential characteristics beyond such logical features. He writes:

> You will note that this philosophy is the philosophy of logical atomism. Every simple entity is an atom. One must not suppose that atoms need persist in time, or that they need occupy space: these atoms are purely logical. (*AR*: 134)
>
> The reason that I call my doctrine *logical* atomism is because the atoms that I wish to arrive at as the sort of last residue in analysis are logical atoms and not physical atoms ... The point is that the atom I wish to arrive at is the atom of logical analysis and not the atom of physical analysis. (*PLA*: 161)

Whatever other features they logically may have, it is atoms in Russell's logical sense that are the atoms of logical atomism.[25]

### 2.3 Russell's Notion of Acquaintance

Russell describes acquaintance as follows[26]:

> I say that I am *acquainted* with an object when I have a direct cognitive relation to that object, i.e. when I am directly aware of the object itself. (*KAKD*: 149)
>
> Acquaintance with objects essentially consists in a relation between the mind and something other than the mind ... (*PoP*: 66–67)

> We shall say that we have *acquaintance* with anything of which we are directly aware, without the intermediary of any process of inference or any knowledge of truths. (*PoP*: 73)

This last point is worth stressing. Russell explicitly denies that acquaintance relations involve judgments.[27] Acquaintance is moreover presupposed in all other cognitive relationships.[28]

The field of acquaintance in Russell's theory includes the following: (1) sense-data, (2) memory-data, (3) introspection-data, including both mental facts like *my seeing the sun* and *my desiring food*, and feelings like those of pain and pleasure, (4) possibly, our own Self, though Russell is unsure of this, and (5) universals, including both sensible properties like whiteness and blackness, and abstract properties like diversity. Russell writes:

> We have acquaintance [1] in sensation with the data of the outer senses, and [2] in introspection with the data of what may be called the inner sense—thoughts, feelings, desires, etc.; we have acquaintance [3] in memory with things which have been data either of the outer senses or of the inner sense. Further, [4] it is probable, though not certain, that we have acquaintance with Self, as that which is aware of things or has desires towards things … we also have acquaintance [5] with what we shall call *universals*, that is to say, general ideas, such as *whiteness*, *diversity*, *brotherhood*, and so on. (*PoP*: 80–81, see also 75–77)

There are again many interesting issues involved in Russell's acquaintance epistemology. But we have enough to grasp the argument for the dominant reading of logical atomism.

### 2.4 *The Argument for the Dominant Reading*

Logical atomism as the search for acquaintance-complexes, then, is the philosophical program of searching for complexes composed of entities with which we are acquainted. J. O. Urmson aptly summarizes this interpretation of logical atomism:

> In the period from 1905 to 1919 Russell attempted to give a reductionist account of empirical knowledge; the basic materials were sense data; the methodological maxim was to replace inferred entities by logical constructions whenever possible, and the theory of descriptions was the main logical tool. (Urmson 1969: 510)

On this reading, Russell's logical atomism depends on both an epistemology with a theory of acquaintance and an ontology including entities with which we metaphysically can stand in acquaintance relations. Let us consider these two doctrines in turn.

A logical construction, as the name suggests, is constructed from something. Russell's language is frequently suggestive of constructing logical complexes from logical atoms.[29] This is why Urmson mentions "basic materials". Now the materials are entities, but what makes "basic" is also their epistemological status. This is where logical atoms enter into logical atomism on the dominant interpretation on that reading, logical atoms are both the epistemological and logical foundation, really the essence, of logical atomism.

Followers of the dominant interpretation, following Russell's remark above, thus understand logical atomism as committed to logical atoms, namely, particulars, qualities, and relations (Pears 1985: 2). D. F. Pears writes, "[Logical atomism's] central claim is that everything we ever experience can be analyzed into logical atoms" (Pears 1985: 1). This commitment arises for epistemological and logical reasons.

The epistemological reasons are as follows. The scope of experience is limited to objects and facts with which we have *acquaintance* in Russell's sense.[30] This interpretative claim is based on Russell's formulation of a "fundamental principle" of analysis[31]:

> All analysis is only possible in regard to what is complex, and it always depends, in the last analysis, upon direct acquaintance with the objects which are the meanings of certain simple symbols. (*PLA*: 173)

The fundamental principle forces upon us an identification of logical atoms and epistemological atoms. This brings us to Russell's further claim about the structures of both logical analysis and epistemological justification:

> All our knowledge, both knowledge of things and knowledge of truths, rests upon acquaintance as its foundation. (*PoP*: 73; see also 175–176)

> All analysis is only possible in regard to what is complex, and it always depends, in the last analysis, upon direct acquaintance with the objects which are the meanings of certain simple symbols. (*PLA*: 173)

In *Problems*, Russell holds an *acquaintance epistemology*, the view that non-inferential justification is given by acquaintance, acquaintance being a relation of direct awareness that is not conceptual, judgmental, or representational.[32] Russell also affirms *epistemological foundationalism*, the view that all justification chains terminate in non-inferential justification. These two claims require epistemological atoms—particulars and relations for Russell—with which we have acquaintance at the base of our justificatory structure. Russell also claims that logical analysis necessarily terminates, if it does, with logical atoms. Crucially, these logical atoms must be objects of acquaintance. We thus have an identification of logical atoms and epistemological atoms: logical atoms are just the entities and facts upon which all knowledge epistemologically depends, and epistemological atoms are just the logically independent entities that are the constituents of facts. Thus Russell adopted logical atomism through an independently motivated "fundamental principle" that in turn led to a foundationalism about analysis: this manifested itself in the view that all words for logical atoms comprehended by a speaker mean objects of acquaintance—such a word being "a symbol whose parts are not symbols" (*PLA*: 173). Pears writes:

> So when analysis could proceed no further, he appealed to acquaintance or direct experience ... Russellian analyses proceed by way of definitions, terminate with indefinables, and, at that point, base themselves upon acquaintance. (Pears 1985: 8)

This is the first half of logical atomism understood as the search for acquaintance-complexes. The other half is an ontology with objects of acquaintance upon which we can ground empirical knowledge. These are to be the atoms of logical atomism. The epistemology and the ontology go hand-in-hand on the dominant interpretation, underscoring the need for atoms on this reading. Pears writes:

> An atom is something indivisible or not further analyzable. A logical atomist, therefore, needs to show not only that the divisions traceable in logic correspond to real divisions in the nature of things, but also that the two corresponding processes of analysis do not continue indefinitely. If Russell is right, there must be a point at which words and things will be found to be not further analyzable. (Pears 1985: 2)

The need for atoms is imposed by the need for the analysis of words to terminate along with the thesis that there is a close correspondence between the analysis of the world and the analysis of language.[33] As we saw above, Russell claims analysis depends on a complex being capable of separation into components. The components of facts will moreover be logically independent of each other. Russell writes:

> It is quite clear that in that sense there is a possibility of cutting up a fact into component parts, of which one component may be altered without altering the others, and one component may occur in certain other facts though not in all other facts. I want to make it clear, to begin with, that there is a sense in which facts can be analyzed. (*PLA*: 172)

And as we saw, in his 1911 "Analytic Realism", he says complexes presuppose logical atoms, but not conversely (*AR*: 134). And Russell holds that a logical atom is logically independent of every other one. Indeed, each atom logically could be the only entity that exists.[34] And expressions whose meaning is some atom will be simple symbols whose meanings can be understood independently of understanding the meaning of any other word.[35] So for a logical atomist on the dominant reading, there is necessarily a close correspondence between words and objects:

> We may lay down the following provisional definitions:
>
> That the components of a proposition are the symbols we must understand in order to understand the proposition;
>
> That the components of the fact which makes a proposition true or false, as the case may be, are the *meanings* of the symbols which we must understand in order to understand the proposition. (*PLA*: 175)

Moreover, words standing for atoms will be understood, necessarily and sufficiently, by acquaintance.[36] Russell's example of such a word is the word "red": he argues "red", unlike complex symbols, cannot be understood except through acquaintance with red objects.[37] And by the close correspondence of language and the world, given that there are such words for atoms like "red", there are logical atoms that are the meaning of such simple symbols.[38]

So one might interpret the text of the lectures as requiring an ontology of logical atoms. There are a number of problems with that argument that have been much-discussed.[39] But the point that matters for present purposes is the sketch of the dominant interpretation of logical atomism as the search for acquaintance-complexes. This reading proposes an intriguing union of epistemological theory of acquaintance and of an ontological theory with simples, all interwoven with a theory of meaning on which the structure of all facts would be made logically perspicuous in an ideal language. It is a captivating idea that remains the dominant reading of logical atomism among philosophers. Yet I argue that this reading of logical atomism does not make logic central to logical atomism: consequently, it poorly fits the history of logical atomism and the text in the 1918 lectures.

## 3 The Logico-Mathematical Origins of Logical Atomism

Viewing Russell's logical atomism as the search for acquaintance-complexes poorly fits the historical record. In particular, it fails to explain three historical facts: (1) nineteenth-century mathematical developments, especially the logical work of Peano and Frege and the work of Cantor, are central to logical atomism; (2) logic is central to logical atomism; and (3) logical atomism is supposed to be what Russell calls a "scientific philosophy". Let us start with (1).

Russell's intellectual autobiography, *My Philosophical Development*, begins as follows:

> There is one major division in my philosophical work: in the years 1899–1900 I adopted the philosophy of logical atomism and the technique of Peano in mathematical logic. (*MPD*: 9)

There is good reason to be suspicious of Russell's remark. For starters, the date is debatable.[40] We have no evidence Russell used the phrase "logical atomism" before 1911: indeed, Russell's first use of "logical atomism" was in French.[41] So we should critically examine what led Russell in 1959 to call himself a "logical atomist" over a decade before he first used the French equivalent of the phrase "logical atomism".

A clue to Russell's meaning comes from the date plus the phrase "the technique of Peano in mathematical logic".[42] Russell in 1901 wrote an essay explaining the importance to philosophy of recent work on mathematics, especially work on mathematical logic by Peano and others. This essay was his 1901 "Recent Work in the Philosophy of Mathematics".[43]

Most of the essay concerns philosophers continuing longstanding controversies about, among other things, the nature of continuity, space, time, infinity, points, and number. Nineteenth-century mathematicians in contrast undertook technical work that dissolved many of these controversies. Philosophers contented themselves with a priori refutations of mathematics. Mathematicians rolled up their sleeves and labored to clarify mathematical notions so as to make mathematics intelligible. Russell stresses that the mathematicians were in the right.

The piece consists in a pattern repeated throughout the essay. The pattern is a discussion of some philosopher's argument against a mathematical notion, followed by a summary of some mathematician's new definition that shows the argument was founded on a faulty and confused view of that notion. Russell's discussion of infinity nicely illustrates the pattern. Let us consider it.

As Russell tells the story, philosophers had for centuries held infinite numbers "were self-contradictory"; but it seemed obvious that there are infinite numbers like the number of whole numbers (*RWPM*: 372). The purported self-contradiction arose from the following fact: an infinitely large collection is sometimes equal in size to an infinitely large proper sub-collection (*RWPM*: 373). For example, the collection of whole numbers is equal in cardinality to the collection of even whole numbers even though the second collection is a proper sub-collection of the first. This contradicts the seemingly obvious thesis that a sub-collection *s* is smaller than any collection *S* of which *s* is a proper part (*RWPM*: 373). But mathematicians like Cantor and Dedekind showed that the violation of this thesis that the proper part has a smaller size than the whole, far from being contradictory, can actually be used to define infinite numbers (*RWPM*: 372–373).[44]

Russell reconsiders philosophers' past discussions of infinity in light of mathematicians' recent work, especially Cantor's. He notes metaphysicists had failed to solve conceptual difficulties associated with the infinite. Mathematicians in contrast dissolved those difficulties:

> Thus on the subject of infinity it is impossible to avoid conclusions which at first sight appear paradoxical, and this is the reason why so many philosophers have supposed that there were inherent contradictions in the infinite. But a little practice enables one to grasp the true principles of Cantor's doctrine, and to acquire new and better instincts as to the true and the false. The oddities then become no odder than the people at the antipodes, who used to be thought impossible because they would find it so inconvenient to stand on their heads. (*RWPM*: 376)

So first we have purported contradictions that plagued the philosophy of infinity for centuries. Then we have technical, mathematical work in the nineteenth century dissolving the purported contradictions. Facility with the new mathematics of infinity was sufficient to dissolve entirely philosophical debate over infinity, at least among those familiar with the new mathematics of infinity. Russell believed that this fruitful pattern was typical of a general development in all philosophy of mathematics:

> In the whole philosophy of mathematics, which used to be as full of doubt as any other part of philosophy, order and certainty have replaced the confusion and hesitation which formerly reigned. Philosophers, of course, have not yet discovered this fact, and continue to write on such subjects in the old way. But mathematicians, at least in Italy [Peano and his school], have now the power of treating the principles of mathematics in an exact and masterly manner, by means of which the certainty of mathematics extends also to mathematical philosophy. Hence many of the topics which used to be placed among the great mysteries—for example, the natures of infinity, of continuity, of space, time and motion—are now no longer in any degree open to doubt or discussion. Those who wish to know the nature of these things need only read the works of such men as Peano or Georg Cantor; they will find there exact and indubitable expositions of all these quondam mysteries. (*RWPM*, 369)

Russell further believes that the introduction of "order and certainty" in the philosophy of infinity can be utilized not just in the philosophy of mathematics, but in all of philosophy. He hopes that the spread of logical and mathematical techniques in philosophy over other areas of controversy will spread order and certainty throughout philosophy. His 1901 essay ends with this recommendation to let logic develop as freely as possible with the strong conviction that logic stands to bring "exactitude and certainty" to all of philosophy:

> What is now required is to give the greatest possible development to mathematical logic, to allow to the full the importance of relations, and then to found upon this secure basis a new philosophical logic, which may hope to borrow some of the exactitude and certainty of its mathematical foundation. If this can be successfully accomplished, there is every reason to hope that the near future will be as great an epoch in pure philosophy as the immediate past has been in the principles of mathematics. (*RWPM*: 379)

His belief in the resolving power of logic for longstanding philosophical problems is then recorded quite early. What has been neglected is that nineteenth-century mathematics was the inspiration for logical atomism. Indeed, his concluding recommendation in the 1901 essay animates Russell's philosophical works thereafter: he urges giving the widest scope to mathematical logic and the logic of relations in the hopes that such development will produce a new logic; a new logic in turn will induce a "great epoch" in philosophy brought about by the emergence of what Russell calls "scientific philosophy". What Russell meant in 1959 by connecting logical atomism to "the technique of Peano" was this: nineteenth-century mathematics—its development of piecemeal technical work within a powerful logic, followed by the dissolution of philosophical problems—was the model for logical atomist philosophy.

The centrality of nineteenth-century mathematics to logical atomism is the historical fact (1) above. And I argue below that this historical fact is unexplained by the dominant reading of logical atomism. But let us first consider the impact of nineteenth-century mathematics on Russell, which brings us to what I denoted by (2) and (3) above: after his encounter with the work of Peano, (2) logic became central to his logical atomist philosophy and (3) logical atomism was to be a "scientific philosophy". Both (2) and (3) occur throughout Russell's writings after 1901, and especially from 1911 to 1924. Here is a collection of occurrences of (2) and (3) in Russell's works from 1911 to 1945:

> The true method, in philosophy as in science, should be inductive, meticulous, respectful of detail, and should reject the belief that it is the duty of each philosopher to solve all problems by himself. It is this method which has inspired analytic realism [that is, "logical atomism" (*AR*: 135)], and it is the only method, if I am not mistaken, with which philosophy will succeed in obtaining results as solid as those obtained in science. (*AR*: 139)

> Philosophy, from the earliest times, has made greater claims, and achieved fewer results, than any other branch of learning … I believe that the time has now arrived when this unsatisfactory state of things can be brought to an end … The problems and the method of philosophy have, I believe, been misconceived by all schools, many of its traditional problems being insoluble with our means of knowledge, while other more neglected but not less important problems can, by a more patient and adequate method, be solved with all the precision and certainty to which the most advanced sciences have attained … [Logical atomism] represents, I believe, the same kind of

advance as was introduced by Galileo: the substitution of piecemeal, detailed, and verifiable results for large untested generalities recommended only by a certain appeal to imagination. (*OKEW*: 3–4)

The philosophy, therefore, which is to be genuinely inspired by the scientific spirit, must deal with somewhat dry and abstract matters, and must not hope to find an answer to the practical problems of life. (*OKEW*: 29)[45]

It is in this way that the study of logic becomes the central study in philosophy: it gives the method of research in philosophy, just as mathematics gives the method in physics. (*OKEW*: 239)

It is not results, but *methods*, that can be transferred with profit from the sphere of the special sciences to the sphere of philosophy. (*SMP*: 57)[46]

First, the detailed scientific investigation of nature does not *presuppose* any such general laws as its results are found to verify. Apart from particular observations, science need presuppose nothing except the general principles of logic, and these principles are not laws of nature, for they are merely hypothetical, and apply not only to the actual world but to whatever is *possible*. (*SMP*: 61)

A scientific philosophy such as I wish to recommend will be piecemeal and tentative like other sciences; above all, it will be able to invent hypotheses which, even if they are not wholly true, will remain fruitful after the necessary corrections have been made. The possibility of successive approximations to the truth is, more than anything else, the source of the triumphs of science, and to transfer this possibility to philosophy is to ensure a progress in method whose importance it would be almost impossible to exaggerate. (*SMP*: 66)[47]

The adoption of scientific method in philosophy, if I am not mistaken, compels us to abandon the hope of solving many of the more ambitious and humanly interesting problems in philosophy. Some of these it relegates, though with little expectation of a successful solution, to special sciences, others it shows to be such as our capacities are incapable of solving. But there remain a large number of the recognized problems in philosophy in regard to which the method advocated gives all the advantages of division into distinct questions, of tentative, partial, and progressive advance, and of appeal to principles with which, independently of temperament, all competent students must agree. The failure of philosophy hitherto has been due in the main to haste and ambition: patience and modesty, here as in other sciences, will open the road to solid and durable progress. (*SMP*: 73)

Philosophical knowledge, if what we have been saying is correct, does not differ essentially from scientific knowledge … and the results of obtained by philosophy are not radically different from those reached in science. (*OoP*: 308)[48]

> I have no doubt that, in so far as philosophical knowledge is possible, it is by such methods [as logical analysis] that it must be sought … (*HWP*: 862)[49]

Russell here takes logic to be central to philosophy, logical atomism. Note that his suggestions for the development of logic are modeled on the development of nineteenth-century mathematics as described in his 1901 "Recent Work on the Philosophy of Mathematics". And when Russell describes the scientific philosophy, that is, philosophy as distinguished from empirical science, what he says describes just logic:

> … certain characteristics may be noted as distinguishing the province of philosophy from that of the special sciences.
>
> In the first place a philosophical proposition must be general … I do believe that a philosophical proposition must be applicable to everything that exists or may exist … What I do maintain is that there are general propositions which may be asserted of each individual thing, such as the propositions of logic … The philosophy which I wish to advocate may be called logical atomism or absolute pluralism, because, while maintaining that there are many things, it denies that there is a whole composed of those things. We shall see, therefore, that philosophical propositions, instead of being concerned with the whole of things collectively, are concerned with all things distributively; and not only must they be concerned with all things, but they must be concerned with such properties of all things as do not depend upon the accidental nature of the things that there happen to be, but are true of any possible world, independently of such facts as can only be discovered by our senses.
>
> This brings us to a second characteristic of philosophical propositions, namely, that they must be a priori. A philosophical proposition must be such as can be neither proved nor disproved by empirical evidence. […]
>
> We may sum up these two characteristics of philosophical propositions by saying that *philosophy is the science of the possible*. But this statement unexplained is liable to be misleading, since it may be thought that the possible is something other than the general, whereas in fact the two are indistinguishable.
>
> Philosophy, if what has been said is correct, becomes indistinguishable from logic as that word has now come to be used. (*SMP*: 64–65)

Now one might argue Russell's emphasis on mathematical logic as the way to make philosophy scientific is a general feature of Russell's philosophy after his discovery of Peano. That need not imply that logic is central to logical atomism. So logic need not be central to logical atomism, except in that logical atomism is part of philosophy. But the text flatly contradicts that suggestion. Russell rather makes logic the essence of logical atomism. He says explicitly that logical atomism emerged from the new mathematical logic both in his logical atomism lectures and in his 1924 "Logical Atomism":

> The kind of philosophy that I wish to advocate, which I call Logical Atomism, is one which has forced itself upon me in the course of thinking about the philosophy of mathematics, although I should find it hard to say exactly how far there is a definite logical connexion between the two ... In the present lectures, I shall try to set forth in a sort of outline, rather briefly and rather unsatisfactorily, a kind of logical doctrine which seems to me to result from the philosophy of mathematics—not exactly logically, but as what emerges as one reflects: a certain kind of logical doctrine, and on the basis of this a certain kind of metaphysic. (*PLA*: 160)

> Also I found myself driven to pluralism. Analysis of mathematical propositions persuaded me that they could not be explained as even partial truths unless one admitted pluralism and the reality of relations ...I began to think it probably that philosophy had erred in adopting heroic remedies for intellectual difficulties, and that solutions were to be found merely by greater care and accuracy. (*LA*: 162–163)

Russell says that logical atomism, as he understands his own view, is inspired by the positive achievements of Peano, Cantor, Weierstrass, Dedekind, and Frege using the new mathematical logic. Russell sees logical atomism as a patient, precise, and dry method of philosophizing, and to identify the method of sound philosophizing with the new mathematical logic:

> It [the new logic] has, in my opinion, introduced the same kind of advance into philosophy as Galileo introduced into physics, making it possible at last to see what kinds of problems may be capable of solution, and what kinds must be abandoned as beyond human powers. And where a solution appears possible, the new logic provides a method which enables us to obtain results that do not merely embody personal idiosyncrasies, but must command the assent of all who are competent to form an opinion. (*OKEW*: 59)

> Therefore every advance in knowledge robs philosophy of some of the problems which formerly it had, and if there is any truth, if there is any value in the kind of procedure of mathematical logic, it will follow that a number of problems which had belonged to philosophy will have ceased to belong to philosophy and will belong to science ... It [applying mathematical logic to philosophical problems] makes it [philosophy] dry, precise, methodical ... (*PLA*: 243)

Let us summarize what we have found. Despite having only used the phrase “logical atomism” in March 1911, some historical facts are accurately reported by Russell’s remark that he became a logical atomist in 1899–1900. First, he learned the new logic. He immediately applied it, and in a sense used logical atomism’s method—the new logic—before he ever coined the term “logical atomism”. This makes good sense if logic is central to logical atomism. And Russell indeed claims logic is central to logical atomism. This is historical fact (2) above.

Second, the centrality of logic is critical to understanding why Russell believes the method of logical atomism—the new logic—can help philosophy become scientific. Taking his cue from the revolutionary impact of nineteenth-century mathematics, Russell believed that making logic central to philosophizing would make philosophy itself scientific, and distinctly scientific in a way that was not feasible without making logic central. This is historical fact (3) above.

Now interpreting logical atomism as a search for acquaintance-complexes fits these two historical facts about logical atomism quite poorly. Consider fact (2). One could argue for the existence of atomic facts, an ontology including logical simples, and acquaintance epistemology without applying mathematical logic in, and making logic central to, philosophy.[50] One does not need to engage in much logic at all to advocate for these views. And a search for acquaintance-complexes is in no sense “precise, dry, methodical” as the new logic is. A search for acquaintance-complexes is not a “*science of the possible*”. And it does not make philosophy “indistinguishable from logic as that word has come to be used”. One cannot be a logical atomist, according to Russell, without applying logic in, and making logic central to, philosophy. But you could be a logical atomist without doing this on the dominant interpretation of logical atomism. The dominant view thus insufficiently accounts for the logical methods of logical atomism and for the general use of logic in philosophy, which Russell insists is vital to logical atomism.

Now consider fact (3). The dominant interpretation poorly explains the supposedly scientific aspect of logical atomism that was inspired by its logico-mathematical origins, including its genesis from nineteenth-century mathematical work that Russell encountered in 1899–1900 before he ever adopted the acquaintance theory and views about simples that we find in the 1910s. Indeed, logical atomism is older than Russell’s acquaintance epistemology and ontology of sense-data and relations, and it survives their demise. What does not change in this period is Russell’s firm belief in the vital importance of logic for making philosophy "scientific". Russell held that nineteenth-century mathematics did make philosophy of mathematics scientific, and logical atomism was supposed to make all philosophy similarly "dry, precise, methodical", as logic is. This makes sense of Russell's claim to be a logical atomist from 1899–1900 onward. And making logic central to logical atomism accounts for why Russell thought that the introduction of the new logic was an "advance" in philosophy rivaling Galileo's advance in physics. On the other hand, viewing logical atomism as a search for acquaintance-complexes produces a philosophy that is not at all scientific in Russell's sense. On the dominant interpretation, logical atomism is not piecemeal or progressive: it is a typical traditional philosophy that does not merit any of the remarkable claims about progress Russell asserts for it.

Viewing logical atomism as a search for acquaintance-complexes does not fit the actual development of logical atomism from 1899–1900 onward through various texts. Even limiting ourselves to Russell's works after his 1911 "Analytic Realism", where he first used the phrase "logical atomism", helps very little: Russell repeatedly points to nineteenth-century mathematical work in discussing what led him to the philosophy of logical atomism.

Understanding logical atomism as the search for acquaintance-complexes fails to explain why Russell traced the origins of logical atomism to mathematical logic. Simply put, the age of logical atomism and its reliance on mathematical logic on this interpretation are inexplicably accidental.[51] As I hope the foregoing summary shows, this reading conflicts with Russell's explicit statements in his 1918 logical atomism lectures, in his 1924 "Logical Atomism", in his 1914 *Our Knowledge of the External World*, in his 1914 "On Scientific Method in Philosophy", and in his 1911 "Analytic Realism", all of which are logical atomist works, and with Russell's intellectual autobiography. It thus conflicts with an abundance of evidence as to the logico-mathematical origins of logical atomism. The dominant interpretation is a bad history.

## 4 Direct Textual Evidence Against the Dominant Reading

Interpreting logical atomism as a search for acquaintance-complexes also conflicts directly with the text of the lectures. Near the beginning of Lecture IV Russell remarks:

> I think one might describe philosophical logic, the philosophical portion of logic which is the portion that I am concerned with in these lectures since Christmas [1917], as an inventory, or if you like a more humble word, a "Zoo" containing all the different forms that facts may have … In logic you are concerned with the forms of facts, with getting hold of the different sorts of facts, different *logical* sorts of facts, that there are in the world. (*PLA*: 191)[52]

Now I submit that Russell does exactly logic, as he describes logic above, in the 1918 lectures on logical atomism. Just look at the table of contents for the logical atomism lectures[53]:

I. Facts and Propositions
II. Particulars, Predicates, and Relations
III. Atomic and Molecular Propositions
IV. Propositions and Facts with More than One Verb; Beliefs, Etc.
V. General Propositions and Existence
VI. Descriptions and Incomplete Symbols
VII. The Theory of Types and Symbolism: Classes
VIII. Excursus into Metaphysics: What There Is (*PLA*: 155)

To see that the logical atomism lectures are a work of logic, let us expand on its table of contents. Lecture I introduces facts and the notion of forms of facts.[54] Lecture II analyzes atomic facts and atomic propositions.[55] Lecture III discusses the purported need for molecular facts.[56] Lecture IV treats the need for belief-facts.[57] Lecture V concerns existence facts and general facts.[58] Lectures VI and VII deal with logic itself. Lecture VI deals with incomplete symbols.[59] Lecture VII concerns type theory.[60] Thus, all but his last lecture deal with logical forms of facts or with logic itself. That is why Russell styles Lecture VIII as an "excursus" and also why he begins Lecture VIII by remarking that it breaks from the themes of the seven earlier lectures:

> I come now to the last lecture of this course, and I propose briefly to point to a few of the morals that are to be gathered from what has gone before, in the way of suggesting the bearing of the doctrines that I have been advocating upon various problems of metaphysics. I have dwelt hitherto upon what one may call philosophical grammar, and I am afraid I have had to take you through a good many very dry and dusty regions in the course of that investigation, but I think the importance of philosophical grammar is very much greater than it is generally thought to be. (*PLA*: 234)

Logic, or "philosophical grammar", is again critical to scientific philosophy as Russell understands it: that is why he spends seven of eight lectures on logical atomism dealing with logic in the sense we saw above. This point is worth emphasizing. The dominant reading would suggest that logical atomists are concerned with discovering the logical simples and arguing for an acquaintance epistemology. Russell's practice in these lectures, save for a one-lecture "excursus", is instead to consider what the logical forms of facts are. Indeed, his discussion of particulars and relations in Lecture II is explicitly concerned with the logical form of atomic facts, and Russell tells us that he is not interested in what particulars there are, if any, but only in the logical characterization of constituents of atomic facts. Rather than search for acquaintance-complexes, as we would expect on the dominant interpretation, Russell's practice in the logical atomism lectures is to do logic. The dominant interpretation gets logical atomist practice wrong.

It may help to clarify this point to consider why Russell believes that mathematical logic is critical to scientific philosophy as Russell understands it. Russell holds that one cannot form the requisite stockpile of logical forms to attack philosophical problems without a robust logic:

> Now I want to say that if you wish to test such a theory as that of neutral monism, if you wish to discover whether it is true or false, you cannot hope to get any distance with your problem unless you have at your fingers' end the theory of logic that I have been talking of. (*PLA*: 242, see also 235)

Russell is not equivocal on this issue: mathematical logic is critical to logical atomism. The reason is that we can hardly make philosophical progress in examining a view without a large stockpile of logical forms or without a powerful logic. The logical examination of a theory is impossibly hindered by a limited logic.

The answer to this limitation is to create a stockpile of logical forms of facts using the powerful new logic. And this is precisely what Russell does, as he tells us when he describes his subject in the logical atomism lectures as an inventory of logical forms that facts may have. The logical atomism lectures are a work describing a search for logical forms, not a search for acquaintance-complexes. It is no accident that Russell calls the logical atomism lectures his "logic lectures" both in correspondence.[61] Even in his appointment diary, a page of which is pictured in Fig. 1.1, Russell refers to the logical atomism lectures as "LL" for "logic lectures".[62]

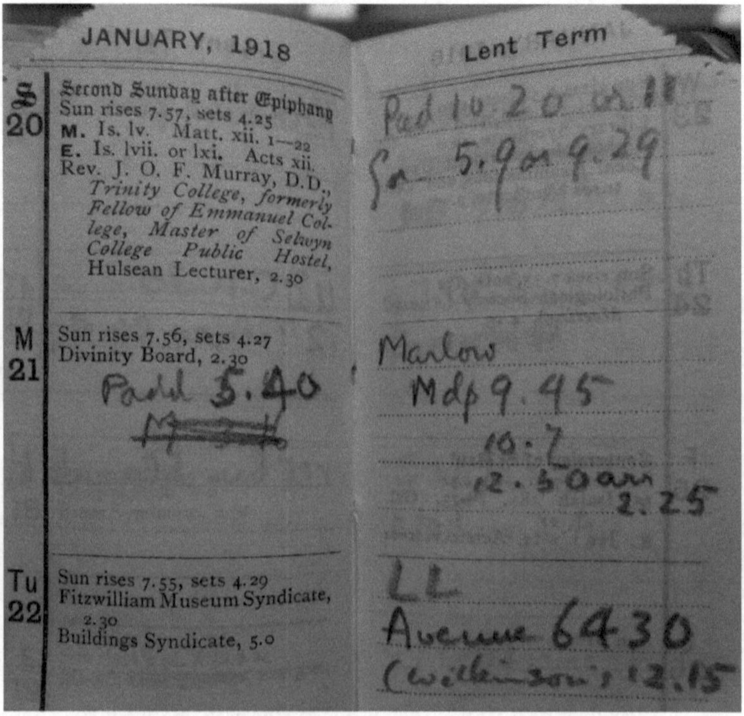


**Fig. 1.1** Russell's appointment diary for 20–22 January 1918

There is no denying what is firmly and uniformly supported by documents contemporaneous with and the text of the logical atomism lectures: the logical atomism lectures are a logic book.[63]

Above all, the text of the logical atomism lectures focuses on, and is organized around, the search for logical forms: this is the stockpiling of logical forms of facts that Russell playfully describes as the inventory of a logical zoo. The text thus conflicts with understanding logical atomism as a search for acquaintance-complexes, even when we confine ourselves to the text and ignore the wider historical context. Having a robust logic, on the dominant reading, is at best just a helpful means for more effectively rendering the truth-conditions of statements in terms of acquaintance-complexes. But this is contrary to the logical atomist practice that we find in the lectures. What Russell in fact makes the center of logical atomism, namely, a robust logic as a means of stockpiling logical forms of facts and logical theses for use in testing possibilities, is exhibited in his practice of searching for logical forms of facts. He does not search for acquaintance-complexes. Again, on the dominant interpretation, the connection of logic to logical atomism is, if fortuitous, accidental. But what we find in the text is logic being used throughout the essential work of classifying logical forms.

Defenders of the dominant interpretation have actually seen this as a virtue of their view. Urmson, for example, says in his 1956 treatment of logical atomism that we can neglect all the "advanced and difficult" parts of Russell's logic:

> Since the metaphysics is thus dependent on the logic, it is clearly indispensable to have some sort of understanding of what sort of logic it is and of the concepts it employs if we are to understand the metaphysics. For our purposes the more advanced and difficult parts of the logic are luckily less important than the most simple and no reference need be made to them. (Urmson 1956: 7–8)

For Urmson, the only necessary parts are those that help us "to understand the metaphysics", and not any parts, if any, that are specially necessary to understand the logical form.

Urmson then discusses truth-functional connectives and truth-functional tautologies and contradictions, omitting quantification theory (Urmson 1956: §2.B). He then says that logicians speak "indifferently about all statements whatever their content or structure" (Urmson 1956: 8). This suggests his view of logical atomism confines the logic of logical atomism to a system of propositional logic in which logical truths are all tautologies, and all non-logical truths are atomic statements, unlike, say, the statement "all humans are mortal". Urmson writes:

> … the formal logician regards himself as supplied with an indefinite number of propositional variables, *p*, *q*, *r*, &c.; since he does not inquire into their structure we may say that they are simple relative to his system. Since he has no logical means of determining the truth or falsehood of the constituents [statements] he limits his interest to those of the functions which can be recognized to be true or false by logical methods—the tautologies and contradictions. Since it is the tautologies which are of importance for research into the foundations of mathematics, this suits him perfectly … So much must suffice as an explanation of the ideas of elementary logic which were adapted for use in the metaphysics of logical atomism. (Urmson 1956: 11; see also 14)

This conflicts with the fact that Russell's logic is far richer than this, and the fact that logical atomist inventory of logical forms goes beyond tautologies and contradictions. Russell writes:

> The technical methods of mathematical logic, as developed in this book [*Principia*], seem to me very powerful, and capable of providing a new instrument for the discussion of many problems that have hitherto remained subject to philosophical vagueness. (*LA*: 163)

The elimination of "philosophical vagueness" is aided by the fact that the language of *Principia* shows "at a glance" the logical form of the facts involved in a statement being true or false (*PLA*: 176). But confining ourselves to a logic as weak as propositional logic, as Urmson does, or even to a first-order logic, destroys all the examples from *Principia* of logical atomism's success. Russell cites, as examples of logical atomism's success, the analyses of number (*PLA*: 234), classes (*PLA*: 228), matter (*PLA*: 235–236), definite descriptions (*LA*: 165–166), series and ordinals (*LA*: 166), ordinary objects (*PLA*: 236–237), points and instants (*LA*: 166), matter (*LA*: 166–167), and mind (*LA*: 167). Propositional logic is far too weak to recover the analyses given in *Principia* of purely mathematical notions like number and class. And a system too inadequate to recover the analyses in *Principia* cannot be taken what Russell meant in describing logic as central to logical atomism. This weighs heavily against Urmson's indication of propositional logic as giving the essentials of what Russell had in mind. And it more generally suggests that the dominant interpretation, which at best accounts for Russell's examples of logical atomist analyses as happy accidents only loosely connected with the search for acquaintance-complexes, has gotten something deeply wrong in its interpretation of logical atomism.

And, for someone who is, supposedly, deeply committed to searching for acquaintance-complexes, Russell is remarkably unconcerned with justifying the existence of logical atoms or with sketching a foundationally structured justification tree for human knowledge.[64] Russell is far more concerned in the logical atomism lectures with classifying logical forms of facts. The dominant interpretation fails to explain why Russell focuses on logical forms throughout the lectures. The dominant interpretation does not explain why Russell views logic as "what is fundamental in philosophy" or why Russell thinks logical atomism will make philosophy scientific.[65] It does not even fit Russell's description of these lectures as his "logic lectures".

The dominant interpretation of logical atomism as a search for acquaintance-complexes, though alluring, is the sort of traditional philosophical view that Russell explicitly contrasts with his own. It conflicts with the logical focus, especially the logical atomist practice of searching for logical forms, that we find in the text of the 1918 logical atomism lectures. The dominant interpretation is a historically inaccurate and textually inadequate view of logical atomism.

## 5 Russell's Logical Atomism

Interpreting logical atomism as a search for acquaintance-complexes composed of ontological simples has inspired rich scholarly discussion. But the dominant interpretation:

1. fails to connect its reading to Russell's remarks about logical atomism and its origins;
2. fails to incorporate Russell's view that a powerful logic is critical to the view;
3. fails to match Russell's logical atomist practice in, and the content of, the lectures.

This historical evidence shows the dominant reading is wrong. What, then, is logical atomism?

Logical atomism is what we might call a "logic-first and logic-last" philosophy (*SMP*: 65). A logical atomist starts by giving a logical system (*LA*: 162). A vital test of a proposed logical system will be its adequacy to certain data that we accept as true (*LA*: 163). A philosophically fruitful logical atomism will require an expressively adequate logic to generate a large variety of logical forms (*OKEW*: 42–43). Logical atomism thus requires a higher-order logical framework. It is the logical atomist's use of a powerful logic that makes logical atomist philosophy distinctively scientific in Russell's sense (*AR*: 139). Where logic does not settle a philosophical issue, the possibilities are open (*SMP*: 72–73). And for traditional philosophers that find Russell's logical atomism and its logic-heavy practice disappointing or difficult, Russell has some unsympathetic advice: "acquire a taste for mathematics" (*PLA*: 243–244).

## Notes

1. Bostock rightly considers the logical atomist period beyond the 1918 lectures (Bostock 2012: vi–vii). Linsky's recent work on logical constructions closely follows the connections between earlier and later works, and he rightly links logical constructions to Russell's logical atomism (Linsky 2003: 372, 2014: §1).
2. Both Landini (2007: §2.1, 2011: 162–163) and Maclean (2014: Chap. 8) criticize the dominant interpretation of logical atomism.
3. Galaugher (2013: Chaps. 1–2) provides a critical historical context for Russell's rejection of the doctrine of internal relations.
4. "We will give the name 'a *complex*' to any such object as '*a* in the relation *R* to *b*' or '*a* having the quality *q*' or '*a* and *b* and *c* standing in the relation *S*.' Broadly speaking, a *complex* is anything which occurs in the universe and is not simple" (*PM*$_2$: 47).
5. "When I speak of a 'fact,' I do not mean one of the simple things of the world; I mean that a certain thing has a certain quality, or that certain things have a certain relation" (*OKEW*: 51).
6. A fact need not make a statement true. Arguably, there would still be logical facts even if there were no truths. This does not change that facts are the sort of thing that could make something true.
7. "The fact itself is objective, and independent of our thought or opinion about it; but the assertion is something which involves thought, and may be either true or false" (*OKEW*: 52).
8. "Thus atomic facts are what determine whether atomic propositions are to be asserted or denied" (*OKEW*: 52).
9. "Now a fact, in this sense, is never simple, but always has two or more constituents" (*OKEW*: 51).
10. "Atomic propositions, although, like facts, they may have any one of an infinite number of forms, are only one kind of propositions. All other kinds are more complicated" (*OKEW*: 51).
11. His examples in 1914 are *this is red*, *this is before that*, *Napoleon was ambitious*, *Napoleon married Josephine*, *A is jealous of B on account of C*, *Charles I was executed*, and *Socrates is a man* (*OKEW*: 51, 53, 57).
12. "The constituents of facts, in the sense in which we are using the word 'fact,' are not other facts, but things and qualities or relations" (*OKEW*: 51). The sense of "fact" here is specifically atomic facts.
13. I once thought that Russell flatly posited them in his 1918 lectures. Perović (this volume) has shown that the issue is more complicated than I had originally supposed. I thank Perović for changing my mind on this point.
14. Now Russell does posit negative facts in his 1919 "On Propositions": "Thus facts, and forms of facts, have two opposite qualities, positive and

negative" (*OP*: 280). But negative facts are not endorsed in his 1924 "Logical Atomism", and he argues against them in the 1940s.

15. "Russell sometimes uses 'monadic relation' for quality, and he sometimes uses 'predicate' for quality; he is explicit about this practice" (*PLA*: 177).
16. "*Particulars* have the purely logical properties of substances, but do not have their metaphysical properties. That is to say, particulars can only be either the subjects of predicates or the terms of relations" (*AR*: 135).
17. "It remains to be investigated what particulars you actually can find in the world, if any. The whole question of what particulars you actually find in the world is a purely empirical one which does not interest the logician as such" (*PLA*: 177).
18. "From the logical point of view, any simple existence is independent of any other, and the only dependence is that of the complex on the simple" (*AR*: 135).
19. In the passage quoted, Russell actually uses the word 'simple'. A *simple* in Russell's logical sense is what has no parts. Note that a simple need not be concrete. Now Russell's sense-data are given as examples of logical atoms, despite having parts. So we can distinguish logical simples from *logical atoms*, which are taken to be simple relative to a given logical construction and are accordingly picked out by, as Russell says, "a simple symbol". For Russell allows that we might analyze them further at a later stage, and yet that we can pick them out by simple symbols in a given logical construction. So logical atoms are not simple in an absolute, construction-independent sense as simples, if there are any, are. The distinction between logical atoms and simples is not critical to my argument here, but it is a vital one to understanding Russell's meaning.
20. "It is analytic, because it claims that the existence of the complex depends on the existence of the simple, and not vice versa, and that the constituent of a complex, taken as a constituent, is absolutely identical with itself as it is when we do not consider its relations. This philosophy is therefore an atomic philosophy" (*AR*: 133).
21. "The only way, so far as I know, in which one thing can be *logically* dependent upon another is when the other is *part* of the one" (*OKEW*: 74).
22. "**Woodrow Wyatt**: What kind of philosopher would you say you are? **Russell**: Well, the only label I've ever given myself is logical atomist, but I'm not very keen on the label. I've rather avoided labels. **Wyatt**: What does that mean? A logical atomist. **Russell**: It means, in my mind, that the way to get at the nature of any subject matter you're looking at is analysis—and that you can analyze until you get to things that can't be analyzed any further and those would be logical atoms. I call them logical atoms because they're not little bits of matter. They're the ideas, so to speak, ideas out of which a thing is built up" (Wyatt 1960: 15). This forty-second

interview segment is viewable on the Internet Archive at the 5:55-6:35 mark: https://archive.org/details/BertrandRussellDiscussesPhilosophy.

23. "… particulars which are known are called *sense-data*" (*AR*: 135).
24. "Logically a sense-datum is an object, a particular of which the subject is aware. It does not contain the subject as a part, as for example beliefs and volitions do. The existence of the sense-datum is therefore not logically dependent upon that of the subject …" (*RSDP*: 9).
25. "You will note that this philosophy is the philosophy of logical atomism. Every simple entity is an atom. One must not suppose that atoms need persist in time, or that they need occupy space: these atoms are purely logical" (*AR*: 135).
26. I cite *Problems* because Russell's descriptions there are far more detailed than his description in the logical atomism lectures. Russell abandoned acquaintance relations by 1919 (*OP*: 294–295; *LA*: 167).
27. "When I speak of a cognitive relation here, I do not mean the sort of relation which constitutes judgment, but the sort which constitutes presentation. In fact, I think the relation of subject and object which I call acquaintance is simply the converse of the relation of object and subject which constitutes presentation" (*KAKD*: 148).
28. "All cognitive relations—attention, sensation, memory, imagination, believing, disbelieving, etc.—presuppose acquaintance" (*CPBR 7*: 5).
29. "In a philosophy of logical atomism one might suppose that the first thing to do would be to discover the kinds of atoms out of which logical structures are composed. But I do not think that is quite the first thing; it is one of the early things, but not quite the first" (*PLA*: 169). "I have been speaking hitherto of what it is *not* necessary to assume as part of the ultimate constituents of the world. But logical constructions, like all other constructions, require materials, and it is time to turn to the positive question, as to what these materials are to be" (*LA*: 169).
30. "Russellian analyses proceed by way of definitions, terminate with indefinables, and, at that point, base themselves on acquaintance" (Pears 1985: 9).
31. "The fundamental principle in the analysis of propositions containing descriptions is this: *Every proposition which we can understand must be composed wholly of constituents with which we are acquainted*" (*PoP*: 91).
32. Paul J. Hager's analysis diagrams are useful in grasping the options here; Hager's foundationalist diagram captures the dominant interpretation's notion of Russellian analysis (Hager 1994: 48, Fig. 4.3).
33. "The theoretical reason for postulating simple particulars is that, when a complex singular expression is fully analyzed, there must be one or more particulars to carry the qualities and relations mentioned in its analysis, and these particulars will be simple because all qualities and relations will have been stripped from them" (Pears 1972: 37).

34. "That is to say, each particular that there is in the world does not in any way logically depend upon any other particulars. Each one might happen to be the whole universe; it is a merely empirical fact that this is not the case. There is no reason why you should not have a universe consisting of one particular and nothing else. That is a peculiarity of particulars" (*PLA*: 179, see also 181).
35. "The acquaintance with the simpler is presupposed in the understanding of the more complex, but the logic that I should wish to combat maintains that in order thoroughly to know any one thing, you must know all its relations and all its qualities, all the propositions in fact in which that thing is mentioned; and you deduce from that that the world is an interdependent whole. It is on a basis of that sort that the logic of monism develops" (*PLA*: 181).
36. "In the same way, in order to understand a name for a particular, the only thing necessary is to be acquainted with that particular. When you are acquainted with that particular, you have a full, adequate, and complete understanding of the name, and no further information is required. No further information as to the facts that are true of that particular would enable you to have a fuller understanding of the meaning of the name" (*PLA*: 179).
37. "This characteristic, that you can understand a proposition through the understanding of its component words, is absent from the component words when those words express something simple. Take the word 'red', for example ... You cannot understand the meaning of the word 'red' except through seeing red things" (*PLA*: 173).
38. "Russell used the empirical argument and claimed, in the spirit of Hume, that, when we find that we cannot push the analysis of words any further, we can plant a flag recording the discovery of genuine logical atoms" (Pears 1985: 5).
39. Confer (Jager 1972: §6.14; Sainsbury 1979: §II.3; Pears 1985: 3–4; Hager 1994: Chap. 4; Linsky 2003: 384–386; Bostock 2012: §14.1). Russell's response to a question from H. Wildon Carr at the end of Lecture II indicates Russell is open to analysis having no end (*PLA*: 180). Likewise Russell's 1924 essay "Logical Atomism" allows for the same (*LA*: 173–174). In both cases, Russell notes that that his considered view is that complexes are composed of simples, and that positing simples is inessential to logical atomism.
40. Russell broke with Neo-Hegelian philosophy in mid-1898 (Griffin 1991: 181). Perhaps Russell means to date his firm adoption of logical atomism to the 1899 publication of Moore's "The Nature of Judgment" in *Mind*. More likely, as we see below he means to include his subsequent adoption of the doctrine of external relations and of Peano's logic.
41. So far as we know Russell coined the phrase "logical atomism" in his 1911 "Analytic Realism", which first appeared in English in 1992 (*AR*: 135).

So strictly speaking, as the article was originally published in French, he first used the phrase "atomisme logique" (*AR*: 412).

42. The other conjunct relates to his rejecting the doctrine of internal relations: "Moreover, by the rejection of à priori constructions the way is opened for philosophy to become inductive, and to begin the patient cooperative accumulation of results by which the triumphs of science have been achieved" (*BoR*: 131). Klement concisely details the importance of this rejection (Klement 2016: §2.1). The importance of this is also missed by reading logical atomism as the search for acquaintance-complexes, so a parallel criticism could be made that the traditional interpretation misses the importance of external relations for Russell's logical atomism. I ignore that here due to space constraints and focus narrowly on the importance of logic.
43. He retitled the essay "Mathematics and the Metaphysicians" in a 1917 reprinting: "The essay 'Mathematics and the Metaphysicians' was written in 1901, and appeared in an American magazine, *The International Monthly*, under the title 'Recent Work in Philosophy of Mathematics'" (*MaL*: v). Another 1901 essay covering similar ground, "Recent Italian Work on the Foundations of Mathematics", remained unpublished until 1993 (*RIWFM*: 350–351).
44. Cantor and Dedekind did not mean *concrete* part. Cantor and Dedekind used the idea of bijections (one-to-one correspondences). Using this idea, we can say a collection is *infinite* means there exists a bijection from itself to a proper sub-collection of itself.
45. Russell, at least by 1914, denied that ethics belonged to scientific philosophy: "Human ethical notions, as Chuang Tzu perceived, are essentially anthropocentric, and involve, when used in metaphysics, an attempt, however veiled, to legislate for the universe on the basis of the present desires of men. In this way they interfere with that receptivity to fact which is the essence of the scientific attitude towards the world" (*SMP*: 63). Russell admits, "the importance or value, within its own sphere", of ethically inspired philosophy, he concludes, "The scientific philosophy, therefore, which only aims at understanding the world and not directly at any other improvement of human life, cannot take account of ethical notions without being turned aside from that submission to fact which is the essence of the scientific temper" (*SMP*: 64). Thus he held that ethical philosophy is disjoint from logical atomist philosophy. Here I leave open the consistency of ethical philosophy with logical atomist philosophy, and so of logical atomist ethics.
46. In this work, Russell affirms logical atomism: "The philosophy which I wish to advocate may be called logical atomism or absolute pluralism …" (*SMP*: 65).

47. Confer also his 1904 review of Moore's *Principia Ethica*: "… philosophy will never advance, until the notion is dispelled, that sweeping general principles can excuse the patient attention to detail which, here as elsewhere, can alone lead to the discovery of truth" (*TMG*: 575).
48. Russell earlier states, "… I call the philosophy which I advocate 'logical atomism'" (*OoP*: 259).
49. I think Russell is referring to his logical atomist philosophy here. He has just illustrated on the previous page "the utility of philosophical syntax" using his theory of definite descriptions (*HWP*: 859).
50. Pears held that Wittgenstein did just that in reasoning a priori for positing simples (Pears 1985: 5–6).
51. Urmson, for instance, says "this new, rich logic" merely "suggested" logical atomism (Urmson 1956: 7).
52. Russell goes even further in his 1914 *Our Knowledge*: "Logic, we may say, consists of two parts. The first part investigates what propositions are and what forms they may have; this part enumerates the different kinds of atomic propositions, of molecular propositions, of general propositions, and so on. The second part consists of certain supremely general propositions, which assert the truth of all propositions of certain forms. This second part merges into pure mathematics, whose propositions all turn out, on analysis, to be such general formal truths. The first part, which merely enumerates forms, is the more difficult, and philosophically the more important; and it is the recent progress in this first part, more than anything else, that has rendered a truly scientific discussion of many philosophical problems possible" (*OKEW*: 57–58; see also *SMP*: 65–66).
53. Here I am taking inspiration from the tree readers of the *Tractatus* (Bazzocchi 2014: IV–VII).
54. "There are a great many different kinds of facts, and we shall be concerned in later lectures with a certain amount of classification of facts" (*PLA*: 164).
55. "I propose to begin today the analysis of facts and propositions, for in a way the chief thesis that I have to maintain is the legitimacy of analysis …" (*PLA*: 169).
56. "I do not see any reason to suppose that there is a complexity in the facts corresponding to these molecular propositions …" (*PLA*: 187).
57. "Today we have to deal with a new form of fact … Now I want to point out today that the facts that occur when one believes or wishes or wills have a different logical form from the atomic facts containing a single verb which I dealt with in my second lecture" (*PLA*: 191).
58. "We have such propositions as 'All men are mortal' and 'Some men are Greeks.' But you have not only such *propositions*; you also have such *facts*, and that, of course, is where you get back to the inventory of the world: that, in addition to particular facts … there are also general facts and existence-facts …" (*PLA*: 206).

59. "I am proposing to deal this time with the subject of descriptions, and what I call 'incomplete symbols', and the existence of described individuals" (*PLA*: 211).
60. "I come now to the proper subject of my lecture, but shall have to deal with it rather hastily. It was to explain the theory of types and the definition of classes" (*PLA*: 226).
61. In a 21 May 1918 letter, Russell writes, "*To P. Jourdain* ... Is he going to print 2 of my logic lectures in July and 2 each subsequent quarter? I hope so" (Griffin 2001: #313). Jourdain was then editor of *The Monist*, where the logical atomism lectures were published.
62. Thanks to the Bertrand Russell Archives in the William Ready Division of Research Collections, McMaster University Library, for permission to use this photograph.
63. In a 17 April 1918 letter, Russell writes, "I wish to write two works concurrently, one to be called 'Introduction to Modern Logic' or some such title, more or less on the lines of the lectures I gave you before and after Christmas ..." (Thompson 1975: 18).
64. A case in point is Russell's casual suggestion that there may not be logical simples (*PLA*: 180). Pears explains this away as Russell's being confused (Pears 1985: 4). Pears thus squares what Russell actually believed with the text. But more to the point is that this interpretation poorly fits the text: for if logical atomism is crucially committed to an ontology with logical simples (knowable by acquaintance), Russell is either unaware of this fact or far too casual in entertaining an ontology without logical simples.
65. "I hold that logic is what is fundamental in philosophy, and that schools should be characterized rather by their logic than by their metaphysic" (*LA*: 162).

## References

### Works by Other Authors

Ayer, A. J. (1972). *Bertrand Russell.* New York: The Viking Press.

Bazzocchi, Luciano (2014). *The* Tractatus *According to Its Own Form:* Supplements *and Other Shavings.* Raleigh, North Carolina: Lulu.

Bostock, David (2012). *Russell's Logical Atomism.* Oxford: Oxford University Press.

Galaugher, Jolen (2013). *Russell's Philosophy of Logical Analysis: 1897–1905.* History of Analytic Philosophy series. New York: Palgrave Macmillan.

Grayling, A. C. (1996). *Russell.* Oxford: Oxford University Press.

Griffin, Nicholas (1991). *Russell's Idealist Apprenticeship.* Oxford: Clarendon Press.

Griffin, Nicholas (2001). *The Selected Letters of Bertrand Russell, Volume 2: The Public Years (1914—1970)*. New York: Routledge.
Hager, Paul J. (1994). *Continuity and Change in the Development of Russell's Philosophy*. Nijhoff International Philosophy series. Dordrecht: Kluwer Academic Publishers.
Jager, Ronald (1972). *The Development of Bertrand Russell's Philosophy*. Muirhead Library of Philosophy. London: George Allen & Unwin LTD.
Klement, Kevin C. (2016). "Russell's Logical Atomism." *Stanford Encyclopedia of Philosophy*: Summer 2016. https://plato.stanford.edu/archives/spr2016/entries/logical-atomism/
Koç Maclean, Gülberk (2014). *Bertrand Russell's Bundle Theory of Particulars*. New York: Bloomsbury Academic.
Landini, Gregory (2007). *Wittgenstein's Apprenticeship with Russell*. Cambridge: Cambridge University Press.
Landini, Gregory (2011). *Russell*. London and New York: Routledge.
Linsky, Bernard (2003). "The Metaphysics of Logical Atomism." In *The Cambridge Companion to Bertrand Russell*, ed. Nicholas Griffin. Cambridge: Cambridge University Press. Chapter 11: 371–391.
Linsky, Bernard (2014). "Logical Constructions." *Stanford Encyclopedia of Philosophy*. Fall 2014. https://plato.stanford.edu/archives/fall2014/entries/logical-construction/
Pears, D. F. (1972). "Russell's Logical Atomism." In *Bertrand Russell: A Collection of Critical Essays*, ed. D. F. Pears. New York: Anchor Books. Chapter 2: 23–51.
Pears, D. F. (1985). "Introduction." In *The Philosophy of Logical Atomism*, Bertrand Russell, ed. D. F. Pears. Open court: La Salle, Ill. 1–34.
Russell, Bertrand, and Woodrow Wyatt. (1960). *Bertrand Russell speaks his mind*. London: A. Baker. URL https://archive.org/details/BertrandRussellDiscusses Philosophy.
Sainsbury, R. M. (1979). *Russell*. London: Routledge & Kegan Paul.
Soames, Scott (2014). *The Analytic Tradition in Philosophy*, vol. 1, *The Founding Giants*. Princeton: Princeton University Press.
Thompson, Michael (1975). "Some Letters of Bertrand Russell to Herbert Wildon Carr." *Coranto*, Vol. 10, No. 1: 7–19.
Urmson, J. O. (1956). *Philosophical Analysis: Its Development Between Two World Wars*. Oxford: Oxford University Press.
Urmson, J. O. (1969). "Russell on Acquaintance with the Past." *The Philosophical Review*, Vol. 74, No. 4: 510–515.

CHAPTER 2

# Logical Atomism's Necessity

*Gregory Landini*

## 1 Introduction: The *Principia* Era Versus the Neutral Monist Era

Russell's Philosophy of Logical Atomism was first articulated in 1911 with his paper "Analytic Realism," while the second of the projected four volumes of Whitehead and Russell's *Principia Mathematica* was at the press. It is a philosophy that can be found in a subdued form in Russell's book *The Problems of Philosophy* (1912). But its best articulation appeared in the book *Our Knowledge of the External World as a Field for Scientific Method in Philosophy* (1914) where Russell proclaims that *logic* is the essence of his new research program for a uniquely scientific philosophy.

The logic Russell has in mind is the synthetic *a priori* ${}^{cp}$Logic (impredicatve comprehension principle logic) of *Principia Mathematica* which proclaims that logic is the study of all the kinds of relational structures there are, subsuming every branch of mathematics (including Euclidean and non-Euclidean geometries). *Principia's* ${}^{cp}$Logic endeavors to eliminate abstract particulars (numbers, sets/classes, spatial figures) from every branch of mathematics and even from logic itself while at the same time enabling the

G. Landini (✉)
University of Iowa, Iowa City, IA, USA
e-mail: gregory-landini@uiowa.edu



L. D. C. Elkind, G. Landini (eds.), *The Philosophy of Logical Atomism*, History of Analytic Philosophy,
https://doi.org/10.1007/978-3-319-94364-0_2

study of every kind of structure. $^{cp}$Logic reveals entirely new logical forms that are otherwise inconceivable and in this way reveals the fallacies imbedded in metaphysical arguments for the ineliminability of abstract particulars and their special kinds of necessity. Russell writes (*LA*: 162):

> I hold that logic is what is fundamental in philosophy, and that schools should be characterized rather by their logic than by their metaphysics.

The $^{cp}$Logic of *Principia* alone enables the *criticism* that Russell finds distinctive of his scientific philosophy and the source of its great value.

*Our Knowledge of the External World* is the natural sequel to *Problems*. In both works, the epistemic foundation of our synthetic *a priori* knowledge was explained as deriving from our acquaintance with universals and our logical perception of the relations between them that hold independently of whether they are exemplified. In both works, Russell held that the distinctive tool of philosophical *criticism* is its use of the theories of structure exemplified in *Principia*'s mathematical logic. *Principia*'s mathematical logic is privileged and unique in its showing the way forward. Russell explains (*AR*: 139):

> The true method, in philosophy as in science … has inspired analytic realism, and it is the only method, if I am not mistaken, with which philosophy will succeed in obtaining results as solid as those obtained in science.

Those blind to logical form (mathematical logic) have, according to Russell, no legitimacy—and those who invent a non-logical necessity governing abstract particulars are among the most blind.

Once commitment to abstract particulars is purged from mathematical logic as a study of relational structure, the field can be seen to be more sound than *any* metaphysics of abstract particulars and specialized kind of necessity governing them. Of course, *Principia* does not avoid metaphysics—not at least in a realist interpretation of its simple types of properties and relation in intension. *Principia's* $^{cp}$Logic assures the existence of relations in intension and studies all kinds of structures that there are by studying the way relations, exemplified or not, order their fields. It is not metaphysics that is anathema to Russell's scientific philosophy. The metaphysically unacceptable claims Russell targets are precisely those claims that there are special abstract particulars and kinds of non-logical necessity. *Principia's* goal was to reveal that no branch of mathematics (not even the non-Euclidean geometries) is dependent on the assumption of abstract particulars. The metaphysicians with their abstract particulars and unique kinds of necessity introduce "muddles" into mathematics (*HWP*: 829). In his enthusiastic 1901 paper "Mathematics and the Metaphysicians," Russell's speaks of the revolution in mathematics that is due (largely) to Cantor who revealed that the field concerns *relational order*, not quantity (number). He later added a footnote honoring Frege's revolutionary contribution in $^{cp}$Logic—a contribution which, put in contemporary terms, involved the discovery that impredicative comprehension, and not quantification theory, is precisely what makes logic an informative science capable of proving a principle of (mathematical) induction. These two revolutions are necessary for Russell's logicism which holds that the study within $^{cp}$Logic of relational structures subsumes every branch of mathematics. There are no abstract particulars unique to mathematics and no special kind of mathematical necessity. Mathematical necessity is logical necessity. At the same time, there is no change of subject matter. The revolutions revealed, finally, what the subject matter of mathematics is. Mathematicians are doing $^{cp}$Logic when they do mathematics.

*Principia*'s $^{cp}$Logic of relations and the acquaintance epistemology of *Problems* had shown the way toward a new scientific philosophy above the metaphysical civil wars. Its agenda is to eliminate non-logical abstract particulars and their kinds of pseudo-necessity. The great value Russell saw in his new science of philosophy, with *Principia's* logic as its essence, lies in its freeing the mind from the *prisons* produced by dogmatisms parading as if they were necessities. Let me quote at length from Russell's 1945 book *A History of Western Philosophy* to illustrate the view and to show that Russell never abandoned it (at least in spirit). We find (*HWP*: 835):

> Intellectually, the effect of mistaken moral considerations upon philosophy has been to impede progress to an extraordinary extent. I do not myself belief that philosophy can either prove or disprove the truth of religious dogmas, but ever since Plato most philosophers have considered it part of their business to produce "proofs" of immortality and the existence of God. … In order to make their proofs seem valid, they have had to falsify logic, to make mathematics mystical, and to pretend that deep-seated prejudices were heaven-sent intuitions. … All this is rejected by the philosophers who make logical analysis the main business of philosophy. … For this renunciation they have been rewarded by the discovery that many questions, formerly

> obscured by the fog of metaphysics, can be answered with precision, and by objective methods which introduce nothing of the philosopher's temperament except the desire to understand. Take such questions as: What is number? What are space and time? What is mind, and what is matter? I do not say that we have here and now give definitive answers to all these ancient questions, but I do say that a method has been discovered by which, as in science, we can make successive approximations to the truth.

Russell's original research program of scientific philosophy is "atomistic" only in the sense that its analytic approach is piecemeal, rejecting metaphysical grandiose systems (Aristotle, Kant, Hegel, Marx, Bradley, Bergson, Spencer, etc.) that are built upon non-logical necessities—pseudo-necessities that in *Problems* and *Our Knowledge* (as well as in "A Free Man's Worship," and "the Essence of Religion") Russell characterized as being tailored to human hopes and aspirations that metaphysical reality should favor good over evil, life over death, and mankind *über alles*. The only *logical* atoms of Russell's original logical atomism are universals which exist independently of being exemplified. But any contingent concrete particulars (such as sense-data), whose existence is warranted *a posteriori* by an empirical science such as physics, can be appealed to as atoms in one's efforts to undermine a metaphysician's argument that distinctive abstract particulars and distinctive necessities governing them are ineliminable. Russell's scientific philosophy aims to dissolve philosophical problems concerning necessity wherever they may lie (in mathematics, physics, ethics, biology, and metaphysics generally). *Principia* showed the way in mathematics, but it takes a good deal of empirical *a posteriori* knowledge in physical science to dissolve indispensability arguments for abstract particulars that metaphysicians impose upon physics, chemistry, biology, psychology, and the like. Russell's scientific philosophy is partly synthetic *a priori* (in using mathematical logic) and partly *a posteriori* (in using current results in empirical science). The new scientific approach affords new hopes for solving *all* philosophical problems—problems arising from metaphysicians offering indispensability arguments for special kinds of necessity and abstract particulars governed by them.

In his paper, "On Scientific Method in Philosophy," Russell explains that "Philosophy is the science of the *possible*" (*SPM*: 111). His student and protégé Ludwig Wittgenstein expressed his *agreement* by proclaiming (*TLP*$_2$: 6.37, 6.375) that logical necessity is the only necessity. Russell eschewed such a characterization because he thought the notion of "necessity" to be itself confused. In any case, both held that it is the business of his scientific philosophy to set matters straight, analyzing away non-logical notions of "necessity" and providing a *privileged* status for mathematical logic. Both Russell and Wittgenstein believed that philosophy must be quite different in kind from the fully empirical sciences of nature. The logical atomist philosopher should be ready on the scene when, and only when, some ineliminability argument is being given for some special non-logical necessity and its abstract particulars. The logical atomist philosopher does not study what there is, but what there *has* to be.

Analysis in Russell's scientific philosophy directs us to find and apply the proper kinds of structures needed to unravel the non-logical pseudo-necessities that generate philosophical problems. Russell's $^{cp}$Logic is informative and its truths include truths *about* relational structures—as are found in the synthetic *a priori* science of mathematics. Russell makes the point salient when he writes: "Philosophy is a study apart from other sciences: its results cannot be established by the other sciences, and conversely must not be such as some other science might conceivably contradict" (*OKEW*: 236). Wittgenstein agreed in saying that "Philosophy is not one of the natural sciences" ($TLP_2$: 4111).

It is quite important to realize that Russell's $^{cp}$Logic does not simply offer a *catalogue* of different linguistic forms for quantification theory with identity. At times, *Our Knowledge* lends itself to such a misinterpretation. Russell writes (*OKEW*: 57):

> Logic, we may say, consists of two parts. The first part investigates what propositions there are and what forms they may have ... The second part merges into pure mathematics, whose propositions all turn out, on analysis, to be such general formal truths. The first part, which merely enumerates forms, is the more difficult, and philosophically the more important; and it is the recent progress in this first part, more than anything else, that has rendered truly scientific discussions of many philosophical problems possible.

Generating a catalog of *logical forms* studied by $^{cp}$Logic would require nothing less than the completion of mathematics itself, since *every* distinct kind of mathematical structure is itself a candidate for a new *logical form*. The notion of finding the correct logical form concerns getting the truth-conditions right and this may well involve significant *a priori* knowledge of mathematical logic as well as significant *a posteriori* knowledge of the given field of study to which mathematical logic is to be applied.

As we can see, the notion of "logical form" involved in Russell's new science of philosophy is not the logical positivist notion of a linguistic form (given by the logical particles of quantification theory with identity) and neither is it the notion of the *form of a fact*. The notion of the *form of a fact* comes largely from Wittgenstein who thought that every theory of *truth* must be impossible. Wittgenstein's Tractarian idea was that in a logically perfect language, an atomic linguistic form *shows* the form of the fact that is its would-be corresponding truth-maker. According to Wittgenstein, the fact that is the truth-bearer and the fact that is its truth-maker, if there is one, have the same logical form. This sameness of form, Wittgenstein tells us, is ineffable and can only be *shown* and not *said*. When his multiple-relation theory of truth collapsed, Russell came to entertain Wittgenstein's notion of the *form of a fact*. But this is a radical change of heart. *Principia's* $^{\varphi}$Logic is about kinds of form—that is, about the kinds of structures that can be exemplified by relations ordering their field. It is not about forms of facts.

Admittedly, this account of the history is out of sorts with the traditional account according to which Russell's logical atomism is an empiricistic program making its first appearance in a series of lectures of 1918 called "The Philosophy of Logical Atomism." This view associates Russell's logical atomism with variants of positions Russell attributed to Wittgenstein, holding Russell hostage to a language whose expressions mirror the forms of the facts, positive, negative, general of their would-be truth-makers. The atoms become sense-data, the epistemology methodologically solipsistic, foundational, phenomenalist. The traditional history wants correction, and the pivotal component of this correction lies in how to properly understand the circumstances of Russell abandoning his 1913 book *Theory of Knowledge*.

It is useful to introduce a division between the acquaintance epistemology of Russell's *Principles* era (1903–1908) and his *Principia* era (1910–1916) and the *noticing* epistemology of his Neutral Monist era (1918–), which was influenced by behaviorist ideas. The *Principles* era is centered on Russell's thesis that a mind may be acquainted with logical abstract particulars that are *propositions*, some among which have the primitive property of being true. The acquaintance epistemology of the *Principia* era abandons propositions and is based on a fundamental relation of acquaintance that a *subject* can have to an object, be it a universal or a transient concrete particular. Brentano's *Principle of Intentionality* is accepted and built into the relation of acquaintance, which is identified with its converse relation presentation and involves a *subject* engaged in an act of selective attention to *this* presented to it. The same point applies to the selective attention of the *subject* involved in the relation of sensation. The relations of acquaintance and sensation are thus central to our empirical knowledge, but it is our acquaintance with universals that is the foundation of our synthetic *a priori* knowledge—our knowledge of the mathematical logic of *Principia*.[1]

The era that followed the *Principia* era is the Neutral Monist era where Russell embarks on a new epistemological tack in sympathy with behaviorism. Brentano's Principle of Intentionality is abandoned. With the *subject* rejected, the relation of acquaintance is replaced by the behavioristically defined *noticing*, which is an "appropriate" reaction, evolved or conditioned, to a stimulus. In his Neutral Monist era, Russell wrote books as replacements for the books that belong to his *Principia* era. There is a nice match up. *Problems* is replaced by *An Outline of Philosophy* (1927). The unfinished *Theory of Knowledge* is replaced by Russell's *The Analysis of Mind* (1921), where thermometers are said to be *sentient* and knowing (even knowing mathematics) consists in appropriate engagements triggered by habits. Russell rewrote *Our Knowledge* twice, altering the original text by addition and subtraction to make later editions compatible with his neutral monism. The rewriting encourages the misleading impression that *Our Knowledge* is out of the *Principia* era and antithetical to *Theory of Knowledge*. In fact, it was originally a part of *Theory of Knowledge* itself. It is, therefore, a very natural sequel to *Problems* and it is best to read only its original edition. Russell's *The Analysis of Matter* (1927) is its neural monist replacement.

The abandonment of *Theory of Knowledge* marks the end of the *Principia era*. It did not come in 1913 and I very much doubt that it came with Wittgenstein's infamous stormy criticisms—criticisms which Russell said he could not understand but left him "paralyzed" because he felt in his bones that he might be onto something. It came well after *Our Knowledge*—and probably not before 1916. The publication of *Our Knowledge* as the sequel of *Problems* is consistent with this. Russell had simply reorganized his research program, publishing early parts of *Theory of Knowledge* on acquaintance in the *Monist*, and publishing the later parts as the book *Our Knowledge*. This, he hoped, would afford time to resolve the difficulty he reached in considering how it is that acquaintance provides the foundation of our knowledge of ᶜᵖLogic itself. In *Theory Knowledge*, Russell had tentatively introduced *logical forms* which are abstract particulars acquaintance with which grounds our understanding of logical quantifiers and particles "and," "or," and "not." This seemed needed to account for the compositionality of quantificational thinking, believing, for example, that $(x)(Fx \vee \sim Gx)$. But the epistemology of acquaintance, so beautifully sketched in *Problems*, collapsed. The question is: *Why*?

The answer is found by understanding Russell's scientific philosophy. The 1918 lectures on the Philosophy of Logical Atomism mark a transition between Russell's *Principia* era and his Neutral Monist era. This is important. The logical atomism which originally belonged to the *Principia* era is properly identified with Russell's science of philosophy. The logical atomism of the Neutral Monist era takes on a quite different epistemological orientation. Russell is in transition and the new behaviorist empirical science of Watson is taking hold on him. He is sympathetic to a behaviorist response to his arguments (published in the *Monist*) that indexicals militate against embracing neutral monism's abandonment of the *subject*. Russell came to believe that behaviorist ideas could undermine his earlier view that the acquaintance relation (or more exactly its converse, the relation of presentation) is precisely what shows that neutral monism's abandonment of the *subject* is mistaken. In *The Analysis of Mind*, Russell goes so far as to imagine the contracting coil of a thermometer to be a "sensation," and *knowing* becomes reacting "appropriately" to an environmental stimulus (*AMi*: 260, 254, respectively). Russell even abandoned his long-held view that universals are themselves individuals. If universals can only occur as exemplified, then they cannot stand as an object to a *subject* in the dyadic relation of acquaintance. In the lectures, Russell categorically denies that the logical particles "and," "or," "not," "all," and "some" indicate any constituents of facts, but he nonetheless entertains general facts, and even negative facts as truth-makers. The 1918 lectures belong to neither era. We must take care not to be misled by the often-tentative statements of Russell's 1918 lectures which do not accurately represent the views of either era. The more coherent positions of his 1924 monograph "Logical Atomism" are better representative of the logical atomism of Russell's Neutral Monist era replacing his fumbling 1918 lectures.

The original logical atomism of the *Principia* era is a program whose agenda is to solve all philosophical problems arising from metaphysical muddles over the nature of abstract particulars and their specialized kinds of necessity. Metaphysicians introduce special kinds of necessity (e.g., mathematical, physical, biological, ethical, metaphysical) and abstract particulars grounding them. In the case of mathematics, the metaphysicians invoke special kinds of necessity governing abstract particulars that are geometric figures and abstract particulars that are numbers (natural, rational, real, imaginary, etc.) or classes/sets. When the non-Euclidean geometers found that it is not a truth of logic that right triangles obey the Pythagorean

Theorem, the metaphysician demands that we accept the abstract particular that is the Euclidean triangle and a special geometric necessity securing its Pythagorean nature. *Principia's* agenda was to reveal that there are no abstract particulars in any branch of mathematics. The branches are studies of relational structures. The only necessity in mathematics is that of logical necessity. Russell's scientific philosophy extends the result. It endeavors to analyze away any kind of metaphysical necessity, using the $^{cp}$Logic of *Principia* together with the best empirical science to reveal that the only necessity is logical necessity. For this to be viable as a research program, Russell's scientific philosophy requires both that $^{cp}$Logic and its epistemology avoid abstract particulars. Only then can it deserve a lofty status above every metaphysics of abstract particulars.

It is possible that Russell had hoped Wittgenstein might make progress in avoiding the *logical forms* of his *Theory of Knowledge*—that is, the abstract particulars acquaintance with which grounds our understanding of logic. Wittgenstein had sharply criticized Russell's *Theory of Knowledge* on just this issue. In *Our Knowledge*, there is a wonderful passage that corroborates this view (*OKEW*: 208):

> If the theory that classes are merely symbolic is accepted, it follows that numbers are not actual entities ... This is in fact the case with all the apparent objects of logic and mathematics. Such words as *or*, *not*, *if*, *there is*, *identity*, *greater*, *plus*, *nothing*, *everything*, *function*, and so on, are not names of definite objects ... All of them are *formal* ... This fact has a very important bearing on all logic and philosophy, since it shows how they differ from the special sciences.

In a footnote, Russell remarks:

> In the above remarks, I am making use of unpublished work by my friend Ludwig Wittgenstein.

In point of fact, however, the only addition to the list that is not *clearly* due to Russell himself is that of *identity*, which Wittgenstein hoped to have been the central achievement of his *Tractarian* reorienting of the foundation of arithmetic *and* logic in terms of the combinatorics of recursive functions. In *Wittgenstein's Apprenticeship with Russell (2007)*, I have argued in detail that Wittgenstein's 1913 *Notes on Logic* and letters reveal that his *Tractarian* Doctrine of Showing began in 1913 with an earnest quest to help Russell in his endeavor to find an epistemology for mathematical logic that preserves its privileged status as the essence of the unique science of philosophy. But the extreme demands of *Showing* vs *Saying* require that mathematics and logic not consist of bodies of truths, but of practices of elucidation. Wittgenstein had remarked to Russell in a 1913 letter from Norway that "identity is the very Devil and *immensely important*" and must be eliminated.[2] Looking back, Russell recalls in *My Philosophical Development* that for a short time he had been convinced by Wittgenstein's concern to eliminate *identity* but he soon rejected the thesis since the implications of the elimination of *identity* make mathematical logic altogether impossible (*MPD*: 115*ff*). Sheffer (1913) and Wittgenstein were both working on a system of logic with only one primitive, so Russell may have had their work in mind too. But the key point is that in the passage Russell remarks that the "apparent objects" of logic and mathematics are not actual entities. These are the halcyon days of Russell's original logical atomism.

Russell's behavioristic logical atomism of the 1920's and after can hardly sustain his thesis that logic is the essence of philosophy. As Dewey was in earnest to reveal, the story of the evolution of cognitive behavior protocols by natural selection changes the very *contents* of logical and mathematical studies. Russell resisted, but he never could recover from Dewey's point. Russell's original logical atomism is viable, only if we return to its original rationalist theory that acquaintance with universals is the foundation of our synthetic *a priori* knowledge of mathematical logic.

In what follows, I shall sketch a solution to the impasse Russell encountered in his unfinished book *Theory of Knowledge*. That is, I shall answer the question as to how Russell's acquaintance epistemology can ground our understanding of logic. The repair opens the way for a return to Russell's original Logical Atomism—a form that need not reject non-logical alethic modality and "rival" logics popular in contemporary philosophy. These kinds of "necessities" can be subsumed into logic just as readily as were the kinds of *faux* "necessities" of non-Euclidean geometries to be subsumed into *Principia's* logicism. They are merely studies conducted *within Principia* of different structures. We shall find that Lewis's counterpart theory (though not his ontological commitments) are the very thing that shows the way!

## 2 *Theory of Knowledge* Solved the Direction Problem

It is not well understood that Russell's definition of "truth" in terms of correspondence with fact was to have been a direct application of his theory of definite descriptions. This is hinted at in a footnote in "On the Nature of Truth" (1906) where we find (*ONT*: 453):

> This is an extension of the principle applied in my article, "On Denoting" (Mind, October, 1905), where it is pointed out that such propositions as "the King of France is bald" contain no constituent corresponding to the phrase "the King of France."

The connection between "truth" as correspondence and the theory of descriptions is alluded to in *Principia* (*$PM_1$*, Vol. 1: 44), and stated rather explicitly in *Theory of Knowledge* (*TK*: 147). It is also stated in "On Propositions: What They Are and How They Mean," where Russell proclaims without argument that one cannot regard the fact that exists if a propositions is true as the "meaning" of the proposition because "… this assimilates propositions too much to names and descriptions" (*OP*: 302). There is a wonderful passage in *Human Knowledge: Its Scope and Limits* (1948) that nicely characterizes the matter. Russell writes (*HK*: 149):

> The difference between a true and a false belief is like that between a wife and a spinster: in the case of a true belief there is a fact to which it has a certain relation, but in the case of a false belief there is no such fact. To complete our definition of "truth" and "falsehood" we need a description of the fact which would make a given belief true, this description being one which applies to nothing if the belief is false. Given a woman of whom we do not know whether she is married or not, we can frame a description which will apply to her husband if she has one, and to nothing if she is a spinster. … In like manner we want a description of the fact or facts which, if they exist make a belief true. Such fact or facts I call the "verifier" of the belief.

This is striking because it is a return to the recursive definition of "truth" as correspondence first set out in *Principia*. The base case of that recursive definition involved Russell's infamous multiple-relation theory.

In 1918 logical atomism lectures, Russell had abandoned the multiple-relation theory and entertained a Wittgenstein-inspired tracking theory, which, it would seem, abandoned the recursive definition. Russell wrote (*PLA*: 168):

> … There are two different relations … that a proposition may have to a fact: the one relation that you may call being true to the fact, and the other being false to the fact. Both re equally essentially logical relations which may subsist between the two, whereas in the case of a name there is only one relation it can have to what it names.

This Wittgensteinean idea proved to be a will'o-the-wisp that led Russell, though perhaps not Wittgenstein himself, to entertain (in diametric opposition to *Principia's* recursive definition of "truth") negative facts and general facts as truth-makers. In *An Inquiry Into Meaning and Truth* (1940) and in *Human Knowledge*, Russell reversed direction again. He rejects negative facts and general facts as truth-makers. Echoing *Principia*, he writes that there is no one verifier for "All men are mortal" (*INT*: 256). Looking back, in *My Philosophical Development*, Russell explains as follows (*MPD*: 186):

> It is only the simpler sort of statement that has a single verifier; the statement "all men are mortal" has as many verifiers as there are men.

And Russell continues: (*MPD*: 188):

> The complexity of the correspondence grows greater with the introduction of logical words such as "or" and "not" and "all" and "some."

Russell never explains how or whether he overcame the difficulties that beset the multiple-relation theory that formed the base case of just this recursive correspondence theory when he tried to develop it in *Theory of Knowledge*. But help comes if we realize that the problems Russell had encountered were not about what facts as truth-makers are involved where molecular and quantified belief are concerned. His problems concerned questions about what the constituents of the facts are that are *truth-bearers*.

In *Theory of Knowledge*, Russell had solved the problem of forming definite descriptions for permutative facts—that is, facts such that other facts can be formed from the same constituents. The solution was his theory of acquaintance with position relations. Russell explains: (*TK*: 146):

> We may now generalize this solution, without any essential change. Let $\gamma$ be a complex whose constituents are $x_1, x_2, \ldots, x_n$ and a relating relation R. Then each of these constituents has a certain relation to the complex. We may omit the consideration of R, which obviously has a peculiar position. The relations of $x_1, x_2, \ldots, x_n$ to $\gamma$ are their "positions" in the complex; let us call them $C_1, C_2, \ldots, C_n$.
>
> …. Thus our complex $\gamma$ can be described, unambiguously without mentioning R, as simply "the complex $\gamma$ in which $x_1 C_1 \gamma$, $x_2 C_2 \gamma$, …, and $x_n C_n \gamma$."

Russell goes on to offer the symbolic expression as well, namely:

$$(\iota\gamma)(x_1 C_1 \gamma \bullet x_2 C_2 \gamma, \ldots, \bullet x_n C_n \gamma).$$

This is a definite description that purports to refer to a permutative fact $\gamma$. It offers a definite description by appeal to the existence of associated "complexes," namely

$$x_1 C_1 \gamma, x_2 C_2 \gamma, \ldots, x_n C_n \gamma.$$

This solves the problem of finding a definite description for a permutative fact by having the description concern position relations and the existence of associated facts that are not permutative and which involve objects having positions in $\gamma$. It is clear, therefore, that the direction problem is the problem of how to form a definite description of a fact in cases where the definite description purports to refer to a permutative fact.

Let us take a specific case. Consider the following definite description which purports to single out a unique fact of Desdemona's loving Othello. We need not neglect the fact that for Russell both "Desdemona" and "Othello" themselves are definite descriptions. We have:

$$(\exists d)\Big(Dy \equiv_y y = d . \& . (\exists o)\Big(Oy \equiv_y y = o . \& . E!(\iota f)\Big(d\, C_1^{L^2} f \bullet o\, C_2^{L^2} f\Big)\Big)\Big).$$

The structure of the purported fact is characterized by the definite description

$$(\iota f)\Big(d\, C_1^{L^2} f \bullet o\, C_2^{L^2} f\Big)$$

in virtue of the position relations $C_1^{L^2}$ and $C_2^{L^2}$ generated by its relating relation "love." Compare the following definite description:

$$(\iota f)\Big(o\, C_1^{L^2} f \bullet d\, C_2^{L^2} f\Big).$$

This definite description purports to single out the fact of Othello's loving Desdemona. I do not know if there is such a fact, since after all he murders her out of jealousy. The relation loves nonetheless determines the position relations $C_1^{L^2}$ and $C_2^{L^2}$ independently of whether there is any such fact. Given there is a fact which I conveniently name "$o - L^2 - d$," we have both:

$$o\, C_1^{L^2}\, o - L^2 - d$$

$$dC_2^{L^2}\ o - L^2 - d.$$

As we can see, position relations do not occur in the fact $o - L^2 - d$. But if there is a fact of Othello's loving Desdemona, then there is a fact of Othello's bearing a position relation to that fact and also a fact of Desdemona's bearing a position relation to that fact.

Russell's solution relies, of course, on it being the case that the facts unified by the position relations in question are ***not*** themselves permutative. For example, Russell needs to be able to establish that the fact

$$d - C_1^{L^2} - \left(d - L^2 - o\right)$$

is ***not*** permutative. This raises an interesting question as to whether the notion of a fact being ***permutative*** must be defined and if so whether it is a logical matter that a given fact is, or is not, permutative. Moreover, we are inclined to say that a fact is permutative when it is *possible* for another fact to be composed of its constituents. But what is this notion of "possibility"? Is it a logical notion or some other?

My answer is that it is *not*, at least not always, an issue for logic to determine whether or not a given fact is permutative. Perhaps the reason we are so easily led into thinking that it is logical issue is because it *is* a logical issue whether or not a given relation is a *p*-relation (a relation that determines position relations). That is, a relation cannot be accidently a *p*-relation. Indeed, ***being a p-relation*** is not properly understood as a property of a relation. It is, for example, logically necessary for the relation "loves" to be a *p*-relation. Moreover, if we are acquainted with a given relation, we understand immediately not only its ***adicity*** but also whether or not that relation is a *p*-relation. Acquaintance with the relation grounds our understanding of what fact would exist were the relation to be exemplified, and this may well be independent of whether the relation is exemplified. Be this as it may, it is not always a matter for logic to determine whether a given fact is or is not permutative. Acquaintance with the *p*-relation loves reveals that the position relations it determines are not themselves *p*-relations, while logic does not reveal any such thing. This is of utmost importance. The quest for a general definition of what it is for a relation to be a *p*-relation in terms of the ***logical possibility*** of a fact being permutative is hopeless for Russell. Clearly, his analysis of the notion of (logically) ***possible truth*** as applied to first-order *wffs* in term of full second-order existential closure and truth cannot provide it. (See Sect. 4 below.) But no matter.

What has to be understood in the case of the relation loves is modest, namely,

$$\sim\left(\left(d - L^2 - o\right)C_1^{L^2}\, d\right).$$

This is easy to understand since Desdemona is not herself a fact.

There is, however, an interesting argument to the result $\sim\left(\left(d - L^2 - o\right)C_1^{L^2}\, d\right)$ and the argument is generalizable, holding not just for the *p*-relation loves and any two people that may exemplify it but holding for every *p*-relation. Consider what happens if we were to have:

$$\left(a - R^2 - b\right)C_1^{R^2} a.$$

If this were the case, then for some *x* and *y*, we have

$$a = x - R^2 - y.$$

Hence, we have

$$\left(a - R^2 - b\right)C_1^{R^2}\left(x - R^2 - y\right).$$

But this implies

$$\left(a - R^2 - b\right) = x.$$

Thus, for some *x* and *y*, we have:

$$a = \left(a - R^2 - b\right) - R^2 - y$$

$$\left(\left(x - R^2 - y\right) - R^2 - b\right) = x.$$

We have an entity *a* that is equal to a fact that contains the entity *a* as one among its distinct constituents. This is an extremely untoward consequence that, in making definite descriptions of permutative facts we often know, does not occur. Indeed, it may well never happen (though it seems not to be logically impossible). Russell's definite description for a permutative fact will be adequate for any such case.

## 3 The Real Difficulty of *Theory of Knowledge*: Compositionality

Intentionality is presupposed in thinking by means of the quantificational apparatus of definite descriptions. That the theory of descriptions is directly involved in Russell's solution of the direction problem reveals clearly enough that his intent was not to explain the *nature* of intentionality. Belief-facts as truth-bears are not about anything. Intentionality is not to be found by an examination of a belief-fact. What then makes a belief-fact truth-apt (appropriate as a truth-bearer)? There is no special *unity* that makes a belief-fact truth-apt, so that it is *about* (or *points to*, or *represents*) its would-be truth-maker. No fact is about anything. A belief-fact, one of whose relata is a *subject*, is truth-apt because it is an artifact of the intentionality of a mind engaged in believing. Hence, the problem facing Russell's multiple-relation theory is not the problem of philosophically explaining the *nature* of intentionality and representation and truth-aptness. The problem of direction concerns the determinate directedness of intentionality, not its *nature*. That problem was addressed, and solved, in *Theory of Knowledge*.

What then created the impasse which shelved the acquaintance epistemology of *Theory of Knowledge*? Russell explains (*TK*: 154):

> Where permutative complexes are concerned, our process of obtaining associated non-permutative complexes was rather elaborate, and no doubt open to objection. One special objection is that, in order to regard the associated complex as non-permutative, we have to regard its atomic constituents, $x_1 C_1 \gamma$, $x_2 C_2 \gamma$, etc., as really its constituents, and what is more we have to regard the corresponding propositions as constituents of the proposition "there is a complex $\gamma$ such that $x_1 C_1 \gamma$, $x_2 C_2 \gamma$, etc." This seems to demand a mode of analyzing molecular propositions which required the admission that they may contain false atomic propositions as constituent, and therefore to demand that admission of false propositions in an objective sense. This is a real difficulty, but as it belongs to the theory of molecular propositions we will not consider it further at present.

As I see it, the "real difficulty," as Russell puts it, concerns compositionality. A recursive definition of a *wff* using the logical signs "•," "~," and "∨" together with variables and quantifiers defines a potential infinity of syntactically distinct expressions. Minds can (at least in principle) understand any among them. Variable binding introduces complicated syntactic constructions as well. Without such compositionality, cognition would be impossible. A theory of the truth-bearers involved in general and molecular belief must accommodate compositionality. But how? What belief-fact is the truth-bearer when a mind believes that there is a complex $\gamma$ such that $x_1 C_1 \gamma$, $x_2 C_2 \gamma$, etc.? Even in cases of logical equivalents, the distinctions must be respected. One cannot hold, for example, that believing that $(x)Fx \bullet (y)Gy$ is the same as believing that $(x)(\sim\sim Fx \bullet Gx)$. Russell could not find a way forward. He gave up—abandoning *Theory of Knowledge* and its multiple-relation theory. He never found a replacement in his appeal to Neutral Monism, with or without its alliance with behaviorism.

A repair on behalf of *Theory of Knowledge* is needed. I offer a repair by first denying the existence of singular (indexical) belief and denying the existence of molecular quantifier-free belief. All belief (and generally all discursive thinking) is quantificational. Indexicals are never involved in soliloquy, but play an essential role in communication. For that, the Russell-Reichenbach token-reflexive theory is certainly adequate. Second, I use the ideas of Quine's paper "Variables Explained Away" to help find the belief-facts that are to play the role of truth-bearers for Russell's multiple-relation theory. Consider $M$ believing that $(\exists y)(x)(R^2xy \vee S^2xy)$. Transform the *wff* using our Quinean technique as follows:

$$m - B^7 - \begin{cases} \exists \\ \forall \\ ijp \ltimes T^4 \\ \vee \\ R^2 \\ S^2 \end{cases}$$

$$(\exists y)(x)\left(R^2xy \vee S^2xy\right)$$

$$(\exists y)(x)\left(\vee R^2S^2\right)(xyxy)$$

$$(\exists y)(x)\left(\ltimes \vee R^2S^2\right)(xyyx)$$

$$(\exists y)(x)\left(p \ltimes \vee R^2S^2\right)(yyxx)$$

$$(\exists y)(x)\left(ijp \ltimes \vee R^2S^2\right)(yx)$$

$$\left(\forall ijp \ltimes \vee R^2S^2\right)y$$

$$\exists\forall ijp \ltimes \vee R^2S^2$$

There is no need to be concerned about the transforms:

$$R^2 xy \vee S^2 xy$$

$$(\vee R^2 S^2)(xyxy)$$

$$(ijp \ltimes \vee R^2 S^2)(yx)$$

These are just converses of a relation, as when we move from "*x* gives *y* to *z*" to "*y* was given by *x* to *z*" and to "*z* received gift *y* from *x*" and so forth. They are readily presented to anyone who understands the relation in question. By sequencing and identifying variables in this way, we can find the belief-fact that is the truth-bearer. The belief-fact is, of course, permutative. But there is no difficulty in finding a definite description for it using the technique of *Theory of Knowledge*, which appeals to position relations. A definite description for it is:

$$(\iota f)\left(m C_1^{B^7} f \bullet \exists C_2^{B^7} f \bullet \forall C_3^{B^7} f \bullet ijp \ltimes^4 C_4^{B^7} f \bullet \vee C_5^{B^7} R^2 C_6^{B^7} f \bullet S^2 C_7^{B^7} f\right).$$

The belief-fact is true if and only if

$$(\exists y)(x)\left(E!(\iota f)\left(x C_1^{R^2} f \bullet y C_2^{R^2} f\right) \vee E!(\iota f)\left(x C_1^{S^2} f \bullet y C_2^{S^2} f\right)\right).$$

This accords with the recursive definition of "truth" in *Principia*.

Finding a consecutive ordering of units for the syntactic composition of *wffs*, however, is only part of our solution to the problem of compositionality. We have to face the problem of what, ontologically, these separate units are. They cannot be universals, for then they would inhere in facts as truth-makers for the belief-facts and, according to *Principia*, the truth-conditions are to be recursive. A quantified statement is not made true by one fact, but by perhaps as many as there are values of the variables. They cannot be abstract particulars, since Russell's research program for philosophy as a unique science would collapse into metaphysics (on a par with rival theories of abstract particulars and their specialized necessities). We cannot move *L-forms* to the belief-relation as follows:

$$m - \left(\exists \forall \delta \vee \varphi^2 \psi^2\right) B^3 - \begin{cases} R^2 \\ S^2 \end{cases}.$$

This would reintroduce the direction problem!

A good part of the solution of this vexing problem is found by applying Quine's technique to the case of binding predicate variables. Not only does this eliminate predicate variables, it shows the way for the elimination of *L-forms* as abstract particulars of acquaintance. Consider again, *M* believes that $(\exists y)(x)(R^2xy \vee S^2xy)$. This time let us offer our version of Quine's methods to treat $R^2$ and $S^2$ as bindable predicate variables. We have:

$$m-\left(\exists\forall\delta \vee e^3e^3\right)B^3-\begin{cases}R^2\\ S^2\end{cases}.$$

$$\left(\exists y\right)\left(x\right)\left(R^2xy \vee S^2xy\right)$$

$$\left(\exists y\right)\left(x\right)\left(\left(\vee e^3e^3\right)\left(R^2, S^2, x, y, x, y\right)\right)$$

$$\left(\exists y\right)\left(x\right)\left(\left(\delta \vee e^3e^3\right)\left(R^2, S^2, y, x\right)\right)$$

$$\left(\exists y\right)\left(\forall\delta \vee e^3e^3\left(R^2, S^2, y\right)\right)$$

$$\left(\exists\forall\delta \vee e^3e^3\right)\left(R^2, S^2\right)$$

This removes the *L-forms* to the belief-relation itself. Hence, they are not to be construed (as Russell had imagined in *Theory of Knowledge*) as abstract particulars. It is *innate* acquaintance with the multiple "belief" relation(s) *themselves* that leads us to imagine separable *L-forms.* In truth, they are inseparable from one another and show up as features of the different belief-relations involved when a mind generates a belief-fact as an artifact of believing.

When a mind *M* believes a logical truth such as $(R)(x)(R^2xx \vee {\sim}R^2xx)$, the belief-relation (as it were) is involved is monadic! That is, we have the following:

$$m-\left(\forall\forall ji^3p\S p \vee e^3\overline{e}^3\right)B.$$

In the case of *M* believing $(R)(x)(R^2xx \vee {\sim}R^2xx)$, we may speak of *M* believing a logical truth. But this way of speaking gives rise to the misleading impression that there is a logical truth (a logical necessary abstract fact)—the truth-maker for his belief—about which *M* believes. This commits a Meinongian fallacy. Abstract logical facts, though they ground logic, are not truth-makers for belief-facts.

## 4 Logical Atomism's Necessary Truth

Russell's *Theory of Knowledge* enables the separation of the question as to what is the ontological *ground* for mathematical logic from the question as to what truth-makers of the truth-bearers are produced when an understanding of the *wffs* of mathematical logic generates belief-facts. Truth is a matter of correspondence relations that depend on the truth-bearers (e.g., belief-facts) that are brought into existence by minds. There is nothing in Russell's multiple-relation theory that offers any help in ascertaining ontological ground(s) of logic. This is as it should be.

When logical necessity concerns the ground of $^{cp}$Logic it comes apart from *truth*.

The facts that ground logical necessity are rarely truth-makers. What truth-makers there are depends on what truth-bearers there are, and what truth-bearers there are depends on what universals a mind may be acquainted with. Only those facts whose universals are objects of acquaintance are truth-makers, and acquaintance with a universal is never assured by acquaintance with the *wff* that renders its exemplification conditions. Now Russell's original logical atomism, the atomism of his *Principia* era, accepts *Principia's* simple-type theory which embodies impredicative comprehension axiom schemas[3]:

$$*12.1(\exists\psi)(x)(\psi!x \equiv \varphi x)$$
$$*12.11(\exists\psi)(x, y)(\psi!xy \equiv \varphi xy).$$

In spite of Russell's many attempts to *emulate* simple (impredicative) comprehension without a blanket ontological commitment to simple types of universals in intention, we must bite the bullet here and embrace a realism of simple types of attributes (universals) in intension. We must, however, form something of a compromise with Russell's struggle to emulate impredicative comprehension. Realism does not require that we hold that we have acquaintance with the universals that we understand to exist by understanding instances of *Principia's* impredicative comprehension axiom schemas. We understand such universals by understanding the *wffs* that render their exemplification conditions, and this in turn requires our use of the *L-forms* that supply our understanding of logic. Acquaintance with a universal assures that the universal is simple-type free. Understanding that a universal exists via our understanding the *wff* that renders its exemplification conditions does not.

Logical necessity, on this view, is not an *invariance* notion and cannot be understood in terms of logically necessary *truth*. For example, that the relation of "identity" is logically symmetric is grounded in the fact *Sym*(=) of the relation identity exemplifying the relation of *being symmetric*. This fact, however, is never a truth-maker for any truth-bearer. The reason is that our minds are not acquainted with the relation *Sym* of symmetry. We understand it via our understanding of the quantification involved in understanding the following instance of comprehension:

$$(\exists S)(R)\big(S(R) \equiv (x,y)(Rxy \supset Ryx)\big).$$

The ontological ground of *Principia's* ᶜᵖLogic is not to be found by any consideration of *truth*.

It is clear from Russell's papers "Necessity and Possibility" (1905), "On the Notion of Cause," (1913), and the logical atomism lectures (1918) that Russell holds that *logical necessary truth* (at least restricted to first-order *wffs*) consists of truth together with full universal generality in the language of *Principia*. Consider the following:

> It is a logically necessary truth that if the inventor of bifocals exists, then he invented bifocals.

> *i.e.*, $\Box\ \big(E!(\iota x I x) \supset [\iota x I x][Ix]\big)$,
> *i.e.*, $(\psi)\big(E!(\iota x \psi!x) \supset [\iota x \psi!x][\psi!x]\big)$, by Russel's transformation:

Where first-order *wffs* are concerned, and where *Principia's* comprehension principles are accepted, Russell's conception of logically necessary truth can be shown to coincide with Tarski's semantic definition of logical truth. Tarski's semantic definition finds "$E!(\iota x I x) \supset [\iota x I x][Ix]$" to be logically true since it is true in every interpretation of "*I*" over any non-empty domain. Clearly, for every such Tarski interpretation of *I*, *Principia's* comprehension assures that there is a property $\psi!$.

It should be observed that a *de re* notion of logically necessary truth (and logically possible truth) is entirely out of sorts with Russell's thesis that these notions, as with mathematics generally, concern structures given by the way relations order their fields (independently of whether the relations are exemplified). Consider the formula:

$$[\iota x I x][\sim\Box I x]$$

This purports to say that some unique inventor of bifocals is such that it is not a necessary truth that he invented bifocals. Clearly the notion of "necessity" here is *not* a notion of logical necessity concerning structure. It smuggles in a quite different notion of "necessity" with perhaps a metaphysical or causal or biological or etc., connotation. The expression "$[\iota x I x]$ $[\sim\Box I x]$" is syntactically horrifying to a Russellian because it seems incompatible with the logical atomist thesis that logical necessity is the only necessity and that logically necessary truth concerns structure. Interestingly, one can rectify matters *semantically*. This is the remarkable technique of Nino Cocchiarella (1975) who shows that if every model is an *anti-essentialist model*, Russell's stand concerning logically necessary truth for first-order *wffs* is vindicated and it semantically coincides with Tarski's notion of logical truth. In such models, *de re* and *de dicto* semantically collapse.

Russell's scientific philosophy—his *original* logical atomism—has been thought to be the enemy of non-logical alethic *de re* modality. And so it is indeed the enemy of metaphysically possible world approaches to logical necessity. This has privileged status as the essence of philosophy itself and it is not an invariance notion. But Russell's original logical atomism, which couched *Principia's* logic as its essence, can accommodate rival "logics" (alethic modal, relevant, intuitionistic, etc.) precisely because they can be considered invariance notions which can be captured quantificationally as studies of different structures. In virtue of this, the *L-forms* that provide our *a priori* grasp of quantification also provide for our understanding of all rival "logics." The logical atomist can explore the different alethic modal system offering *faux* necessities akin to those appealed to in geometries (Euclidian vs non-Euclidean).

There is *only* the cognitive apparatus of *L-forms* which are lurking behind all discursive—that is, quantificational—thinking. There is no meaning to the question of whether quantificational thinking is "classical." All the different formal studies of structure that parade themselves as if they were rival non-logical necessities and "logics" (relevant logics, paraconsistent logics, modal logics, etc.) can now come under the same cognitive roof—the same *quantificational* cognitive apparatus that enables our understanding of *Principia*. Many of the transcriptions are quite familiar. We find:

$$(\Box p)^{w} = df\,(\alpha)(wR\alpha \supset p^{w}) \quad ((x)\varphi x)^{w} = df\,(x)(xIw \supset (\varphi x)^{w})$$
$$(\Diamond p)^{w} = df\,(\exists\alpha)(wR\alpha \bullet p^{w}) \quad (p \vee q)^{w} = df\; p^{w} \vee q^{w}$$
$$(\sim p)^{w} = df \sim (p^{w})$$

It then is easy to prove results *a priori* such as

$$\text{Reflex}(\text{R}) \vdash \Box p \supset p$$
$$\text{Trans}(\text{R}) \vdash \Box p \supset \Box\Box p$$
$$\text{Trans}(\text{R}), \text{Symm}(\text{R}) \vdash <> p \supset \Box <> p$$

All this is known *a priori* since quantification theory is known *a priori*. The proofs involve the infamous rule of necessitation:

$$\text{From} \vdash p \text{ infer} \vdash \Box p.$$

But the rule of necessitation is just the rule of universal generalization, once the non-logical alethic □ is understood as a universal quantifier. It is therefore knowable *a priori* just as surely as is the rule of universal generalization. We can go on to accept David Lewis's approach to *were-would* subjunctives:

$$(p \Box\!\!\rightarrow q)^{i} = df$$
$$(\exists w)(p^{w} \text{ is accessible to } i) \supset (\exists w)\left((p \bullet q)^{w}\right) \bullet (w')\left((p \bullet \sim q)^{w'} \supset w \leq_{i} w'\right).$$

Such logical forms are well *within* the original Russellian logical atomist scientific philosophy (though their common modern interpretations are not.)

Of course, Russell's science of philosophy requires that the notation "$p^{w}$" be also eliminated. Russell's science of philosophy cannot accept "proposition *p* obtains *at* world *w*." Russell rejects propositions altogether. David Lewis (1968) has shown the way to find an analysis congenial to Russellians. He offers the following:

$$(\Box Fx)^{w} = df\; xIw \bullet (\alpha)(wR\alpha \,.\supset.\, (y)(yCx \bullet yI\alpha \,.\supset.\, Fy))$$
$$(\Diamond Fx)^{w} = df\; xIw \bullet (\exists\alpha)(wR\alpha \bullet (\exists y)(yCx \bullet yI\alpha \,.\bullet.\, Fy)).$$

Lewis's logical forms are congenial to Russellian atomists precisely because they eliminate "$p^w$." There has been some concern that Lewis's counterpart transcriptions are inadequate to modal statements such as "All who are rich might have been poor." Clearly, it won't do to put:

$$(x)(Rx \supset <> Px).$$

This is inadequate since the original is to be interpreted not to mean that for each person who is rich there is some world in which that person is poor, but rather that all who are rich might have been poor *together*—that is, in the same possible world. One might naturally try to capture it with

$$(x_1,\ldots,x_n)((Rx_1\bullet,\ldots,\bullet Rx_n) \supset \Diamond(Px_1\bullet,\ldots,\bullet Px_n)).$$

This works since there are naturally only finitely many of those who are rich. But in cases where such an assumption cannot be made, there is still no problem. The Russellian logical atomists with *Principia* at their side need only to bind predicate variables and capture this *non-logical* notion of necessity with the following:

$$(\exists F)((z)(Fz \equiv Rz)\bullet(z)(<> Fz \supset \Box Fz).\bullet. <> (x)(Fx \supset Px)).$$

The clause $(z)(<>Fz \supset \Box Fz)$ assures that the property $F$ is *rigid*, keeping the same extension in all worlds.

Confronted with a desire to avoid such predicate quantification, however, some think we are compelled to add an operator $@[p]$ to our first-order quantified modal language and then write

$$\Diamond(x)(@[Rx] \supset Px).$$

Some objectors to Lewis's counterparts have appealed to a need for an actuality operator in quantified modal logic, pointing out that any such *actuality* operator is out of sorts with there being a general recipe for a Lewisian counterpart transcription.[4] This metaphysical contention, as we can readily see from the above perspective afforded by *Principia's* logic which binds predicate variables, is a result of an impoverished conception of logical form (as if logic must be limited to a first-order quantification theory)—just as Russell's atomism predicts. The baroqueness of a metaphysical ontology is inversely proportional to the comprehensiveness of the logical forms of one's mathematical logic.

Admittedly, it remains an ongoing research project to find interpretations of the Lewisian counterpart logical forms that are consistent with Russell's robust sense of reality. But that project works entirely *within* Russell's original research program of scientific philosophy—his original logical atomism. Lewisian counterpart relation is a relation of comparative similarity and a world might be construed as simply a relevant similarity class. When the parameters of "similarity" are constrained by specific conceptions of necessity (e.g., physical, biological, psychological, metaphysical), one can investigate the relevant constraints on similarity. Unfortunately, ignorance in the field reigns and this ignorance is what makes it seem as if truth-conditions for such claims of non-logical necessity and possibility (as invariance notions) require worlds and entities. In investigating whether, say, a three-tailed kangaroo is biologically possible, the relevant counterparts are not kangaroos in possible worlds, but may well be genes all of which are active in actual organisms of an appropriate relevance class (i.e., a three-tailed kangaroo's being biologically possible may have its truth-conditions in its relevant similarity to a three-tailed lizard, or a three-toed bird). Different kinds of considerations of relevance apply to different conceptions of non-logical necessity, but the appeal to the logical forms of Lewis's counterparts and relevance is clearly the way forward for a Russellian atomist. This is a bare sketch of a research method, but there is little space to do more in this chapter than offer hints. For the present, let us go on to illustrate how one may generalize the philosophical method of appropriating the semantics as the syntax for analyzing rival "logics" from within Russell's *original* logical atomism.

Quite obviously, logical atomism can study alethic modal structures. Intuitionistic "logic" readily succumbs to the method. That is because intuitionistic logic is isomorphic to S4 modal logic. Thus, intuitionism is not a rival logic. It can be wholly subsumed and studied within *Principia* as one among kinds of structures. It is no more a rival than the study of a Euclidean metric is a rival of the study of a Riemannian. *Principia's* $^{cp}$Logic readily affords the study of both.

Relevant "logic" is more difficult because its complete analysis involves us in the philosophy of mind itself. It involves the philosophy of mind because the natural interpretation is that it takes $(p)^x$ to mean that $p$ is information at information state $x$. An information state may well be inconsistent and incomplete. No such information state is a possible world, since by definition a "possible world" is consistent and complete. The notion of "information" and an "information state" concerns intentionality of mind. The philosophy of mind (which engages with empirical fields such as cognitive psychology and neuroscience) is not sufficiently developed to enable logical atomistic constructions. But the Routley-Meyer semantics take us most of the way toward a logical atomist analysis of relevant informational structures. We have:

$$\left(A\vee^{r}B\right)^{z}=df\ A^{z}\vee B^{z}$$

$$\left(A\bullet^{r}B\right)^{z}=df\ A^{z}\bullet B^{z}$$

$$\left(A\supset^{r}B\right)^{z}=df\left(\forall x,y\right)\left(Rzxy\,.\supset.A^{x}\supset B^{y}\right)$$

$$\left(\sim^{r}A\right)^{z}=\sim\left(A^{z^{*}}\right)$$

Of course, for relevant entailment, one needs different axioms governing its triadic accessibility relation $R$ for different systems.[5] Now in relevant entailment, we get both:

$$\sim^{r}\sim^{r}A\supset^{r}A$$

$$A\supset^{r}\sim^{r}\sim^{r}A.$$

But without the classical axiom $(x)Roox$, we do not get:

$$A\bullet^{r}\sim^{r}A\,.\supset^{r}.\,B$$

$$A\vee^{r}B\,.\supset^{r}.\,\sim^{r}A\supset^{r}B.$$

While ***relevant disjunctive addition*** is innocuous, so that from $A$ we can get $A \vee^r B$, the *wff* $B$ might be quite irrelevant to the premise. Hence, relevant logicians reject

$$A\vee^{r}B,\sim^{r}A\vdash B.$$

But this is not incompatible with accepting a classical disjunctive argument as applied to information states:

$$(A \vee B)^x, \sim (A^x) \vdash B^x.$$

This should be of no surprise whatever.

The correct thing to say in response to the relevant logician or the intuitionistic logician or any mathematical logician is *not* that *Principia's* classical "⊃" and "~" are the *correct* logical connectives characterizing the *correct* relational structure. The notion of a *correct* relational structure is misguided. There can only be correctness or incorrectness with respect to the application of a given structure. (Euclidean geometry is not "correct," while a non-Euclidean "incorrect." But it might be correct as applied to a physical local spatial metric.) The proper thing to say is that there is no conflict between relevant "logic" and classical "logic," and intuitionistic "logic" and any other "logic." *None* are logic. The analogy with geometry is apt. They are each important studies of distinct structures, just as are Euclidian metrics *vs* non-Euclidean. All can be fully understood and studied by the same quantificational cognitive apparatus of *L-forms* that enables our understanding of the *wffs* of *Principia's* $^{\text{cp}}$Logic. In introducing our Quinean apparatus for quantification, inference cognitively proceeds without any variables, bound or free. There is no cognitive analog of quantifier-free inference. All inference is quantificational because quantification scaffolds intentionality itself. There is no logical deductive or semantic consequence relation holding between meaning propositions. There is only the homogeneous apparatus of quantification which involves *L-forms*, one and all, and inseparable. There is no molecular (negative, disjunctive, etc.) thinking. All discursive thinking is quantificational. Indeed, there are no cognitive logical relations indicated by our grasp of the so-called logical particles.

Russell's *original* logical atomism—the logical atomism of his *Principia* era—is alive and well. The logical atomism of Russell's Neutral Monist era, or that which was influenced by Wittgenstein, embracing negative and general facts, *forms of facts*, with the impredicative comprehension principles of *Principia* still in hope of being emulated or obviated, is a dead end. It is only the original logical atomism with the full power of *Principia's* $^{\text{cp}}$Logic as its essence that reveals itself wherever a philosopher uses the analytic tools of mathematical logic to look askance at the metaphysical dogmatisms that parade themselves as purported kinds of non-logical necessity with their special abstract particulars. Lewis's own work on the counterpart logical forms for alethic modality and all the important work on formal systems that abound today (para consistent or otherwise) are not lost according to Russell's original, but can be made into its allies. It nonetheless remains wholly antithetical to metaphysical abstract particulars (worlds, *propositions*, numbers, geometric spatial figures, times, sets/classes, etc.) and their non-logical necessities. Russell observes that "[I]n the welter of conflicting fanaticism, one the few unifying forces is scientific truthfulness" (*HWP*: 836). Russell's *original* logical atomism still offers the last and best hope for a genuinely scientific philosophy.

## Notes

1. I use the expression "$^{cp}$Logic" to characterize the thesis that logic embodies the *impredicative* comprehension of functions which makes it a genuine informative synthetic *a priori* science.
2. See Wittgenstein "Letter of 29 November 1913", in Wittgenstein (1914), p. 123.
3. As I interpret *Principia*, the syntax is that of simple impredicative type theory. The exclamation sign (!) serves only to distinguish genuine bindable object-language predicate variables (e.g., $\varphi!$, $\psi!$, $f!$, $g!$) from schematic letters ($\varphi$, $\psi$, $f$, $g$) for *wffs*. Followers of Church's interpretation hold that *Principia's* grammar codes ramification into its object-language, so that letters ($\varphi$, $\psi$, $f$, $g$) are for object-language bindable non-predicative variables. This interpretation is quite unhistorical and generates insuperable difficulties for an interpretation of Russell's *Theory of Knowledge*. For all of Russell's trials and tribulations, there is not one mention of ramification in any of his works on the epistemology of *Principia's* logic. Ramification, as I see it, concerned Whitehead and Russell's failed nominalistic semantics for *Principia's* bindable predicate variables. It was never part of its object-language grammar.
4. See Fara and Williamson (2005).
5. The following are the basic axioms, where "o" designates a basis:

   Idempotence: $(x)\mathrm{R}xxx$
   Identity: $(x)\mathrm{R}oxx$
   Monotony: $\mathrm{R}oab \wedge \mathrm{R}bcd .\supset. \mathrm{R}acd$
   Star1: $\mathrm{R}abc \supset \mathrm{R}ac^{*}b^{*}$
   Inheritance: $\mathrm{R}oab \wedge A^{a} .\supset. A^{b}$
   Star2: $b^{**} = b$
   Commutation: $\mathrm{R}abc \supset \mathrm{R}bac$

   If one adds the axiom $(x)\mathrm{R}oox$, the system emulates the classical quantifier-free calculus.

## References

### Works by Other Authors

Cocchiarella, Nino B. (1975). "On the Primary and Secondary Semantics of Logical Necessity." *Journal of Philosophical Logic*, Vol. 4: 13–27.

Fara, Michael and Timothy Williamson (2005). "Counterparts and Actuality." *Mind* Vol. 114: 1–30.

Landini, Gregory (2007). *Wittgenstein's Apprenticeship with Russell.* Cambridge: Cambridge University Press.

Lewis, David (1968). "Counterpart Theory and Quantified Modal Logic." *Journal of Philosophy*, 65: 113–126.

Sheffer, H. (1913). "A Set of Five Independent Postulates for Boolean Algebras with Application Logical Constants." *Transactions of the American Mathematical Society*, Vol. 14: 481–488.

Wittgenstein, Ludwig (1914). *Notebooks 1914–1916*, eds. G. H. von Wright and G. E. M. Anscombe, second edition 1979. Chicago: The University of Chicago Press.

CHAPTER 3

# Logical Atomism in Russell's Later Works

*Gülberk Koç Maclean*

## 1 Introduction

The question I will discuss in this chapter is whether and in what sense Bertrand Russell's later works such as *Inquiry into Meaning and Truth* (1940) and *Human Knowledge: Its Scope and Limits* (1948) embrace the logical atomist philosophy. Even though logical atomism is preserved as a research method in Russell's later metaphysical and epistemological works, I will argue that logical atomism as a metaphysical thesis is not preserved. My argument is based on the following: (1) In Russell's later work, the atoms of reality do not include particulars, whereas it is essential to the logical atomist metaphysics that there are unrepeatable particular atoms as ultimate kinds of reality. (2) The metaphysical structure of simple facts is that universal qualities are in compresence relations; the logical atomist metaphysics, however, requires a metaphysical structure whereby particulars exemplify universal qualities and relations. (3) Furthermore, the metaphysical structure of reality is a causal structure. In contrast, the logical atomist metaphysics requires that the structure of reality is a logical structure. (4) Finally, there are no logical facts in Russell's later works. The logical atomist ontology, on the other hand, requires logical facts to correspond to true negative propositions, general propositions, and existential propositions.

G. Koç Maclean (✉)
Mount Royal University, SW Calgary, AB, Canada



L. D. C. Elkind, G. Landini (eds.), *The Philosophy of Logical Atomism*, History of Analytic Philosophy,
https://doi.org/10.1007/978-3-319-94364-0_3

I should ideally determine exactly which period of Russell's early works (before 1940) is committed to logical atomism as a metaphysical thesis before I give reasons for holding that Russell's ontology is not a logical atomist one in the later period. This, however, is not a simple task, as the other chapter in this volume will attest to it. Since Russell clearly defines his philosophy as logical atomism in "The Philosophy of Logical Atomism" lectures delivered in 1918 and explicitly states that in these lectures he proposes a metaphysics implied by the philosophy of logical atomism (*PLA*: 160), I will adopt the strategy of determining the distinctive features of logical atomism as a metaphysical thesis in *The Philosophy of Logical Atomism* lectures, and thereupon conclude that Russell's later metaphysics does not qualify as such, even though he still employs logical atomism as a method of philosophical inquiry.

## 2 Logical Atomism as a Research Method

According to Russell, logical atomism is the right method of philosophical research. The main components of logical atomism are logical analysis and logical construction.[1] In *The Philosophy of Logical Atomism* lectures, logical analysis is defined as "the view that you can get down in theory, if not in practice, to ultimate simples, out of which the world is built" (*PLA*: 234). This method of logical analysis yields a metaphysics with particulars, and qualities and relations in Russell's logical atomist period (*PLA*: 161).

One of his arguments in support of logical atomism as the right method for philosophical inquiry is based on his argument for analysis being the right method for doing philosophy, since if reality is indeed composed of wholes (complexes), the ultimate simples (atoms) of which will have to be found by analysis, then analysis ought to be the right method for philosophical inquiry (ibid.: 178). Russell defends the method of analysing complex wholes into atoms against the falsification charge, namely, that when we analyse a whole into its parts, the whole loses its causal properties. Russell accepts that this may be the case; he points out, however, that "analysis into 'atoms' is perfectly valid so long as it is not assumed that the causal efficacy of the whole is compounded out of the separate effects of the separate atoms" (*OoP*: 258–259).

Another way in which Russell defends logical atomism as a philosophical method is by defending analysis as the method which enables us to move from what seems, to the Cartesian inquirer, an undeniable datum, nevertheless vague, to its components which are precise. Russell gives the following statement as an example for what seems to be undeniable data:

"There are a number of people in this room at the moment". He urges us to admit that when we start attempting to clarify this statement and define its constituents, such as a person, a room, the relation of something being in something else, we realize how vague the statement is and the process of analysis helps us render precise what is imprecise (*PLA*: 161).

Logical construction, the other key element of logical atomism as a way of doing philosophy, is a method implied by the adoption of the principle of Ockham's razor, according to which we ought to go through all the propositions that describe reality "with a view to finding out ... the smallest apparatus out of which [we] can build up these propositions" (*PLA*: 235). In this logical process, we will find that some metaphysical entities have been assumed to exist dogmatically. Logical construction gives us a way of avoiding this dogmatism. We "construct a logical fiction having the same formal properties to those of the supposed metaphysical entity and itself composed of empirically given things, and ... logical fiction can be substituted for [our] supposed metaphysical entity"[2] (*PLA*: 236). For instance, Russell employs this method of logical construction to avoid commitment to the existence of persisting substances, which we are not given in experience. An ordinary object, such as a desk persisting through time, is taken as a logical fiction instead of "a continuous entity persisting through changes" (*PLA*: 235–236). The desk is a "system of correlated particulars, hung one to another by relations of similarity and continuous change and so on" (*PLA*: 237). The particulars referred to here are sense-data, which are transient particulars, that is, "passing particulars of the kind that one is immediately conscious of in sense" (ibid.). The ordinary particular is merely a construction out of transient particulars.

Here are the main reasons why I argued elsewhere[3] that Russell employs logical atomism as a research method in his later works, in particular, *Inquiry* (1940) and *Human Knowledge* (1948). First, we find that logical atomism is still at work as a philosophical method: logical analysis of events (transient particulars), which were provisionally ultimate kinds of entities, yields universal qualities and compresence relations as analysans, not particulars. Second, Russell still believes that logical construction of entities, whenever possible, is an important method in philosophizing. Russell's ordinary particulars are constructions, that is, series of complexes of compresent qualities (Koç Maclean 2014: 63–64).

To elaborate on the first point, transient particulars in *Inquiry* and *Human Knowledge* are complexes of compresent qualities; for instance, when I perceive a red ball, the qualities of redness, roundness, and softness make up one complex. The compresence of these qualities "at a certain space-time location" is a transient particular. Whereas the ball understood as a persisting object is an ordinary particular. The ordinary particular is a mere construction, while the transient particular is not; it is a complex.

Logical analysis of events, which were provisionally taken to be particulars in the beginning sections of *Human Knowledge*, reveals that there really is no proper justification to take events as unrepeatable things, that is, particulars. We could only appeal to the substrata to justify their uniqueness, but the substrata are not experientially given; they are mere philosophical posits to explain individuation. All we find after thorough logical analysis of events are repeatable entities, that is, universals. When universal qualities are in compresence relations, they comprise complexes,[4] which we call "events" or "transient particulars". But these linguistic expressions do not correspond to any ontological kinds, that is, particulars.

The second point is that logical construction is employed in *Human Knowledge* whenever possible. Russell has not ceased to think that this is a useful methodology to appeal to when philosophizing. The "transient particulars" are complexes of compresent qualities, not logical constructions. The "ordinary particulars", however, are logical constructions out of complexes of compresence. We should note here though that when Russell identifies transient particulars with complexes of compresent qualities and ordinary particulars with logical constructions, his identifications are eliminative, not reductive. That is, he eliminates the ontological category of particulars altogether. The significance of this move is that it protects Russell's theory of particulars from objections which arise from an individual's (particular) being identical to a bundle of qualities or a set of bundles in a reductive manner, such as the problem of accidental predication and the problem of change. When there are no particulars (individuals) in one's ontology, the problem of accidental predication can be explained away since there are no individuals which have some of their properties essentially and others accidentally. And there is no need to account for how an individual can change their properties, since there are no individuals that persist through time and go through a change in their properties (Van Cleve 1985: 129).

## 3 Logical Atomism as a Metaphysical Thesis

As a metaphysical thesis, logical atomism is comprised of the following theses: (1)–(6) below. The first (1) is Pluralism: Reality is not one, but many. In response to a question at the end of his first Philosophy of Logical

Atomism lecture as to whether and how Russell justifies his claim that there is a multiplicity of things, Russell points out that the arguments against pluralism are a priori. Bradley's argument against pluralism is that the assumption of plurality leads to a contradiction. If there are many things, then the relations between them will have to be external to them. Each thing is a logically independent entity; its identity has to be independent of its relations to other atoms. Suppose, argues Bradley, there are two things, *A* and *B*, and they are in a relation, *C*. If relations exist external to the things they relate, as Russell wants to maintain, then we cannot explain how *A*, *B*, and a relation, *C*, form a unity because *C*, being another entity, will need a relation, *D*, to relate it to *A* and *B*. But now, since again relations are supposed to be external to what they relate, we will need another relation *E* to relate *A*, *B*, and *C* to *D*. And this positing of relations will go on infinitely without ever being able to explain the initial unity between *A* and *B* via the relation *C*. So, the assumption of plurality leads to its impossibility (Bradley 1893: 18). Russell argues that there is no a priori reason for why there should be either one thing or many things. This question can only be settled a posteriori. So, he claims he has an empirical argument for pluralism; it is simply true based on experience that there are many things (*PLA*: 168; *LA*: 339).

(2) There are two kinds of atoms. Logical analysis of simple propositions yields two kinds of atoms: particular things, and universal properties and relations. "The atoms [Russell arrives at] as the sort of last residue in analysis are … 'particulars'—such things as little patches of colour or sounds, momentary things- … predicates or relations and so on" (*PLA*: 179). Russell's use of *patches* of colour as distinct from *shades* of colour has ontological significance. Patches of colour are particulars, unrepeatable things, whereas shades of colour are universals, repeatable entities. What accounts for the particularity of a patch of colour is its substratum. This understanding of particulars in *The Philosophy of Logical Atomism*, I argue, is a continuation of Russell's endorsement of the substratum theory of particulars, which goes as far back as *The Principles of Mathematics* (1903). My reasons are based on the implications of Russell's understanding of a "term" in *Principles*. A term is the logical subject of a proposition; it is anything that can be thought about. At the same time, a term is an individual, it is an ontological entity; it is something numerically distinct from everything else. A term has other characteristics of "substance" as being unchanging and indestructible. Consider "Socrates is mortal". Socrates is a term. Socrates, as a term, is supposed to be unchanging, but he clearly did change.

This can be accounted for in two ways; either the term is Socrates's essence, that is, his essential properties, and these essential properties remained unchanging, while the accidental ones changed, or the term is a substratum. Now if the term is an essence in an Aristotelian sense, then it cannot explain the numerical distinctness of Socrates when there is another individual which shares the essential properties of Socrates. If the essence is an individual essence, such as the property of being identical to Socrates or being the teacher of Plato, then this would contravene Russell's argument for the existence of relations independently of their terms, since the above properties are relational properties. So, the term has to be a substratum in order that it can meet all of the above criteria for something to be a term (Koç Maclean 2014: 21).

Russell's particulars are no longer sense-data in "On Propositions" (1919); they are sensations or images, since the rejection of the assumption of a mental subject, under the influence of William James, implies that sense-data, which are objects of the cognitive subject, need to give way to sensations. Russell expresses, in "On Propositions", how much he is attracted to William James's theory of neutral monism, according to which the stuff of what we call the mind and the stuff of what we call matter are in fact of the same kind, that is, a kind neutral between mind and matter (*OP*: 299). As of 1921, we find Russell full-heartedly endorsing neutral monism in *The Analysis of Mind* (*AMi*: 287), *The Analysis of Matter* (*AMa*: 10, 402), and *An Outline of Philosophy* (*OoP*: 293, 307). Thus, Russell's particulars in *The Analysis of Mind* and *The Analysis of Matter* are events and "the mind" is to be constructed out of events that obey psychological laws and "the matter" is to be constructed out of events that obey physical laws, while all events are ultimately neutral.

What is significant for our discussion here is that these events are particulars and their particularity is still explained by their possession of an ontological constituent, a substratum. Russell takes a particular in the logical sense to "fulfill the definition of substance" (*AMa*: 277), which is "that which can only enter into a proposition as a subject, never as predicate or relation" (*AMa*: 238). As I explained above, a particular understood as a logical subject needs to be numerically distinct and serve as the holder of properties and the term of relations; therefore, a particular event needs to have a substratum as a component to serve these purposes.

(3) The atoms of the world form complexes, which are facts, and facts "belong to the objective world" (*PLA*: 183).

(4) The metaphysical structure of simple facts is that non-recurring particulars exemplify universal properties and relations.

(5) Whether there is an underlying causal structure of all reality cannot be known; therefore, no assumptions are made with respect to the existence of such a structure.

(6) There are not only positive facts, but also logical facts to correspond to negative propositions[5] (*AMa*: 211), general propositions, and existential propositions so that there are negative facts, general facts, and existence facts (*AMa*: 235).

We find that in Russell's major metaphysical and epistemological works in the later period, most of these metaphysical commitments disappear, namely, (2), (4), (5), and (6), but the rest remain. The commitments which remain, namely, (1) pluralism, that is, the view that reality cannot be a monistic whole, but must be a multiplicity and (3) objectivity of facts, that is, the view that facts that make up reality are independent of the ways we perceive and conceptualize them, are necessary features of logical atomism as a metaphysical thesis.

Russell persists in holding on to the objectivity of facts, though there is a change in the nature of this objective reality in the later period. Reality is no longer physical or mental in its nature; it is neutral between the two. "Mental" and "physical" are not intrinsic qualities of reality; rather they are merely our two ways of describing, and acquiring knowledge about, reality: direct knowledge and inferential knowledge, respectively. Facts are neutral occurrences, some of which are known directly, without inference, which Russell calls "percepts", whereas other facts are known by inference (*IMT*: 284–285; *HK*: 209). Thus, objectivity is based on a neutral reality. When a proposition is true, it corresponds to a fact, which is neutral in itself.

Some philosophers hold that Russell's metaphysics in the later period was logical atomist. Engel claims that "although in 1940 Russell rejects a number of [his earlier logical atomist] views, he keeps the basic structure of logical atomism" (Engel 2006: 143). In support, Engel cites Russell's correspondence theory of truth and objectivity of facts and his commitment to the existence of facts "over and above things and their properties or relations" (ibid.). However, I will argue that commitment to plurality and objectivity of facts are necessary, but not sufficient for a metaphysics to count as logical atomism as espoused in *The Philosophy of Logical Atomism*. The loss of particular atoms, the change in the metaphysical structure of simple facts, and the postulation of a causal structure to reality, along with the disappearance of logical facts, imply that Russell's later metaphysics is no longer *logical atomism*.

## 4 The Ontological Status of Atoms in the Later Period

It was in *Our Knowledge of the External World* that Russell eliminated ordinary particulars, such as myself considered as "persisting" over time, but he still held onto transient particulars, which were sense-data and sensibilia in *Our Knowledge*, and events in *The Analysis of Mind* and *The Analysis of Matter*. So, sense-data, and later events, were particular atoms in this early period. In 1948, Russell in *Human Knowledge* eliminates transient particulars from his ontology as well, ridding his ontology of all particular atoms. He argues that transient particulars or events are not ultimate kinds of reality, since considered apart from their properties, which are universals, they are mere substrata with no properties of their own, which means, according to Russell, that they are not empirically known[6] or verifiable (*HK*: 294).

> It is difficult to see how something so unknowable such as a particular would have to be required for the interpretation of empirical knowledge. The notion of substance as a peg on which to hang predicates is repugnant.

The second reason why Russell holds that particulars are not among logical atoms is the philosophical principle that we ought not to posit the existence of entities unnecessarily.[7] The philosophical problems that lead Russell to infer the existence of particulars, namely, the need to account for individuation of particulars and construction of the space-time series, Russell was now able to explain without appeal to extra entities, i.e., substrata.

The later Russell does not believe that the *logical* possibility, described by Max Black, of a universe with nothing but two spheres that share all their qualities and relations in a symmetrical universe is a philosophical problem that requires explanation. Based on his later ontological framework, which recognizes only universal atoms (qualities and relations), if two bundles share all their properties, they are simply the same bundle of compresent universal qualities occurring twice; they are not two different particular spheres (O'Leary-Hawthorne 1995: 193). In the realm of metaphysical possibilities, taken as narrower than the logical one, if one thing is different from another, this will be explained not by appeal to the different substrata they have, but in terms of a difference in their qualities[8] (*IMT*: 102; *HK*: 306). That is, Russell no longer believes that he has to posit the existence of entities, which cannot be experienced, for the sake of accounting for individuation in logically possible scenarios.

As a corollary, Russell no longer believes we need particulars to ensure that we construct a space-time series with no recurring particulars. He identifies events with incomplete complexes (bundles) of compresence of qualities and employs complete[9] complexes as the elements of the space-time series. These complexes include positional qualities (*HK*: 263) that increase the chances of non-recurrence (*HK*: 198, 265, 294, 295). Russell admits, however, that he can only claim that the complete complexes will not recur as a physical fact, not as a logical fact (*HK*: 83).

Russell's responses above to the problem of recurrence, both in the case of particulars and in the case of the construction of the space-time series, is dismissive of logical possibilities. Earlier, he would have taken seriously the logical possibility that two "things" share all their properties and yet be different. Similarly, earlier he would have been troubled by the logical possibility that his atoms might recur, and therefore endanger the accuracy of his construction of the space-time series. But he does not seem to think these logical possibilities are philosophically problematic anymore. This suggests that logical considerations no longer trump all other philosophical considerations, such as the principle of simplicity and the primacy of experience in ontological commitment.

This is not to say that he does not engage in logical analysis anymore, for he clearly does. He still believes it is a useful philosophical tool to understand the nature of reality. He believes that the atoms that are the relata of the structure of reality have a metaphysical nature to be discovered by logical analysis.

## 5 The Metaphysical Structure of Simple Facts

Logical analyses yield the metaphysical parts of the structure of simple facts, that is, we identify the parts (atoms) of an object and "the ways in which they are interrelated" (*HK*: 250). Since Russell believes in an objectively existing reality, so that atoms of reality exist independently of the ways in which we perceive them, the structural relations between atoms are also objective facts.

Earlier, the structural relation between atoms of reality, that is, particulars and universal properties and relations, was one of exemplification. Particulars exemplified universal properties and relations. The logical analysis of the structure of the simple fact, which corresponds to the proposition "This is red", for instance, would reveal that there would be a transient particular, which exemplifies the universal property of being red.

In the later period, by contrast, the logical analysis shows that there are no transient particulars, in the sense of a substantial entity. Instead, there are only universal properties and relations. The proposition "This is red", when true, corresponds to a fact and this fact is a complex of qualities, including the universal quality red, which are in a relation of compresence with each other. Thus, the metaphysical structure of simple facts is that universal qualities are in compresence relations.

## 6 The Metaphysical Structure of Reality

Russell, in the earlier period, suspends judgement on whether there is a causal structure of reality or spatio-temporal continuity between different temporal slices of an object. For instance, an ordinary object, such as a person persisting over time, Russell takes to be merely a class of sets of particular experiences (*PLA*: 240).

> It does not matter in the least to what we are concerned with, what exactly is the given empirical relation between two experiences that make us say, "These are experiences of the same person. It does not matter precisely what that relation is, because the logical formula for the construction of the person is the same whatever that relation may be, and because the mere fact that you can know that two experiences belong to the same person proves that there is such an empirical relation to be ascertained by analysis".

Thus, the early Russell was agnostic with respect to the question as to whether the metaphysical structure of reality was interwoven by relations of causation and spatio-temporal continuity. He believed that we can give a comprehensive account of the world and our experiences without committing ourselves to the existence of such relations.

The later Russell, however, is no longer agnostic with respect to the metaphysical structure of reality. He maintains that we can know what the structure is *probably* like, though not what it *must* be like. In Russell's later works, logic is not powerful enough to explain all our experiences; logical fictions turn out to be inadequate in explaining why perception occurs, without positing a causal structure to the world (*AMa*: 399; Bostock 2012: 194). As a consequence, in *Human Knowledge*, instead of suspending judgement on the structure of reality, Russell postulates its causal structure. As well as the postulate of causal structure, Russell assumes the following postulates to be true in order to be able to justify our non-demonstrative inferences: quasi-permanence, separable causal lines, spatio-temporal continuity in causal lines,[10] and analogy. Russell does not hold these postulates to be logically necessary, but only probably true (*HK*: 487).

Based on our observation of the interaction of parts of reality, we formulate causal laws. Russell's definition of a causal law is that it is "a general principle in virtue of which, given sufficient data about certain regions of space-time, it is possible to infer something about certain other regions of space-time" (ibid.: 308). In line with his belief in the existence of a neutral world existing independently of our mental or physical descriptions of it, Russell believes that it is because the world has a certain causal structure that we are able to come up with causal laws, principles which allow us to make, for the most part, successful inferences about unobserved parts of reality. But he admits that he cannot justify this belief either a priori or by experience.

In *Human Knowledge*, Russell retracts from his earlier view on causation as invariant or frequent succession, and instead assumes causation as one of his postulates about reality. One reason is that he now believes that taking a cause as an "invariable antecedent" does not quite capture what we mean when we say, "A causes B". It blurs the real distinction between the invariant succession of two events and the causal connection between two events. He gives the example of two clocks running concurrently: "when one points to the hour, the other strikes, but we don't think that the one has 'caused' the other to strike" (*HK*: 315). Russell holds now that "when *A* is caused by *B* the sequence is not merely a fact, but is in some sense necessary" (ibid.).

The reasons Russell has for postulating a causal structure to the world concern the accounts of perception and persistence of material objects (*HK*: 316). Russell needs causal connections to exist independently of us to explain perception. When I perceive a blue ball, for example, there must be something in the world, which functions as the first event in a causal chain of events leading to my sensation of blueness and roundness. With respect to the explanation of persistence, since there are no substances (ordinary particulars), he does not have the tools to account for the persistence of a "thing". If he had ordinary particulars, he would have explained the persistence of an object by holding that each ordinary particular has a substratum that takes on certain properties at one time and replaces them with others at another. Meanwhile, the substratum would stay the same through change of qualities. However, with no substrata to serve the role of individuation, Russell explains that he needs to postulate causal lines. A causal line is one where if you knew some of the events in the line, then you would be able to infer "something about the others without having to know anything about the environment" (*HK*: 316).

Russell gives the following example to explain the point: "If my doors and windows are shut, and at intervals I notice my dog asleep on the hearthrug, I infer that he was there, or at least somewhere in the room, at the times when I was not noticing him" (ibid.). So, in order to be able to make this inference that his dog continues to exist at times it is not being perceived by anyone, Russell needs to accept that there are causal lines between the events involving his dog.

Thus, reality has a certain causal structure, which we need to postulate, since we cannot justify its existence either a priori or by experience. We believe a priori that reality has a causal structure, but we cannot justify this belief a priori. And with respect to the laws that govern this reality, our knowledge comes in degrees of probability.

## 7 Logical Facts

The early Russell's claim that "there is an objective complexity in the world ... [which] is mirrored by the complexity of propositions" (*PLA*: 184), coupled with his admittance of negative facts, existence facts, and general facts to his ontology, misleadingly suggests that the *logical* complexity of a proposition which results from using logical words points at a corresponding *logical* complexity in the world of facts so that simple propositions would correspond to simple facts and any logical word the propositions include would correspond to logical objects in reality.[11]

Russell's atomic propositions correspond to atomic facts. They express the fact that either one particular has a certain property or two or more particulars are in a certain relation (*PLA*: 177). Molecular propositions are ones where atomic ones are combined with logical connectives, "or", "if", "and" (*PLA*: 184). But interestingly "not" is not included in the list. Russell says, "supposing you have the proposition 'Socrates is mortal', either there would be the fact that Socrates is mortal or there would be the fact that Socrates is not mortal" (*PLA*: 185). This suggests that there are negative facts and negative facts will have to be simple facts. Indeed, Russell holds that we should take "negative facts as facts [and] assume that 'Socrates is not alive' is really an objective fact in the same sense in which 'Socrates is human' is a fact". Negative facts are ultimate (*PLA*: 189; *OP*: 288), which means that there is not a specific kind of *logical* structure to be found in negative facts.

The argument Russell gives against Mr. Demos's alternative account of the truth of negative propositions is that this account does not offer a simpler account than Russell's, which accepts the existence of negative facts. Mr. Demos proposes to explain negative propositions in terms of incompatibility relations between propositions so that when we assert *not-p* (Socrates is not alive), we are in fact asserting that there is some other proposition, *q* (Socrates is dead), which is incompatible with *p* (Socrates is alive). Russell's objection to this proposal is that since "no two facts are incompatible" (*PLA*: 189), the incompatibility is between propositions and so what makes *not-p* true will have to be a fact involving propositions. Incompatibilities between propositions will have to be ultimate, objective facts about the world. But it is difficult to accept that they are, more difficult than accepting negative facts because propositions are not real; they are not part of the inventory of the world; they do not have independent being (ibid.).

There are existence facts to correspond to existence propositions, such as "Some men are Greeks", and general facts to correspond to general propositions, such as "All men are mortal" (*PLA*: 206; *OP*: 289). Russell reasons that "when you have enumerated all the atomic facts in the world, it is a further fact about the world that those are all the atomic facts there are about the world, and that is just as much an objective fact about the world as any of them are" (*PLA*: 207). But Russell does not give us an analysis of general facts, which would tell us if general facts are complexes with logical structure, in which case there would be ontological analogues of "all" or "some", or whether general facts are ultimate. He writes, he does "not profess to know what the right analysis of general facts is" (*PLA*: 207).

As we have seen above, Russell does not treat all logical words the same way. The logical words "and", "or", "if", and "then" go into making up molecular propositions and these propositions do not reflect any complexity corresponding to the complexity of the relevant molecular propositions. Molecular propositions are made true or false by correspondence to the simple propositions they connect, and therefore there are no ontological analogues of "or", "if", "then", or "and" (*PLA*: 179, 186–187). Yet, the propositions that include logical words "not", "all", and "some" are not made true or false by correspondence to the simple propositions they are annexed to, but by different facts, namely, negative facts, existence facts, and general facts. And Russell does not seem to think these reflect the complexity of the propositions that express these facts either. So, in what sense is there a mirroring of complexity between our propositions and facts that constitute reality when there are no ontological counterparts of our logical words "or", "not", "if", "then", "all", or "some", even though there exist negative facts, existence facts, and general facts? Could it be that the logical complexity of reality is revealed by the complexity of the logical form of simple propositions, such as "*xRy*", which we arrive at by generalizations from such simple propositions as "Socrates loves Plato"?

Russell explains that these completely general propositions are merely variables in a certain logical form (*PLA*: 208). He accepts that it is not clear what the constituents of a logical proposition could be: "*x* and *R* and *y* are nothing, and they are not constituents" (ibid.: 209). How about the form? Is that a constituent? Russell does not think so. He argues that if we take the form of the dual relation in the simple proposition "Socrates loves Plato" to be a constituent of it, we "would have to have that constituent related to other constituents" (ibid.). Russell points out, however, that the form of a dual relation "may *possibly* be a constituent of general statements about propositions that have that form". Russell says, "it is possible that logical propositions might be interpreted as being about logical forms" (ibid.). So, the question as to whether logical form is an ultimate fact about reality seems to be left open in the logical atomism lectures.

I think that Russell's logical form reveals ontological form. The logical form of propositions, "*xRy*", is indeed reflected in reality because "*xRy*" is the symbolic expression of the metaphysical structure of two particulars exemplifying a universal relation. The logical form of proposition is merely an expression of the metaphysical structure of simple facts; it does not indicate a logical object in reality. That is, logical complexity of the proposition mirrors ontological complexity.

Thus, in the early period, even though there are no logical objects or logical form as ontological constituents of reality, there are negative facts, existence facts, and general facts. By contrast, in the later period, there are no logical facts. In *Inquiry*, Russell shows that logical words, such as "or", "some", and "all" are not necessary in describing objective facts about the world, but they are "indispensable for the description of certain mental facts" (*IMT*: 93). There are no disjunctive facts in the *Inquiry* (84–87), and neither were such facts recognized in *The Philosophy of Logical Atomism* lectures. However, back then, the truth of a disjunctive fact, say *p* or *q*, would require correspondence to two facts, not one fact (*PLA*: 185). In *Inquiry*, disjunctive propositions express a state of the speaker, namely, hesitation or uncertainty, as when I say, "I will have oatmeal or toast". Supposing my use of "or" to be exclusive, when this proposition is true, there is not a disjunctive fact about the world: (my-having-toast-or-oatmeal) or two facts, that is, (my having toast and my not having oatmeal) or (my not having toast and my having oatmeal), but only either one of these facts (my having toast) or (my having oatmeal). Some uses of the logical word "some" are also explained in this manner, as expressing hesitation or uncertainty, as when Russell says, "My book is somewhere in this room" (*IMT*: 89–90). Other uses of "some" are explained by the fact that at times "the generalization is more important to us than the particular instances", as when we say, "Some swans are black" (*IMT*: 88).

Logical words do not express *logical* facts about the world even in the case of the logical word "all", which Russell earlier thought was indispensable to the description of the objective world because after we have enumerated everything in the world, we would still have to add that "this is all that exists" to make the description complete (*PLA*: 206). But in *Inquiry*, Russell argues we need "all" merely to be able to "*state* that the description is complete" (*IMT*: 92) [original italics], which suggests that "all" is needed only for the description of our knowledge, not as part of reality.

However, the logical word "all" in some cases indicates an *extra-logical fact* in *Human Knowledge*. Russell argues that although there does not need to be any logical, or otherwise, facts to correspond to general propositions in the case of tautological general propositions, such as "All pentagons are polygons" or "All widows have had husbands", there must be extra-logical facts about the world when it comes to empirical general propositions, such as "All dogs bark" or "All copper conducts electricity", where the propositions in question have not been proven by complete enumeration. There must be a fact about the relation between the property of being copper and the property of conducting electricity, which *partially* accounts for the truth or falsity of the general proposition, "All copper conducts electricity" (*HK*: 138–140). For induction based on a positive correlation between being copper and conducting electricity is not sufficient to yield the true conclusion that all copper conducts electricity, without there being some fact about these properties that makes it antecedently probable that they would be so correlated (*HK*: 404–405; 435–436). This is pointing to what Goodman (1979) will call "the new riddle of induction". We need to be able to explain the difference between law-like generalizations, such as "All copper conducts electricity", and accidental generalizations, such as "All copper coffee pots in my kitchen are gifts" (Goodman 1979: 73).

Whereas Goodman argues that there is no reality, a neutral world underlying all worlds (Goodman 1968: 175–176), which comes with facts to ground the law-like generalizations, Russell argues that there must be a fact about the world that explains why some generalizations are law-like, that is, they have initial credibility; they are probable even before we find any confirming evidence for them. Russell maintains that we need to postulate a law to this effect in order to be able to justify our scientific knowledge, since it is largely based on non-demonstrative inferences. To this end, Russell accepts "the structural postulate", which states that if there are two identical structures, such as having the molecular structure of copper and the molecular structure which generally allows something to conduct electricity, it is probable that they have one of the following kinds of causal connections: Either they both have a common causal ancestor or "two structures are composed of similar ingredients and there exists a causal law leading such ingredients to arrange themselves in a certain pattern" (*HK*: 468). The latter kind of similarity of causal structure is what explains the antecedent probability of the generalization that all copper conducts electricity. Thus, Russell does not accept the existence of general facts in his later work. However, in place of general facts concerning empirical generalizations, he postulates that reality has a certain causal structure, and therefore accepts the existence of *extra-logical* facts, by which he means facts about the causal structure of reality.[12]

Negative facts are also ultimately purged from Russell's ontology. In *Human Knowledge*, Russell explains that if we think of "Socrates is dead" as an elementary proposition, there are no facts over and above the ones that make positive propositions true. That is, the proposition "Socrates is not dead" will be made false by the fact that Socrates is dead (*HK*: 120–122). When the proposition is negative, it applies to the facts in an indirect manner: "When we say, 'it is not raining', the proposition 'it is raining' is first considered and then rejected. 'Not-*p*' may be defined as "*p*' is false'; it is, in fact, a statement about *p* in quotes" (*IML*: 360). So, the truth of "Socrates is not alive" no longer requires the negative fact that Socrates is not alive; it requires only the fact that Socrates is dead. We consider the proposition, "Socrates is alive" and realize that it is false because it fails to correspond to reality, namely, that there is no such complex as Socrates's being alive in reality, and therefore we reject the proposition that "Socrates is alive" by asserting "Socrates is not alive".

We find that even the Law of the Excluded Middle, that is, each proposition is either true or false, does not express an objective fact about the world. The assertion "*p* or not-*p*" is, according to Russell, to be correctly stated as, "'*p*' is true or '*p*' is false". Assuming "true" already defined, Russell defines "false" as follows: "Some among propositions are true; the remainder, by definition, are to be called 'false'" (*IML*: 360). Russell explains that if we accept this view, then we must accept that "the law of the excluded middle is purely verbal; it is a definition of 'false' or 'not' according to taste" (ibid.). Russell continues (*IML*: 360):

> The proposition 'Queen Anne is dead' is true in virtue of a relation between the words and a certain non-verbal fact. Where such propositions are concerned, 'truth' is not a syntactical concept. But what fact ensures the truth of Queen Anne is dead or not dead? No fact about Queen Anne ... There is however a fact which is relevant; it is a fact about the meanings of the words 'or' and 'not'.

The law of identity faces the same fate. "The Law of Identity appears [to be] a convention of typography" (*IML*: 354). To show that the law of identity does not describe facts but it is merely a convention, Russell gives the following example: Consider the proposition, "*A*, *B*, and *C* were all murdered by the same man" (ibid.). Russell interprets this proposition in such a way as to eliminate the notion of identity: "'There is an entity M such that *M* murdered *A*, *M* murdered *B*, and *M* murdered *C*'". Here "identity is replaced by the repetition of the letter *M*" (ibid.).

As logic loses its ontological status as part of objective reality, mathematics goes down with it, since arithmetic relies on logic (*IML*: 357). Thus, Russell's final thoughts on the ontological statuses of logic and mathematics are that they are both linguistic. He writes (*IML*: 362):

> Our conclusion is that the propositions of mathematics and logic are purely linguistic, and that they are concerned with syntax ... They are assertions as to the correct use of a certain number of small words.

## 8 Concluding Remarks

Even though logical language, as reduced to mere syntax, cannot tell us whether there is a logically necessary structure of reality, Russell argues that it can tell us about the metaphysical nature of the atoms (simple parts) of reality. The fact that the same words are used in propositions when similar properties in reality are found together suggests that qualities and relations in reality are of a recurrent nature, that is, they are universals. Suppose that we all utter the sentence "This is black" while we are standing next to the blackboard. If this proposition is true, what verifies "this" will be a complex of qualities, one of which will be black. The repetition of the word "black" while pointing at a certain region of space-time will show that there is a characteristic of reality in that region which recurs, hence the universal nature of qualities (*IMT*: 343).

In *Inquiry*, Russell supplants his belief in the existence of universals with his infinite regress argument against nominalism, which he had articulated earlier in *The Problems of Philosophy* (*PoP*: 55). Russell holds that the fact that certain things are similar to each other is a fact about the world, not about language (*IMT*: 347). He insists that "the similarity of two things is truly a non-linguistic fact as the yellowness of one thing" (ibid.). Here is how Russell ends his book, *Inquiry* (*IMT*: 347):

> Complete metaphysical agnosticism is not compatible with the maintenance of linguistic propositions. Some modern philosophers hold that we know much about language, but about nothing else. This view forgets that language is an empirical phenomenon like another, and that a man who is metaphysically agnostic must deny that he knows when he uses a word. For my part, I believe that, partly by means of the study of syntax, we can arrive at considerable knowledge concerning the structure of the world.

Thus, in the later period, Russell still maintains that we ought to study the logical, syntactical structure of our propositions describing our observations of the world in order to understand the metaphysical structure of the world and make inferences as to what the atoms are like and what kind of relations they get into. Since we are part of this reality, as is our language which aims to describe it, the ways in which we use language to describe reality may help us understand its underlying nature. If we repeat certain words in our descriptions, there must be a repeatable feature of the world that causes us to reiterate. That is, atoms of reality must be universals. Russell does not believe that the logical structure of our propositions corresponds to a logical structure in reality; nevertheless, they correspond to a metaphysical structure. The ontological structure of simple facts is that universal qualities are in compresence relations. And we aim to capture the ontological structure of the relations between simple facts by formulating laws, which rely on postulates, which enable us to make inductive inferences, but which themselves cannot be justified either by experience or a priori.

To recapitulate my main argument, logical atomism as a philosophical method is employed by the later Russell, but he no longer holds logical atomism as a metaphysical thesis. The later Russell does not maintain that the metaphysical structure of reality is such that particulars exemplify universal properties or relations. He is not agnostic as to whether reality has a causal structure. He does not admit logical facts, such as negative facts, existence facts, or general facts in his ontology. Finally, the metaphysical nature of reality does not include particulars as ultimate kinds of entities.

In fact, Russell explains in "Logical Atomism" (1924) that he is not strictly committed to any specific metaphysics that he arrives at; instead what he is steadfastly committed to is a certain way of doing philosophy (*LA*: 323):

> I don't regard the issue between realists and their opponents as a fundamental one; I could alter my view on this issue without changing my mind as to any of the doctrines upon which I wish to lay stress. I hold that logic is fundamental in philosophy, and that schools should be characterized rather by their logic than by their metaphysic. My own logic is 'atomic' … I prefer to describe my philosophy as 'logical atomism', rather than as 'realism', whether with or without some prefaced adjective.

An interesting question that arises from Russell's abandonment of philosophical atomism as a metaphysical thesis is whether he can coherently sustain the philosophy of logical atomism as a research method in this later period. Since logic is not the underlying structure of the world, but instead merely a combination of syntactical rules, we have the problem of the justification of logical analysis as the right method for philosophical inquiry. If we do not have a priori access to logical truths which underlie the metaphysical structure of the world and which exist in some sense independently of the ways we think of and perceive the world, is the fact that our linguistic structures are part of reality a good enough reason to think that engaging in logical analysis will help us understand the nature of reality?

## Notes

1. Russell defines the task of philosophy in "Logical Atomism" (1924) as "essentially that of logical analysis, followed by logical synthesis" (*LA*: 341).
2. See also *LA*: 326–328.
3. Koç Maclean (2014): 133–145.
4. Some philosophers, such as Engel, have mistakenly taken Russell's bundles to be sets of properties (145), but bundles of properties are not sets of properties, according to Russell. They are complex wholes, composed of qualities (*IMT*: 128–129).
5. Propositions are merely complex symbols (sentences). See *PLA*: 166.

6. In contrast, Edwin Allaire distinguishes between substrata and bare particulars, and argues that bare particulars can be directly experienced. He states that particulars present themselves to us as one in experience and he argues that it is the bare particular of a concrete particular that accounts for its numerical difference (Allaire 1963: 120).
7. In a footnote added in 1955 to his paper "On the Relations of Universals and Particulars" (1911), which argued that the division between universal and particular is "ultimate" Russell wrote (*RUP*: 124): "The argument in the above article in favor of the existence of particulars no longer seems to me valid for reasons which I have explained in *Human Knowledge: its scope and limits*. ... I no longer think that there are any spatial or temporal relations which always and necessarily imply diversity. This does not prove that the theory which asserts particulars is wrong, but only that it cannot be proved to be right. The theory which asserts particulars and the theory which denies them would seem equally tenable. If so, the latter has the merit of logical parsimony".
8. Russell holds that there are positional qualities. These are qualities such as centrality (being at the centre of a visual field) or dexterity (being to the right of the centre of one's visual field) (*HK*: 294, 295, 298).
9. *Incomplete* complexes of compresent qualities are events (transient particulars). However, when a sufficient number of qualities form a complex such that all the qualities in the group are compresent and no quality lying outside the group is compresent with every quality in the group, they form a *complete* complex. It is these complete complexes which are the elements of the space-time series to be constructed (*HK*: 304).
10. As Engel explains (Engels 2006: 147) Russell recognizes even earlier in *Inquiry* that we need "non-demonstrative principles of inference, which cannot be rendered either probable or improbable by any empirical evidence" (*IMT*: 304). Russell mentions a principle, which is similar to the spatio-temporal continuity postulate he articulates later on in *Human Knowledge*: "There is evidently a causal relation between seeing and hearing the explosion; when I am on the spot, they are simultaneous; we therefore assume that when they are not simultaneous, there has been a series of occurrences, which, however, were not perceived, and are therefore not in perceptual space. This point of view is reinforced by the discovery that light, as well as sound, travels with a finite velocity. We may therefore take, as a principle: if, in my experience, an event of kind *A* is always followed, after a finite interval, by an event of kind *B*, there are intermediate events which connect them" (*IMT*: 302–303).
11. I initially thought that Russell's thesis that the logical complexity of propositions is mirrored in the complexity of facts implied that, in the case of negative facts, there would be an ontological constituent of the negation

symbol as well as a positive fact. Soames has also made the same mistake: "~S is true because it corresponds to a fact in which an abstract constituent designated by '~' combines with a non-linguistic complex represented by the sentence S. In this way, we are led to the surprising conclusion that the negation symbol stands for something in the world" (Soames 2003: 187–188). But, as Landini noted in correspondence, this interpretation is mistaken because Russell states that he takes negative facts to be ultimate and fundamental, which means they cannot have an abstract constituent to stand for the negation symbol.

12. My interpretation of extra-logical facts as facts about the causal structure of reality is supported by the following passage in "The Nature of Inference" (1920): "From the premise that 'Socrates was a wise man' we can validly infer that he was a man. Such inferences are considered trivial when they are simple, as in the above instance, yet they cover the whole of pure mathematics. But they do not cover anything else. They do not cover, for example, the sort of inferences which lead to a man's being condemned in a criminal trial. All such inferences rest upon causality and are in some sense extra-logical" (*TNI*: 84–85).

## References

### Works by Other Authors

Allaire, Edwin (1963). "Bare Particulars." In *Metaphysics: Contemporary Readings*, ed. Michael J. Loux. Routledge, 2008: 114–120.

Bostock, David (2012). *Russell's Logical Atomism*. Oxford: Oxford University Press.

Bradley, Francis Herbert (1893). *Appearance and Reality: A Metaphysical Essay*. Second edition (revised), with an Appendix. First Published 1897. London: Swan Sonneschein & Co. Ltd. 1908.

Engel, Pascal (2006). "Bertrand Russell: An Inquiry into Meaning and Truth." *Central Works of Philosophy, Volume 4: The Twentieth Century: From Moore to Popper*, ed. John Shand. Kingston, Canada: McGill-Queens University Press. 134–154.

Goodman, Nelson (1968). "Words, Works, and Worlds," *The Pragmatism Reader: From Peirce through the Present*, eds. Robert B. Talisse and Scott F. Aikin. Princeton University Press, 2011: 174–187.

Goodman, Nelson (1979). "The New Riddle of Induction," in *Fact, Fiction, and Forecast*. Harvard: Harvard University Press, 1983.

Koç Maclean, Gülberk (2014). *Bertrand Russell's Bundle Theory of Particulars*. London: Bloomsbury.

O'Leary-Hawthorne, John (1995). "The Bundle Theory of Substance and the Identity of Indiscernibles." *Analysis* 55.3: 191–196.
Soames, Scott (2003). *The Philosophical Analysis in the 20th Century Volume 1*, Princeton: Princeton University Press.
Van Cleve, James (1985). "Three Versions of the Bundle Theory of Particulars." In *Metaphysics: Contemporary Readings*, ed. Michael J. Loux. Routledge. 2008: 121–130.

PART II

# Influences on Russell's Logical Atomism

CHAPTER 4

# Russell and Frege on the Power of Symbols and the Compositionality of Linguistic Expressions

*Pieranna Garavaso*

## 1 Introduction

In the lectures on *The Philosophy of Logical Atomism*, Russell acknowledges his debt to Ludwig Wittgenstein by mentioning his former student in four places: in the preface to the publication of the lectures in *The Monist*, when discussing relations between facts and beliefs (*PLA*: 167), in the third lecture on atomic and molecular propositions (*PLA*: 182), and when discussing how to describe the logical form of a belief (*PLA*: 199). In Russell's discussion of facts, one can hear loud and clear Wittgenstein's voice; Russell's statement "the world contains facts" (*PLA*: 163) echoes proposition 1.2 of the *Tractatus Logico-Philosophicus*: "The world divides into facts" (*Die Welt zerfällt Tatsachen*).

Gottlob Frege is never mentioned in these lectures of Russell; yet when reading them one hears also Frege's voice. Is this the echo of an actual reading or communication between Russell and Frege, just like the echoes

P. Garavaso (✉)
University of Minnesota, Morris, MN, USA
e-mail: garavapf@morris.umn.edu



L. D. C. Elkind, G. Landini (eds.), *The Philosophy of Logical Atomism*, History of Analytic Philosophy,
https://doi.org/10.1007/978-3-319-94364-0_4

of Wittgenstein's voice? We know that Russell read the first volume of Frege's *Grundgesetze der Arithmetik* (1893) since he refers to it in his 1901 letter to Frege regarding the contradiction he had discovered in Frege's system. The appendix to Russell's *The Principles of Mathematics* attests to the fact that by 1903, Russell had read some parts of Frege's *Begriffsschrift* (1879) and *Die Grundlagen der Arithmetik* (1884), although not when they were originally published but later on when Russell realized that those works of Frege dealt with problems he was also thinking about.[1] The appendix shows that Russell also read some of Frege's early articles such as "*Über Sinn und Bedeutung*," and "On Concept and Object", but did he read also "On the Scientific Justification for a Conceptual Notation?"

Michael Beaney points out many similarities between Russell and Frege in their theoretical interests and many differences in their lives, personalities, and careers. Beaney reclaims for Giuseppe Peano the merit of having drawn attention to Frege's work, although he also admits that "it was Russell who was instrumental in introducing Frege's work to the English-speaking world" (Beaney 2003: 130). The historical evidence seems to indicate that in 1901–1902 Russell was not aware of Frege's comments on symbols and language as expressed in his earlier writings such as *Begriffsschrift* and the articles mentioned above. However, the lack of evidence of an actual exchange between Frege and Russell on some matters does not automatically disqualify our understanding of their work as somehow mutually influential beyond the documented historical interactions between them. Beaney states (Beaney 2003: 139):

> … it was just because Russell had worked his own way through the problems that he was able to recognize Frege's genius, and (given his generosity in acknowledging the work of others) do what he did in making Frege's writings better known. So although it does seem that Russell arrived at his logical results independently of Frege, Frege nevertheless played a role in sharpening their articulation. In turn, it was Russell's deeper involvement in the philosophical debates of the day that brought out the significance of Frege's work, and indeed, later prompted Frege himself to clarify his own philosophical ideas, as their correspondence from 1902, in particular, shows.

I agree with Beaney that the mutual interaction between these two philosophers and logicians might have been more subtle and indirect than can be captured by the historical evidence of actual exchanges and contacts. In contrasting Russell and Frege, Beaney focuses on the better-known and more widely discussed parts of Frege's and Russell's thought, that is, on their work in mathematical logic especially with regard to the concept of number, Russell's contradiction, and the notion of analysis. In this chapter, I explore some other topics on which I perceive some interesting similarities and dissimilarities between Russell's and Frege's positions. In the absence of historical evidence that Frege and Russell directly discussed these matters, for example, in their correspondence following Russell's discovery of the contradiction in Frege's system, each of them may have been expressing ideas that were in some form being discussed by mathematicians, logicians, and philosophers of logic and mathematics who were their contemporaries and that these ideas were simply independently captured by both thinkers. Additionally, I would like to submit that, when Frege and Russell express different ideas, they may have been expressing views that were more deeply their own.

I discuss two *prima facie* unrelated topics of contrast and comparison between Russell and Frege. The first topic is the importance of logical symbolism: this is an example of both harmony and contrast. Both philosophers stress the significance and power of symbols; however, contrary to common opinion, I argue that while Russell's main concerns with regard to symbols were their capacity to "fit" and adequately represent the components and structure of the symbolized reality, Frege's concerns were focused more strongly on the relationships between symbols and the conceptual or informational content they symbolized and on the power of symbolic expressions in engendering the grasping of their contents. I characterize this difference between Russell's and Frege's emphases by labeling Russell's concerns as "metaphysical or ontological" and Frege's as "epistemological or conceptual."

The second theme focuses on the thesis supported by both Russell and Frege that the composition of linguistic expressions representing Russellian facts or Fregean thoughts, respectively, corresponds to the complexity of the thoughts and facts that those linguistic expressions represent. The debate on Frege's views on the composition of thoughts and their alleged inconsistency with Frege's repeated assertions that different sentences express the same thought has generated an extended literature[2]; some of which also include discussions of Frege's logical notation in his more technical writings such as the *Grundgesetze*.[3] This chapter has a narrower scope and focuses on Russell's comments on the composition of propositions as stated in his lectures on *The Philosophy of Logical Atomism* and on Frege's articles from the period after the publication of Begriffsschrift and before the publication of the first volume of the *Grudgesetze*. Russell changed many of his views during his long philosophical career; yet, the examination of Frege's and Russell's views in these specific time frames provides an interesting contrast even if they may not be Frege's and Russell's definitive views.

A query that is at the origin of this chapter and that ties a link between the topics I discuss is the following: As there are Fregean echoes in Russell's lecture on *The Philosophy of Logical Atomism*, might there also be Russellian echoes in the scholarly debate about Frege's Principle of Compositionality? In pursuit of a substantiated answer to this question, I submit that the contrast of emphases between Russell's and Frege's views on the power and significance of symbols, which I outline in the first part of the chapter, provides a useful hypothesis to explain how a Russellian "metaphysical" reading of the complexity of linguistic expressions might have prevented a more accurate "epistemological" interpretation of Frege's remarks on the compositionality of logical and ordinary language.

## 2 The Power and Significance of Symbols

In *The Philosophy of Logical Atomism*, Russell states, "There is a good deal of importance to philosophy in the theory of symbolism, a good deal more than one time I thought" (ibid.). Frege also strongly believed in and stressed the significance of symbols; in his article "On the Scientific Justification of a Conceptual Notation," published in 1882, in which he advocates for, and attempts to divulgate the merits of his Begriffsschrift, Frege states, "Symbols have the same importance for *thinking* that discovering how to use the wind to sail against the wind had for navigation. Thus, let no one despise symbols! A great deal depends upon choosing them properly" (Frege 1972b: 84).[4] For both Russell and Frege, "symbolism" is a broad term that includes different types of signs from words to sentences to mathematical and logical operators and their use is very important in both word languages and mathematical or logical languages (*PLA*: 166):

> Perhaps I ought to say a word or two about what I am understanding by symbolism, because I think some people think you only mean mathematical symbols when you talk about symbolism. I am using it in a sense to include all language of every sort and kind, so that every word is a symbol, and every sentence, and so forth. When I speak of a symbol I simply mean something that "means" something else, and as to what I mean by "meaning" I am not prepared to tell you.

Although Frege does not provide a definition of the meaning of "symbol," his discussion of them always includes examples from mathematics or logic as well as from ordinary language. Thus, for Frege as well as for Russell, symbolism is to be understood broadly.

Furthermore, both Russell and Frege stress the fact that symbols are concrete and sensible: the physical nature of symbols is an essential feature from which their efficacy ensues. For Russell, we "think about the symbols, because they are tangible" (*PLA*: 166); Frege points out the link between the "tangible" and "sensible" nature of symbols and their usefulness as an aid to human thinking (Frege 1972b: 83):

> Even most animals, through their ability to move about, have an influence on their sense-impressions: they can flee some, seek others. And they can even effect changes in things. Now man has this ability to a much greater degree; but nevertheless, the course of our ideas would still not gain its full freedom from this [ability alone]: it would still be limited to that which our hand can fashion, our voice intone, without the great invention of symbols which call to mind that which is absent, invisible, perhaps even beyond the senses. ... And [the value of symbols] is not diminished by the fact that, after long practice, we need no longer produce [external] symbols, we need no longer speak out loud in order to think; for we think in words nevertheless, and if not in word, then in mathematical or other symbols.

For Frege, symbols make it possible for thinking to arise. For Russell, the physical nature of symbols explains why we use them: symbols fit our nature, that is, the fact that our learning and other mental activities rely on sensory experiences. Russell and Frege also agree on the fact that the importance of symbols is to a certain extent "negative," as Russell qualifies it, that is, due to the symbolism's role in avoiding mistakes. Russell warns (*PLA*: 166):

> I think the importance is almost entirely negative, i.e., the importance lies in the fact that unless you are fairly self-conscious about symbols, unless you are fairly aware of the relation of the symbol to what it symbolizes, you will find yourself attributing to the thing properties which only belong to the symbol. That, of course, is especially likely in very abstract studies such as philosophical logic, because the subject-matter that you are supposed to be thinking of is so exceedingly difficult and elusive that any person who has ever tried to think about it knows you do not think about it except perhaps once in six months for half a minute. The rest of the time you think about the symbols, because they are tangible, but the thing you are supposed to be

> thinking about is fearfully difficult and one does not often manage to think about it. The really good philosopher is the one who does once in six months think about it for a minute. Bad philosophers never do.

For Russell, other logicians and philosophers have not given enough attention to symbols or they seldom do so, especially to the symbols' relations to what they symbolize. In the second half of this passage, Russell describes this common mistake as that of attributing properties that belong to the symbols to the objects that are symbolized.[5] An accurate symbolism can prevent mistakes and this is the "negative" importance of symbols. At the very beginning of "On the Scientific Justification," Frege points out (Frege 1972b: 83):

> Time and again, in the more abstract regions of science, the lack of a means of avoiding misunderstanding on the part of others, and also errors in one's own *thinking*, makes itself felt. Both [shortcomings] have their origin in the imperfection of language, for we do have to use sensible symbols to think. (my emphasis)

In "Concept and Object," originally published in 1892, that is, ten years after the publication of "On the Scientific Justification of a Conceptual Notation," Frege still warns that "we must not let ourselves be deceived because language often uses the same word now as a proper name, now as a concept" (Frege 1984: 189). Two of Frege's main concerns with ordinary language are (1) its ambiguity, that is, the fact that the same term might be used to symbolize different things; and (2) the looseness of its grammar that does not align accurately with the formal correctness of logic (Frege 1972b: 84–85):

> Language proves to be deficient … when it comes to protecting *thinking* from error. It does not even meet the first requirement which we must place upon it in this respect; namely, being unambiguous. …
>
> Language is not governed by logical laws in such a way that mere adherence to grammar would guarantee the formal correctness of thought processes.

So far we have found at least three similar features in Frege's and Russell's views concerning symbolism. Symbols are important and must be carefully chosen; they capture our attention and are useful because of their tangible or sensible nature; and they have a vital "negative" role in preventing errors. Despite these commonalities, however, there are also interesting differences in Russell's and Frege's views on the role of symbols.

While for Russell the significance of symbols is mostly negative as it "lies in the fact that unless you are fairly self-conscious about symbols, unless you are fairly aware of the relation of the symbol to what it symbolizes, you will find yourself attributing to the thing properties which only belong to the symbol" (*PLA*: 166), Frege appreciates both the necessity and the positive roles of symbolisms much more readily than Russell:

> [O]ur thinking is closely bound up with language and thereby with the world of the senses. (Frege 1979c: 269).
>
> [W]e have to use symbols to think. …
>
> Also, without symbols we would scarcely lift ourselves to conceptual thinking. …
>
> This does not exhaust the merits of symbols; but it may suffice to demonstrate their indispensability. (Frege 1972b: 83–84)

For Frege, certainly more explicitly than for Russell, symbols are not merely a necessary evil; rather, they play an indispensable role in allowing human thinkers to reach conceptual thinking. Without symbols, we could not generalize and think about the commonalities between present and perceptible objects and objects we perceived in the past. Because of the ambiguity and logical looseness of ordinary language, "We need a system of symbols from which every ambiguity is banned, which has a strict logical form from which the content cannot escape" (Frege 1972b: 85).

Russell acknowledges that symbols have an important epistemological function, which is inextricably connected to the psychological features of each person's apprehension of a cognitive content (*PLA*: 167):

> I will in the course of time enumerate a strictly infinite number of different things that "meaning" may mean but I shall not consider that I have exhausted the discussion by doing that. I think that the notion of meaning is always more or less psychological, and that it is not possible to get a pure logical theory of meaning, nor therefore of symbolism. I think that it is of the very essence of the explanation of what you mean by a symbol to take account of such things as knowing, of cognitive relations, and probably also of association. At any rate I am pretty clear that the theory of symbolism and *the use of symbolism is not a thing that can be explained in pure logic without taking account of the various cognitive relations that you may have to things.* (my emphasis)

In this passage, Russell outlines a psychological notion of meaning that relies on and represents the individual cognitive associations that symbols have with the objects they signify. So, the word "Moon" that symbolizes the Earth's satellite will mean also the psychological associations linked to my experiences of seeing the moon in this and in other countries. The psychological notion of meaning that Russell mentions in the above passage resembles Frege's notion of the idea or mental image (*Vorstellung*) linked to a word that Frege contrasts with the object or referent and the sense or mode of presentation in "Über *Sinn und Bedeutung*" (Frege 1892b: 161–162):

> The meaning of a proper name is the object itself which we designate by using it; the idea which we have in that case is wholly subjective; in between lies the sense, which is indeed no longer subjective like the idea, but is yet not the object itself. The following analogy will perhaps clarify these relationships. Somebody observes the Moon through a telescope. I compare the Moon itself to the meaning; it is the object of the observation, mediated by the real image projected by the object glass in the interior of the telescope, and by the retinal image of the observer. The former I compare to the sense, the latter is like the idea or experience. The optical image in the telescope is indeed one-sided and dependent upon the standpoint of observation; but it is still objective, inasmuch as it can be used by several observers. At any rate it could be arranged for several to use it simultaneously. But each one would have his own retinal image. … [W]hen we say 'the Moon', we do not intend to speak of our idea of the Moon, nor are we satisfied with the sense alone, but we presuppose a meaning. To assume that in the sentence 'The Moon is smaller than the Earth' the idea of the Moon is in question, would be flatly to misunderstand the sense. If this is what the speaker wanted, he would use the phrase 'my idea of the Moon'.

In this famous and useful analogy, Frege depicts a tripartite distinction between three realms that are each related in a different fashion to the use of symbols. At the most subjective level is the idea; this is the level most closely connected with psychological features, for example, the idea that comes to my mind when I hear or use the word "Moon." Frege compares the idea to the retinal image of the Moon seen through the telescope, which is different for each individual person. At the other end of the spectrum, and of the telescope, is the planet Moon, the real entity seen through the lens of the telescope and referred to by the word "Moon" when we use a sentence about that planet. In the middle is the optical image of the Moon which corresponds to the sense or the conception connected to the word "Moon"; this is neither the object nor the idea, rather the content of information we grasp when we hear or understand this word, that is, a content that is intermediate between the subjective ideas and the actual objects. The sense has distinguishing non-subjective and non-psychological features that derive from the particular point of view of the telescope and how it captures an image of the Moon. Senses are specific ways in which the informational content of a word, a symbol, or a sentence is transmitted to those who hear or grasp the word, symbol, or sentence. Thus, for both Russell and Frege, symbols are connected with aspects of meaning that cannot be explained in the mere logical form of language. In the last quoted passage, Russell points out that "the use of symbolism is not a thing that can be explained in pure logic without taking account of the various cognitive relations that you may have to things" (*PLA*: 167). This point was very clear for Frege as well as he distinguishes between what he calls the conception or sense of words, the "logical kernel" of thoughts, on one side, and their "psychological" verbal clothing. Frege even characterizes different types of thinking according to their different level of entanglement with psychological components.[6]

There is a further difference that I would like to point out between Russell's and Frege's approaches to symbolism. Frege worried about the potential ambiguity of symbols, that is, their being often used to name both a concept and an individual, as it happens with the word "horse" (Frege 1972b: 84). Russell worries about ambiguity too: "That is why the theory of symbolism has a certain importance, because otherwise you are so certain to mistake the properties of the symbolism for the properties of the thing" (11). Yet, Russell's concern seems quite different from Frege's (*PLA*: 166):

> [The theory of symbolism] has other interesting sides to it too. There are different kinds of symbols, different kinds of relation between symbol and what is symbolized, and very important fallacies arise from not realizing this. The sort of contradictions about which I shall be speaking in connection with types in a later lecture all arise from mistakes in symbolism, from putting one sort of symbol in the place where another sort of symbol ought to be. Some of the notions that have been thought absolutely fundamental in philosophy have arisen, I believe, entirely through mistakes as to symbolism—e.g. the notion of existence, or, if you like, reality. ... Now my own belief is that as [these mistakes] have occurred in philosophy, they have been entirely the outcome of a muddle about symbolism, and that when you have cleared up that muddle, you find that practically everything that has been said about existence is sheer and simple mistake, and that is all you can say about it. I shall go into that in a later lecture, but it is an example of the way in which symbolism is important.

In this passage, Russell stresses the importance of symbols in a metaphysical sense, that is, in order to keep the type of entities that are symbolized clearly distinct, rather than in an epistemological sense, that is, in order to avoid ambiguities and aid us in correctly identifying what the symbols represent. Russell's metaphysical concern originates not only from the aim of avoiding ambiguity but also from a desire to provide an accurate symbolic representation of different kinds of reality. For Russell, the core goal and significance of language is its aptness to represent reality accurately. Appealing to Frege's tripartite distinction between ideas, senses, and real objects, I would characterize Russell's concern as more intensely focused on the link between symbols and objects, rather than between symbols and either ideas or senses. At least in the articles I have been quoting from, Frege's emphasis was more epistemological than Russell's in the sense that it was focused on the challenge of choosing symbols that are best suited to represent the senses or the content of information connected with the objects or aspects of reality they denote.

In his 1990 article criticizing what he takes to be Gareth Evans' overly "Russellian" interpretation of Frege, David Bell also supports a reading of Frege's early views on language that are substantially different from Russell's. In particular, Bell argues that Frege's concern with the expressivity of language was broader than Russell's as it stressed the different functions of language: "[T]he semantic model presupposed in the Begriffsschrift appeals, not to referential relations between linguistic and non-linguistic items, but to the expressive function of language: language is significant in so far as it successfully expresses our thoughts, judgements, ideas, and reasonings" (Bell 1990: 271). I read Frege and Russell as Bell does: while for Russell the referential relation of symbols to objects, of propositions to facts is crucial and exemplifies the core function of language, for Frege language can be significant, thoughts may have a sense, whether or not their components "hook up" with reality. Bell explains as follows (Bell 1990: 270):

> Frege did not begin with a general assumption that language functions by referring to, or describing reality, nor did he assume that the thoughts expressed by its means are thoughts about reality. His problem was not to explain how language can hook on to items in the world—as part of an attempt to account for the overall possibility of meaningful language as such—but to explain and codify relations of logical entailment that hold between sentences, or between the thoughts they express.

That this is an accurate representation of Frege's view can be seen from the following passage from Frege's "Introduction to Logic" (1906), also quoted by Bell (Frege 1979a, b: 191).

> [A] proper name is a constituent of a sentence, which expresses a thought. Now what has the object got to do with the thought? We have seen from the sentence 'Mont Blanc is over 4000 m high' that it is not part of the thought. Is then the object necessary at all for the sentence to express a thought? … [T]he object designated by a proper name seems to be quite inessential to the thought-content of a sentence which contains it. To the thought-content! For the rest, it goes without saying that it is by no means a matter of indifference to us whether we are operating in the realm of fiction or truth.

Frege, of course, does not think that truth and fiction are the same, but the content of thought expressed by a sentence is not affected by the lack of reference for some of the sentence's parts. And this is because Frege's focus was on the cognitive significance of sentences rather than on their correctly representing reality, that is, on their truth or falsehood.

In conclusion, since I have not found any passages in Frege's early discussions of the importance of symbolism that parallel what I call the metaphysical concerns of Russell's in the *The Philosophy of Logical Atomism*, I submit that their views on symbols were, despite their more superficial similarities, as seen in the quotes cited at the beginning of this section, quite different. As a concern with reference is quite explicit in Russell, I regard Frege's comments on the importance of symbols to be at least more focused on their epistemic and logical/inferential value than Russell's, if not exclusively so.

## 3 The Compositionality of Linguistic Expression and the Complexity of Facts and Thoughts

In the lectures on logical atomism, Russell believed that propositions are symbols and that they have internal complexity (*PLA*: 166):

> A proposition is just a symbol. It is a complex symbol in the sense that it has parts which are also symbols: a symbol may be defined as complex when it has parts that are symbols. In a sentence containing several words, the several words are each symbols, and the sentence comprising them is therefore a complex symbol in that sense.

Russell uses the terms "proposition," "statement," and "sentence" synonymously[7] and, at the end of the following passage, explicitly defines propositions as sentences (*PLA*: 165):

> [S]tatements or propositions or judgments [are] all those things that do have the duality of truth and falsehood. … A proposition, one may say, is a sentence in the indicative, a sentence asserting something, not questioning or commanding or wishing. It may also be a sentence of that sort preceded by the word "that."

For Russell, a proposition is what we usually call a declarative sentence or an objective clause, that is, a sentence that takes the place of the object of verbs that express propositional attitudes such at "to believe" and "to know." Propositions symbolize or express facts: "What I call a fact is the sort of thing that is expressed by a whole sentence, not by a single name like 'Socrates'" (*PLA*: 164) and the facts described make the propositions true or false according to whether or not they obtain (*PLA*: 167):

> [T]here are two propositions corresponding to each fact. Suppose it is a fact that Socrates is dead. You have two propositions: "Socrates is dead" and "Socrates is not dead." And those two propositions corresponding to the same fact, there is one fact in the world which makes one true and one false. That is not accidental, and illustrates how the relation of proposition to fact is a totally different one from the relation of name to the thing named. For each fact there are two propositions, one true and one false and there is nothing in the nature of the symbol to show us which is the true one and which is the false one.

There is a relationship between facts in the world and propositions; propositions are symbols for facts and for every fact there are two propositions relating to it. The true proposition is the one that symbolizes a fact that obtains; the false one is the one that symbolizes a fact that does not obtain.[8] Nothing in the symbols can tell us whether or not a proposition is true.

Russell points out the internal complexity of propositions and facts and their mutual correspondence: "[t]here is an objective complexity in the world, and … it is mirrored by the complexity of propositions" (*PLA*: 176). The complexity of the proposition, that is, of the symbol, correspond to the complexity of the fact asserted by the proposition: "The complexity of a fact is evidenced, to begin with, by the circumstance that the proposition that asserts a fact consists of several words, each of which may occur in other contexts" (*PLA*: 171). The words that compose a proposition are symbols that occur in several different contexts; our grasp of the content of a proposition is thus dependent on our ability to understand the meaning of those components (*PLA*: 174–175):

We may lay down the following provisional definitions:

> That the components of a proposition are the symbols we must understand in order to understand the proposition;
>
> That the components of the fact which makes a proposition true or false, as the case may be, are the *meanings* of the symbols which we must understand in order to understand the proposition.

Russell does declare that he is not prepared to provide a definition of "meaning" in these lectures and also that there is "a strictly infinite number of different things that 'meaning' may mean" (*PLA*: 167). For the purposes of my discussion, the above definition is important because it shows how for Russell, the parts of a proposition, that is, the words that compose the proposition, are related to the parts or components of the fact symbolized by the proposition. Furthermore, Russell points out how we are able to grasp the content of a whole proposition by understanding the individual symbols composing it. Thus, the complexity and articulation of propositions play an epistemological role in allowing us to grasp and understand the informational content provided by these linguistic entities; the content we grasp is the outcome of the combined input that the components provide. I will call this the epistemic significance of the principle of compositionality that asserts that there is a correspondence between the parts of sentences and the parts of what the sentences describe.

Frege echoes Russell's views as expressed in the last quote concerning both the parallel complexity of sentences and what they express and the epistemological function that this complexity plays. Here is one of Frege's statements of the Principle of Compositionality: "The world of thoughts has a model in the world of sentences, expressions, words, signs. To the structure of the thought there corresponds the compounding of words into a sentence" (Frege 1977a: 38). Notice that for Frege, sentences, expressions, words, and signs correspond to thoughts; Frege does not appeal to facts as the content described by symbols. In connection with the thesis that the meaning of a complex symbol or expression is fully determined by the meaning of its components, Frege also states: "[understanding a totally new thought] would be impossible, were we not able to distinguish parts in the thought corresponding to the parts of a sentence, so that the structure of the sentence serves as an image of the structure of the thought" (Frege 1977b: 55). This comment clearly mirrors Russell's claim in the second provisional definition quoted above to the effect that "the components of the fact which makes a proposition true or false ... are the *meanings* of the symbols which we must understand in order to understand the proposition" (*PLA*: 175). Thus, for both Russell and Frege, the internal articulation of whatever verbal representations we use to express facts and thoughts plays a role in our ability to grasp the information conveyed by these verbal expressions.

We can thus preliminarily conclude that Frege and Russell shared the theses that (1) there is a complexity in symbolic expressions such as sentences and propositions that has an epistemic role in our ability to understand new sentences and propositions composed of symbols with which we are already familiar and (2) there is a correspondence between the components of sentences and propositions and what these sentences and propositions symbolize, that is, thoughts and facts, respectively.

I would like now to examine what I believe are some differences between Russell's and Frege's views on the parallel compositionality of facts/thoughts and linguistic representations of facts/thoughts. The first important difference is the difference between a Fregean thought and a Russellian fact; this difference is crucial in understanding Russell's and Frege's goals but it is difficult to make it without risking of misrepresenting Frege's position. For Russell, facts are clearly in the world, "the world contains facts" (*PLA*: 163), facts are what constitute reality. Facts are as real as any of the objects of our perception. Russell is concerned to clarify that the complexity inherent in facts is not psychological or mind dependent; following the above quoted provisional definitions regarding the compositionality of propositions and facts, Russell states (*PLA*: 175)

> I call these definitions preliminary because they start from the complexity of the proposition, which they define psychologically, and proceed to the complexity of the fact, whereas it is quite clear that in an orderly, proper procedure it is the complexity of the fact that you would start from. It is also clear that the complexity of the fact cannot be something merely psychological. If in astronomical fact the earth moves round the sun, that is genuinely complex. It is not that you think it complex, it is a sort of genuine objective complexity, and therefore one ought in a proper, orderly procedure to start from the complexity of the world and arrive at the complexity of the proposition.

It is clear from this passage that for Russell the complexity of a proposition or sentence symbolically expressing a fact is dependent on the original complexity of the fact symbolized by the proposition. This complexity is not psychological: "It might be suggested that complexity is essentially to do with symbols, or that it is essentially psychological. I do not think it would be possible seriously to maintain either of these views" (*PLA*: 175). The use of the term "psychological" to describe the complexity of the proposition is interesting; we have seen earlier how Russell outlines what he calls a psychological notion of meaning based on the individual cognitive associations between symbols and the objects they signify. The articulation of propositions into parts symbolizing the parts of the symbolized facts is guided by and mirrors the actual, objective, and primary articulation of the world; it does not follow the articulations coming from the mind.

Although Frege's thoughts are not Russell's facts, similarly to facts they are not mind-dependent entities in any way. Thoughts are objective for Frege and they do not reduce to individual acts of thinking; the following passages are exemplary of Frege's conception of thoughts that he denoted by the German terms *Gedanke* and *Gedanken* in contrast with the term *Denken* that he used to indicate the psychological or mental activity of thinking: "thoughts are independent of our thinking;" "thoughts are not generated by, but grasped by, thinking" (Frege 1979a, b: 127, 133): "Unlike ideas, thoughts do not belong to the individual mind (they are not subjective), but are independent of our thinking and confront each one of us in the same way (objectively)" (Frege 1979a, b: 127, 133, 148); "Thoughts are not mental entities, and thinking is not an inner generation of such entities but the grasping of thoughts which are already present objectively" (Frege 1980: 67). For Frege, thoughts are objective in the sense that they can be grasped by different knowers, are not mental or mind-dependent entities such as ideas, and they are not characterized or affected by psychological coloring.

There has been, however, extensive debate on whether or not Frege believed that thoughts have unique essential structures. The affirmative interpretation traces back to Dummett's reading of Frege; for Dummett's Frege, analysis is *unique* and "displays ... the 'essential structure' of the sentence, and may be said to uncover the internal structure of the thought it expresses" (Dummett 1981: 272).[9] However, Frege also repeatedly and clearly states that different sentences may express the same thought.[10] As a result, Fregean scholars have been faced with the challenge of resolving an apparent inconsistency between the theses that (i) thoughts have unique internal structures, that (ii) the structure or composition of a sentence mirrors the structure or composition of the thought it expresses, and that (iii) different sentences (with different structures) may express the same thought—which according to (i) has only one essential structure. Dummett's proposal is to distinguish between analysis, which gives us the unique essential structure of a thought, and decomposition, which provides different ways of separating the components of sentences and thoughts.

A second difference between Russell's and Frege's views on the compositionality of sentences expressing facts and propositions, respectively, concerns the domain of relevant linguistic expressions. When Frege discusses the structure of thoughts and the fact that different sentences may express the same thought, he does not draw his examples only from an ideal or perfect language; rather, his examples are taken both from ordinary language and from mathematical discourse. Two of Frege's examples of pairs of sentences expressing the same thoughts are "At Platea the Greeks defeated the Persians" and "At Platea the Persians were defeated by the Greeks" and "There is at least one square root of 4" and "The concept *square root of 4* is realized" (Frege 1972a: 112; 1984: 188). With regard to the compositionality of language, Russell explicitly mentions an ideally perfect logical language (*PLA*: 176):

> [I]n a logically correct symbolism there will always be a certain fundamental identity of structure between a fact and the symbol for it; and … the complexity of the symbol corresponds very closely with the complexity of the facts symbolized by it …
>
> In a logically perfect language the words in a proposition would correspond one by one with the components of the corresponding fact, with the exception of such words as "or", "not", "if", "then" which have a different function.

This difference in the range of symbolic expressions included in the thesis that structure of linguistic expressions mirrors the structure of the contents expressed by the relative symbolic expression is significant in the attempt to resolve the alleged inconsistency in Frege's position. For even solutions that look at Frege's more technical work in order to find an interpretation that eliminated the inconsistency need to be able to handle also the examples of sentences of ordinary language supposedly expressing the same thought.

I would like to conclude this discussion by suggesting that the difference that we found in the first section of this chapter may lead us to see an analogous difference in the way in which Russell sees the correspondence between the structure of facts and the structure of sentences expressing those facts and the way in which Frege's sees the correspondence between the structure of sentences and the structure of thoughts expressed by the relative sentences. For Russell, the structure of a fact is primary and it dictates how the propositions will be articulated. For Frege, it is the compositionality of a sentence that determines how we will grasp a certain thought similarly to how senses are modes of presentation that help us identify objects signified by a certain symbol. Recall that for Russell it was of the utmost importance that symbols were adequate to what they were supposed to symbolize. In contrast, Frege stresses that the choice of symbols is important mostly for their epistemic role in allowing or helping us to grasp what is symbolized. For Russell, the internal articulation of a proposition is an inevitable consequence of the essential internal articulation of the fact that makes that proposition true. As Frege's thoughts are not sensible entities and thus they are not perceivable by the senses, we need sensible symbols to grasp them. For Frege, a symbolic language of any kind is our indispensable means for grasping thoughts and this essential role that language plays in enabling us to grasp thoughts brings about the complexity of the symbolic representations of thoughts. In Frege's view, the articulation of sentences is a core feature of language that has a crucial epistemic function for our understanding what language expresses and for communicating with each other in any linguistic context, whether within a logically correct or an ordinary language. For Russell, the complexity of propositions is an inevitable consequence of the ontological complexity of facts.

Let us recall Frege's Moon analogy once again to better illustrate the contrast between Russell's approach to the complexity of language expressions vis-à-vis reality versus Frege's views on the complexity of sentences expressing thoughts. I like to describe this contrast by saying that while there is an alleged metaphysical feature of reality at the source of Russell's claim that language is compositional, there is an epistemic feature of human ability to grasp thoughts at the origin of Frege's claim that language has a compositional nature. Whereas Frege is focused on the correspondence between the compositionality of linguistic expressions and their ability to reflect the internal structure of the informational content that these expressions convey, Russell is focused on the connection between the complexity of the linguistic expressions vis-à-vis the internal complexity of facts as the portion of reality those linguistic expressions represent. In the tripartite framework of Frege's telescope's analogy, Russell examines the relationships between objects or facts and symbolic expression, whereas Frege's interest is more on the relationships between linguistic expressions and informational contents or senses. Frege's and Russell's emphases can thus be regarded as complementary rather than oppositional; yet it is wrong to take them as analogous.

A remaining well worth pursuing question that exceeds the scope of this chapter is whether Russell's approach to the compositionality of language may not have functioned as a lens through which many scholars have read, in my opinion not fully accurately, Frege's views on the composition of sentences and thoughts. This reading may have been the reason why many scholars have attributed to Frege the belief that thoughts too, just like Russell's facts, have one unique essential internal structure. I submit that my suggestion that this is an inaccurate reading of Frege's comments on the compositionality of language is supported by the differences between Russell's and Frege's views on symbols and their roles, which we explored in the first section.[11]

## Notes

1. In his "Russell and Frege," Michael Beaney points out that James Ward, Russell's tutor, gave him a copy of *Begriffsschrift* in October 1895, when Russell received a fellowship at Oxford.

   Russell however reports in his autobiography that he "possessed the book for years before [he] could make out what it meant. Indeed, I did not understand it until I had myself independently discovered most of what it contained" (cited in Beaney 2003: 130–131); "Russell reports that he only read [Frege's *Grundlagen*] some sixteen years after its publication [i.e., around 1900]" (Beaney 2003: 131).

   See also, "It was not until 1902, after POM (*Principles of Mathematics*) had gone to press, that Russell discovered Frege's work" (Griffin 2003: 21, FN 23); "It will be noted that many such locutions are readily handled by the quantification theory that Frege had developed in the *Begriffsschrift* in 1879 and of which Russell was still unaware" (Griffin 2003: 23).

   Gideon Makin interestingly characterizes Russell and Frege as "theoretically" contemporaries despite Frege's chronological precedence: "[Frege's] work became known to Russell, at least in any detail, only after Russell had completed his *Principles*.

... So although chronology would have made it possible for Russell to be of a subsequent philosophical generation (as, for example, Wittgenstein relates to Russell and Frege), circumstances made them, in effect, into contemporaries whose philosophical developments ran parallel to each other" (Makin 2000: 6).

2. The literature on this topic is quite extended; Bronzo (2017) and Klement (forthcoming) provide useful and updated references.

   Michael Dummett offers a solution of the alleged conflict between Frege's views on the composition of thoughts and his thesis that different sentences can express the same thought based on a distinction between analysis and decomposition.

   Some scholars who discuss Dummett's approach are Hodes (1982), Currie (1985), Garavaso (1991), Bermúdez (2001), Penco (2003), Textor (2009), and Kemmerling (2010).
3. See for example Klement (forthcoming) and Landini (1996, 2012: Chap. 5).
4. In this passage, Frege uses *Denken* not *Gedanke*; I have accordingly altered Bynum's translation "thought" with "thinking" to reflect Frege's choice of terms; in all cases in which I replace "thought" or "thoughts" with "thinking" in standard and commonly cited translations of Frege, I signal the change by italicizing the term "thinking."

   For a defense of the use of "thinking" instead of "thought/thoughts" in translating Frege's relevant passages as well as evidence that Frege's use of these terms was quite deliberate, see Garavaso and Vassallo (2014: 9–13).
5. An example of the misunderstanding Russell had in mind here would be very useful.

   I only surmise that perhaps he was thinking of expressions such as fictional names or definite descriptions that function as grammatical subjects in sentences and thus seem to refer to an object even if there might be no object to be referred to by the symbol or symbolic expression.

   The alleged referential property of the symbol leads to ascribing the property of being an object to what is symbolized.

   The remark about the lack of good philosophers thinking about symbolisms at the end of this passage seems like an odd remark from Russell if one is reminded of the fact that in that period several logicians such as George Boole, Peano, and Frege were quite concerned about, and engaged with, new types of symbolism.
6. For a thorough discussion of different types of thinking in Frege, see Garavaso and Vassallo (2014), especially Chap. 3.
7. In the texts I quote, Russell does not seem to distinguish between sentences or statements as sequences of words and propositions as the content of information conveyed by statements and sentences.

   Neither does he seem to distinguish between types and tokens of sentences.

Although I consider these distinctions important, I don't believe that their neglect creates irredeemable problems in the interpretation of the passages I quote from Russell's lectures.

8. In these lecture on logical atomism, Russell introduces the notion of negative facts: "Another distinction, which is perhaps a little more difficult to make, is between positive facts and negative facts, such as 'Socrates was alive'—a positive fact—and 'Socrates is not alive'—you might say a negative fact" (*PLA*: 165) and discusses it in greater detail in the last pages of the third lecture devoted to atomic and molecular propositions (*PLA*: 187–190).

   As negative facts don't play any significant role in the contrast I am trying to draw between Frege and Russell and since the discussion of Russell's notion of negative facts has generated spirited debates—even riots at Harvard—I leave this topic out of my discussion.
9. Andreas Kemmerling lists several scholars supporting this view: David Bell (1987), Dalia Drai, Michael Dummett, Paul Horwich, Jeffrey King, Christopher Peacocke, Ian Rumfitt, Stephen Schiffer, Robert Stalnaker, and Pavel Tichý (2011: 166).
10. For citations of Frege's examples of different sentences expressing the same thought, see Garavaso (1991, 2013), Janssen (2001), Pelletier (2001).
11. This chapter was written as a contribution to the Summer Seminar 2017 *Bertrand Russell's The Philosophy of Logical Atomism: A Centenary Celebration* organized by Landon Elkind and Gregory Landini at the Obermann Center for Advanced Studies at the University of Iowa. I am grateful to the organizers for the invitation to participate in this seminar and to all the participants for useful feedback on this project.

    I also thank my colleagues Clement Loo and Lory Lemke for helpful discussions and comments.

## References

### Works by Other Authors

Beaney, Michael (2003). "Russell and Frege." In *The Cambridge Companion to Bertrand Russell*, ed. Nicholas Griffin: 129–131. Massachusetts: Cambridge University Press.

Bell, David (1987). "Thoughts." *Notre Dame Journal of Formal Logic*, Vol. 28: 36–50.

Bell, David (1990). "How 'Russellian' Was Frege?" *Mind* no. 99: 267–277.

Bermúdez, José Luis (2001). "Frege on Thoughts and Their Structure." *Logical Analysis and History of Philosophy*, no. 4: 87–105.

Bronzo, Silver (2017). "Frege on Multiple Analyses and the Essential Articulatedness of Thought." *Journal for the History of Analytic Philosophy* 5 no. 10: 1–34.

Currie, Gregory (1985). "The Analysis of Thoughts." *Australasian Journal of Philosophy*, no. 63: 283–298.

Drai, Dalia (2002). "The Slingshot Argument: An Improved Version." *Ratio*, Vol. 15, No. 2: 194–204.

Dummett, Michael (1981). *The Interpretation of Frege's Philosophy*. London: Duckworth.

Frege, Gottlob (1879). *Begriffsschrift, eine der arithmetischen nachgebildete Formelspache des reinen Denkens*. In van Heijenoort, ed. (1967a): 1–82; 1st Published. Halle: Niebert.

Frege, Gottlob (1892a). "On Concept and Object." *Collected Papers on Mathematics, Logic and Philosophy*, tr. Max Black, V. H. Dudman, Peter Geach, Hans Kaal, H. H. W. Kluge, Brian McGuinness, and R. H. Stoothoff, 182–194. Oxford: Basil Blackwell, 1984.

Frege, Gottlob (1893). *Grundgesetze der Arithemtic*, Volume 1, Jena: Verlag Herman Pohle.

Frege, Gottlob (1972a). *Conceptual Notation and Related Articles*, tr. T. W. Bynum. Oxford: Clarendon Press [1879].

Frege, Gottlob (1972b). "On the Scientific Justification of a Conceptual Notation." *Conceptual Notation and Related Articles*, tr. T. W. Bynum, 83–89. Oxford: Clarendon Press [1882].

Frege, Gottlob (1977a). "Negation." *Logical Investigations*. New Haven: Yale University Press: 31–53.

Frege, Gottlob (1977b). "Compound Thoughts." *Logical Investigations*. New Haven: Yale University Press: 55–77.

Frege, Gottlob (1979a). "Introduction to Logic." In *Posthumous Writings:* 185–196.

Frege, Gottlob (1979b). *Posthumous Writings*, ed. Hermes, Kambartel, Oxford: Basil Blackwell.

Frege, Gottlob (1979c). "Sources of Knowledge of Mathematics and the Mathematical Natural Sciences." In *Posthumous Writings*: 267–274.

Frege, Gottlob (1980). "Frege to Husserl." In *Philosophical and Mathematical Correspondence*. Chicago: University of Chicago Press [1906]. 66–70.

Frege, Gottlob (1984). "Über Sinn und Bedeutung." *Collected Papers on Mathematics, Logic and Philosophy*, transl. Max Black et al., 157–181. Oxford: Basil Blackwell [1892].

Garavaso, Pieranna (1991). "Frege and the Analysis of Thoughts," *History and Philosophy of Logic*, no. 12: 195–210.

Garavaso, Pieranna (2013). "Four Theses in Frege." *Paradigmi* no. 31: 43–59.

Garavaso, Pieranna, and N. Vassallo (2014). *Frege on Thinking and its Epistemic Significance*. Lanham, MD: Lexington Books.
Griffin, Nicholas (2003). *The Cambridge Companion to Bertrand Russell.* Massachusetts: Cambridge University Press.
Hodes, Harold T. (1982). "The Composition of Fregean Thoughts." *Philosophical Studies*, no. 41: 161–178.
Janssen, Theo M. V. (2001). "Frege, Contextuality and Compositionality." *Journal of Logic, Language, and Information*, no. 10: 115–136.
Kemmerling, Andreas (2010). "Thoughts Without Parts: Frege's Doctrine." *Grazer Philosophische Studien* 82 no. 1: 165–188.
Klement, Kevin C. (forthcoming). "*Grundgesetze* and the Sense/Reference Distinction." *Essays on Frege's Grundgesetze der Arithmetik*, eds. M. Rossberg and P. Ebert.
Landini, Gregory (1996). "Decomposition and Analysis in Frege's *Grundgesetze*." *History and Philosophy of Logic*, no. 17: 121–139.
Landini, Gregory (2012). *Frege's Notations What They are and How They Mean.* London: Palgrave Macmillan.
Makin, Gideon (2000). *The Metaphysicians of Meaning. Russell and Frege on sense and denotation.* London and New York: Routledge.
Pelletier, Francis Jeffry (2001). "Did Frege Believe Frege's Principle?" *Journal of Logic, Language, and Information*, no. 10: 87–114.
Penco, Carlo (2003). "Frege: Two Theses, Two Senses." *History and Philosophy of Logic*, 24 no. 2: 87–109.
Textor, Mark (2009). "A Repair of Frege's Theory of Thoughts." *Synthese*, no. 167: 105–123.

CHAPTER 5

# Russell's and Wittgenstein's Logical Atomisms

*David G. Stern*

## 1 What Did Russell Learn from His "friend and former pupil Ludwig Wittgenstein"?

Russell's preface to the published text of his lectures on logical atomism informs the reader that those eight lectures, given in early 1918

> are very largely concerned with explaining certain ideas which I learnt from my friend and former pupil Ludwig Wittgenstein. I have had no opportunity of knowing his views since August 1914, and I do not even know whether he is alive or dead. He has therefore no responsibility for what is said in these lectures beyond that of having originally supplied many of the theories contained in them. (*PLA*: 160)

Despite this clear and explicit introductory disclaimer, it has proven surprisingly difficult to say precisely which ideas and theories Russell was referring to in this passage.

D. G. Stern (✉)
University of Iowa, Iowa City, IA, USA
e-mail: david-stern@uiowa.edu



L. D. C. Elkind, G. Landini (eds.), *The Philosophy of Logical Atomism*, History of Analytic Philosophy,
https://doi.org/10.1007/978-3-319-94364-0_5

If we take Russell at his word in the passage quoted from his preface, the ideas and theories in question are ones that Wittgenstein communicated to Russell shortly before the start of the First World War, and that Russell set out in those lectures given during the last year of that war. One might expect, then, that identifying them would be a matter of comparing and contrasting what was said in Russell's lectures, on the one hand, with what Wittgenstein wrote while he was a student of Russell's, on the other. However, Wittgenstein wrote very little before August 1914. Furthermore, in a letter written to Russell in March 1919, Wittgenstein said that his *Tractatus* contained "all of my work of the last six years" and that he believed he'd "solved our problems finally". Consequently, until quite recently, almost all work on the relationship between Russell's and Wittgenstein's logical atomisms has turned on the assumption that Wittgenstein's *Tractatus* is our best source for the views that he had previously worked out while engaged in close conversation with Russell in 1912 and 1913. In other words, most work on the relationship between was said in Russell's lectures and Wittgenstein's own views has taken the form of a debate over three closely related issues. First, which ideas and theories make up Russell's logical atomism, as set out in *The Philosophy of Logical Atomism*? Second, which ideas and theories make up Wittgenstein's logical atomism, as set out in *Tractatus*? Third, how should we understand the relationship between their views? On this approach, the principal evidence for our answers to the first two questions, and thus the basis for our answer to the third, dates from 1918, the year in which Russell gave his lectures and Wittgenstein assembled *Tractatus* in its almost final form.

As an example of the way that this approach has served as a common framework for otherwise very different interpretations of Russell and Wittgenstein, consider the disagreement between David Pears and Gregory Landini over the relationship between Russell's logical atomism and Wittgenstein's. In his introduction to the Open Court edition of *The Philosophy of Logical Atomism*, Pears described the two philosophers as pursuing related, but fundamentally different, kinds of logical analysis. On his account, Russell developed an empiricist analysis of sense data, while Wittgenstein provided a rationalist theory of meaning. According to Pears, while both philosophers agreed that philosophical analysis must lead to minimal units of meaning, "logical atoms", they adopted different approaches that led them to very different conceptions of those atoms. Russell's philosophy of language was largely motivated by epistemological concerns about the nature of acquaintance and the basis of reference, with the result that his logical atoms were sensory givens. On the other hand, Wittgenstein's philosophy of language was supposedly motivated by ontological concerns about what the world must be like for meaning to be possible, and so his logical atoms were the result of an a priori argument about the nature of analysis.

In *Wittgenstein's Apprenticeship with Russell*, Landini rejects Pears's interpretation. He argued that Pears's definition of logical atomism is misconceived, for Russell's empiricist epistemology is not really part of his logical atomism, or at the very least, not an essential part of it. Furthermore, Wittgenstein could not have taught Russell a theory of meaning, for Wittgenstein was developing a theory of meaning he learnt from Russell. In accordance with the interpretive template I have just described, neither Pears nor Landini pays close attention to the interaction between Russell and Wittgenstein during the pre-war period. Instead, they each concentrate on the (very different) systematic commitments they identify in *The Philosophy of Logical Atomism* and *Tractatus*. Indeed, until recently, it has usually been taken for granted that this was not only the best way of identifying Russell's and Wittgenstein's logical atomisms, but that there was no realistic alternative, given the available evidence. For instance, Landini glosses our opening passage from Russell's preface to the lectures on logical atomism in the following terms:

> Russell was eager to launch his new program for philosophy as the science of logical form and to introduce Wittgenstein. But in his efforts to establish a reputation for Wittgenstein, he embellished matters greatly. Wittgenstein's ideas were at a very immature stage during his conversations with Russell. (Landini 2007: 24)

On this way of approaching the Russell-Wittgenstein relationship, there can be no possibility of a detailed account of a two-way conversation over an extended period. Once we accept that Wittgenstein's ideas really were at such an "immature stage" during his conversations with Russell, it is natural to approach their working relationship along the following lines. During the time that they worked closely together in 1912 and 1913, their conversation must have been primarily a matter of Wittgenstein's learning from Russell, and then working out those ideas in a more systematic way later on.

Pears builds his interpretation of the Wittgenstein-Russell relationship in terms that appear, at first sight, to be diametrically opposed to Landini's, arguing that Wittgenstein offered deep and far-reaching criticisms of Russell. However, like Landini, rather than attending to their correspondence, or other evidence of their relationship at the time, Pears also concentrated on a reading of Russell's 1918 lectures and *Tractatus* when answering the question what Russell learnt from Wittgenstein.[1]

A shared difficulty for both Pears's and Landini's accounts is that they each have to maintain that Russell's preface is seriously mistaken when Russell describes the relationship between his own logical atomism and Wittgenstein's. If Pears is right, their logical atomisms are very different. If Landini is right, they are very similar, but Russell learned very little from Wittgenstein.

In the remainder of this chapter, I do not aim to intervene directly in the debate between Pears and Landini. Instead, I explore the proposal that neither of them is able to take what Russell says in his preface at face value because they both take it take for granted that Wittgenstein was committed to a relatively stable system of ideas during the years in question, a logical atomism that can be systematically defined. However, a number of very different trains of thought can be identified even if one restricts one's attention to what Russell said in his eight lectures on logical atomism, or on what Wittgenstein wrote in *Tractatus*, let alone the radical changes that took place over any longer period one cares to identify. As a result, any unitary account of either of those key texts, let alone the process of development that led to each of them, requires some substantive and inevitably quite controversial principles of selection. Furthermore, even before dealing with the widely recognized difficulties associated with the "synchronic" problem of identifying a definitive statement of logical atomism from their writings in 1918, there is a prior and much less well-known "diachronic" difficulty. Russell's and Wittgenstein's philosophical outlooks, and their views on many of the specific issues they discussed, changed dramatically and repeatedly, both during the two-year period beginning in late 1911 when they were in close contact, and during the war years, when they had no philosophical communication at all.

The case for the alternative account of the relationship between Wittgenstein's views and Russell's discussion of those views in those lectures he gave in early 1918 that I will outline in the remainder of this essay can be summed up as follows. Wittgenstein first met Russell in October 1911, and visited his lectures during the last term of 1911, but he was only a student in Cambridge for five terms, from the beginning of February 1912 to the end of May 1913. That period of close contact was itself relatively short, and even during it, they repeatedly disagreed or simply failed to understand each other. Their views evolved and repeatedly changed both during those pre-war years, when they were in close contact, and even more so afterwards, when they had no opportunity for philosophical discussion. Consequently, it is methodologically inadvisable, to put it mildly, to begin work on the very idea of "logical atomism" in the confident expectation of identifying a shared set of views common to the pre-war period in which they collaborated, let alone both Russell's 1918 lectures and Wittgenstein's *Tractatus*. The plural in my title—"logical atomisms", rather than "logical atomism"—is thus meant not only to convey that the two philosophers' logical atomisms were significantly different, but also that each of them repeatedly changed their views concerning central philosophical questions.

After their period of close collaboration during 1912 and early 1913, Wittgenstein's only personal contact with Russell was a series of meetings in Cambridge in early October 1913. During that period, he dictated the "Notes on Logic", summarizing the main results of his work to date. Consequently, the best starting point for an understanding of the views of Wittgenstein's that are discussed in Russell's lectures on logical atomism are the "Notes on Logic". The "Notes" amount to a telegraphic summary of Wittgenstein's philosophical outlook immediately before his departure for Norway, which proved to be the end of his period of close collaboration with Russell. Although Russell subsequently saw the notes dictated to Moore six months later, which include important further developments in Wittgenstein's path towards the system set out in the *Tractatus*, these left little impression on him: almost without exception, the views of Wittgenstein's which Russell discusses in his lectures are set out in the "Notes on Logic".

In other words, my main aim here is to argue that before we can be in a position to identify, let alone settle the questions on which they differ, we first need a much more fine-grained account of the development of Wittgenstein's early philosophy in general, and of a number of distinct stages in his interactions with Russell, in particular.

Before we can be in a position to give a satisfactory answer to the question with which we began, about what Russell learned from Wittgenstein, we need to first give much more consideration to their interaction in the pre-war years, and the ways in which each of them changed and developed after they were no longer collaborating with each other, rather than directing our attention to the polished and summative works they produced at the end of this period. If we start from work of Wittgenstein's that is more directly in dialogue with Russell's pre-war work than the *Tractatus*, and see which ideas and theories are present there, and which only emerge in the course of later work, we will be much better able to appreciate what they had in common and where they disagreed. In other words, we need to replace the overly crude and simple question with which we began—"What did Russell learn from Wittgenstein?"—with a revised version of that question that explicitly acknowledges that their collaboration took place over a number of years, namely: "When did Russell learn what from Wittgenstein?"

## 2 When Did Russell Learn What from Wittgenstein?

### *2.1 February 1912 to August 1913: Four Episodes*

Most of the information about the development of Wittgenstein's views that I will review in this section is already well known. However, its full significance has rarely been properly appreciated. Because it is usually approached as evidence of the emergence of a single unitary "logical atomist" view, or alternatively, a clearly contrasted pair of views, the tensions between Russell's and Wittgenstein's views at the time, or over time, are only too often smoothed out in order to facilitate that account on which they were all along travelling towards their respective destinations.

We can identify a number of clearly defined stages in Wittgenstein's philosophical development during the 1910s. While a student in Manchester, Wittgenstein develops an initial interest in working on the philosophy of logic and mathematics. During October–December 1911, he is a visitor at Russell's lectures in Cambridge. In February 1912, Wittgenstein begins his studies with Russell. As Potter observes, during the first of his two years in Cambridge, "Wittgenstein came to see himself as engaged in a joint project with Russell" (Potter 2009: 20).

From an early stage, Wittgenstein's attention was focused on the question of the nature of logic. There is relatively little documentary evidence of the details of their discussions, but several key disagreements are recorded in the correspondence with Russell. A few points in this first stage of his journey stand out.

(1) In June 1912, Wittgenstein writes to Russell:

> Logic is still in the melting-pot but one thing gets more and more obvious to me: The prop[osition]s of Logic contain ONLY APPARENT variables and whatever may turn out to be the proper explanation of apparent variables, its consequence *must* be that there are NO *logical* constants.

> Logic must turn out to be of a TOTALLY different kind than any other science. (Wittgenstein 2012: 30)

This is the first statement of what Wittgenstein called his "fundamental thought" in the *Tractatus*: "that the 'logical constants' do not represent".[2] As Potter argues, what is fundamental to Wittgenstein's conception of logic is not so much this specific claim, but the reasons that lead him to make it (Potter 2009: §5.7). For it turns on making a sharp contrast between logic, which is not about anything, and so is "empty", and thus not informative, and any empirical investigation. It also leads to a sharp distinction between the logical relations between propositions, and the logical structure of the propositions themselves. It is also Wittgenstein's first major challenge to Russell's conception of logic as the most general science.

(2) In January 1913, Wittgenstein proposes, in another letter to Russell, that

> all theory of types must be done away with by a theory of symbolism showing that what seem to be *different kinds of things* are symbolised by different kinds of symbols which *cannot* possibly be substituted in one another's places. (Wittgenstein 2012: 38)

This is part of Wittgenstein's response to what he had called the "Complex Problem" in his previous letters, the question of how to analyse statements about complexes—including both the problem of the nature of logical connectives and the problem of the meaning of quantifiers—and turns on the denial that the complex is a further entity over and above the components.

Here we also have a foreshadowing of Wittgenstein's later conception of analysis, which aims to replace Russellian theory construction by an appropriate notation with the aim of clarifying our current use of language. Russell did not fully appreciate the depth of this disagreement. On the first page of his introduction to the *Tractatus*, he declared that "In order to understand Mr. Wittgenstein's book, it is necessary to realize what is the problem with which he is concerned. In the part of his theory which deals with Symbolism he is concerned with the conditions which would have to be fulfilled by a logically perfect language". (Wittgenstein 2016: Introduction)

(3) In a letter written to Russell in the first half of June 1913, Wittgenstein criticizes Russell's multiple relation theory of judgement. He wrote:

> I can now express my objection to your theory of judgment exactly: I believe it is obvious that, from the prop[osition] "A judges that (say) a is in the Rel[ation] R to b", if correctly analysed, the prop[osition] "aRb.∨.~aRb" must follow directly *without the use of any other premiss*. This condition is not fulfilled by your theory. (Wittgenstein 2012: 40)

The following month, Wittgenstein wrote to Russell that his work was going well and that he felt "rather hopeful":

> All my progress comes out of the idea that the *indefinables* of Logic are of the general kind (in the same way as the so called *Definitions* of Logic are general) and this again comes from the abolition of the real variable. Perhaps you laugh at me for feeling so sanguine at present; but although I have not solved *one* of my problems I feel very, very much nearer to the solution of them all than I ever felt before .... I am very sorry to hear that my objection to your theory of judgment paralyses you. I think it can only be removed by a correct theory of propositions. (Wittgenstein 2012: 42)

Three years later, in a letter to Ottoline Morrell, Russell later described Wittgenstein's criticism as "an event of first-rate importance in my life ... I saw he was right, and I saw that I could not hope ever again to do fundamental work in philosophy" (*Aut*: 66).

There is an extraordinarily large literature on Wittgenstein's objection to Russell, not only because of Russell's acknowledgement that it was an important turning point in their philosophical relationship, but also because we have so little documentary evidence of precisely what Wittgenstein's "paralysing" objection to Russell's theory of judgement in May and June 1913 consisted in. Interpreters have usually regarded two crucial passages in the "Notes on Logic" as providing further information about the nature of this objection:

> Every right theory of judgment must make it impossible for me to judge that this table penholders the book. Russell's theory does not satisfy this requirement. (Wittgenstein 1979: 103)

> The proper theory of judgment must make it impossible to judge nonsense. (Wittgenstein 1979: 95)[3]

Following Jinho Kang (2017), I believe this is a mistake, and that these two passages set out additional and rather different objections to Russell's theory. For obvious reasons, Kang calls the first of these two further objections the "illegitimate combination" objection, and the second, the "nonsense" objection. Both of these objections are much more general than the initial "paralysing" objection. That objection, namely that from a proper analysis of the judgement that aRb " 'aRb.∨.~aRb' must follow directly *without the use of any other premiss*", involves a particular requirement internal to Russell's own views at the time. The other two, later objections, have to do with different versions of the requirement that a theory of judgement be constructed in such a way that it rule out certain forms of impossible judgement.

(4) According to David Pinsent's notes of his conversations with Wittgenstein, it was only later in the summer of 1913 that Wittgenstein's work began to crystallize. His diary for 25 August, the first day they saw each other since the middle of June, records that Wittgenstein explained to Pinsent

> his latest discoveries in Logic. They are truly amazing and have solved all the problems on which he has been working unsatisfactorily for the last year. He always has explained to me what he has been working at, and it is exceedingly interesting to see how he has gradually developed his work, each idea suggesting a new suggestion, and finally leading to the system he has just discovered—which is wonderfully simple and ingenious and seems to clear up everything. … It is like the transition from Alchemy to Chemistry. (Flowers and Ground 2016: 216)

Unfortunately, these tantalizing remarks are not followed up by a philosophically informative summary of the most recent discoveries in logic, the solution of the problems Wittgenstein had been working on, or the system Wittgenstein had discovered. Clearly, Wittgenstein convinced Pinsent that he had made a breakthrough, but we are left none the wiser about any of the details. However, in early October 1913, Wittgenstein dictated the "Notes on Logic" for Russell, which Russell clearly studied closely. In a letter to Ottoline Morrell, dated 3 October 1913, Russell described this work as "as good as anything that has ever been done in logic" (Potter 2009: 263) and he discussed it and distributed copies of it to his graduate seminar at Harvard the following spring.

All of these four episodes in the evolution of Russell and Wittgenstein's philosophical relationship are striking, but we know very little more about the details of each of them, and a great deal of ink has been spilled over speculation about how best to fill in the gaps. The overall outcome of the extensive literature on these supposed "turning points" is ultimately indecisive and disappointing: it turns out that with sufficient ingenuity and determination, it is possible to come up with a bewildering variety of mutually incompatible interpretations of these events, despite the fact that at first sight they seem so promising. Instead of concentrating on the first three of these much-debated events—the emergence of Wittgenstein's conception of logic, and of analysis, his criticism of the multiple relation theory of judgement—we should instead focus our attention on the outcome of this period of interaction between Russell and Wittgenstein, the "Notes on Logic", if we wish to identify what Russell took from Wittgenstein in his 1918 lectures. Similarly, if we want to identify those parts of *Tractatus* that are not the direct result of interaction with Russell, we should look for the ideas found there that are not present in the "Notes on Logic" and that first emerge in Wittgenstein's writing during 1914–1917.

### 2.2 *October 1913: Wittgenstein's "Notes on Logic"*

Thanks to Michael Potter's close reading of the "Notes on Logic", we can now see, not only that Wittgenstein's ideas were remarkably well-developed by the end of his conversations with Russell, but also how to map out the stages by which Wittgenstein moved from those views to the ones he sets out in the *Tractatus.* This is why I believe we need not only to focus on the "Notes on Logic", rather than the *Tractatus*, in order to understand the relationship between Wittgenstein's logical atomism and Russell's, but also to look much more closely at the complex and intricate process that first led up to Wittgenstein's writing the "Notes on Logic", and then further developing and changing his philosophical outlook over the next several years. That process is recorded in his subsequent work: dictations to Moore in the spring of 1914, his 1914–1916 notebooks, and "Core Prototractatus" and "Proto-Protractatus", early versions of *Prototractatus* that Wittgenstein drafted during the first two years of the war, to name some of the most significant later stages in the path that led to the composition of *Tractatus.*

Until recently, when Wittgenstein's "Notes on Logic" have been the focus of scholarly attention, they have usually been mined for clues about the nature of Wittgenstein's position in *Tractatus.* However, Potter's book on the "Notes" has shown that they can also be read as setting out a surprisingly systematic philosophy of logic, or to put it another way, a significant fraction of the Tractarian system. They are Wittgenstein's first attempt at a full statement of his philosophy, written at a time when he still thought of himself as collaborating with Russell on "our problems", as he termed them (Wittgenstein 2012: 35).

Furthermore, there was almost no philosophical contact between Russell and Wittgenstein after this point. Although Russell did read the notes Wittgenstein dictated to Moore in Norway in the spring of 1914 a year later, he got little from them, and the developments recorded in them that postdate the 1913 notes for Russell are not reflected in Russell's lectures on logical atomism. In particular, the Moore notes begin with an extended discussion of the show/say distinction, and its implications for a number of central topics in the philosophy of logic. As that topic is not discussed in the "Notes on Logic", it seems clear that Wittgenstein regarded it as a principal result of his first few months of work in Norway. However, Russell does not mention it or its implications in the lectures on logical atomism, and there is no other sign, as far as I can see, that he made use of the Moore notes in that work. Indeed, this is underlined by Wittgenstein's conviction in his initial reply to Russell's first letter about the *Tractatus*, written from the prisoner of war camp in August 1919, that

> you haven't really got hold of my main contention, to which the whole business of logical prop[osition]s is only a corollary. The main point is the theory of what can be expressed (gesagt) by prop[osition]s—i.e. by language—(and, which comes to the same, what can be *thought*) and what can not be expressed by prop[osition]s, but only shown (gezeigt); which, I believe, is the cardinal problem of philosophy. (Wittgenstein 2012: 98)

To sum up: as the philosophical positions attributed to Wittgenstein in the logical atomism lectures are so close to the ones found in the 1913 notes, we should regard the "Notes on Logic" as the principal source of Wittgenstein's contribution to Russell's logical atomism. However, it is important to note that Russell's choice of topics from the wide-ranging menu that Wittgenstein provided was quite selective. Russell's focus as a working philosopher was, quite naturally, on the issues and ideas that he found most fruitful, and he had no interest in providing a balanced report on his student's work as a whole. Potter sums up the relationship between Russell's 1918 lectures on logical atomism and Wittgenstein's 1913 *Notes on Logic* in the following terms, which provide constructive guidance as to the extent to which we can rely on the former for guidance in reconstructing Wittgenstein's views when he wrote the latter:

> In the lectures, especially, he was more interested in recollecting what Wittgenstein's views had been than in developing his own. This is relevant to us here, because there had been, as we have noted, very little philosophical contact between them since Wittgenstein's departure for Norway in October 1913. The views of Wittgenstein's which Russell was recollecting are therefore overwhelmingly those of the *Notes on Logic*. For this reason Russell's lectures represent a significant additional source for the interpretation of these notes. (Potter 2009: 135)

Here, then, is a telegraphic summary of some of the main points to be found in the "Notes on Logic", drawing on Potter's reading of that material. These include Wittgenstein's conception of analysis, the "central task of explaining the unity of the proposition" (Potter 2009: 109), his "fundamental thought" that there are no logical constants (Potter 2009: Chap. 4; Wittgenstein 2016: 4.0312), and what Potter calls the "symbolic turn", namely moving to a conception of the proposition "as symbolizing what it expresses, rather than being identical with it" (Potter 2009: 78). Related ideas include Wittgenstein's view that "simple objects are whatever is represented by the simple symbols" (Potter 2009: 69), his critique of Fregean senses for names (Potter 2009: Chap. 7), his view that the theory of types is superfluous, for the reason that such a theory would be a misguided attempt to determine, rather than describe which combinations make sense (Potter 2009: 83), his critique of Frege's views that a proposition is a name of a truth value (Potter 2009: Chap. 9), his rejection of Russell's distinction between an asserted and an unasserted proposition (Potter 2009: 97), and his conception of the meaning of a proposition as a complex, not a fact, and relying in his account of propositional meaning on the model of the relationship between a name and its meaning (Potter 2009: 135). A key point in these criticisms of Russell is Wittgenstein's insistence on the distinction between complex and fact, and the correlative conception of a world of facts, not of things (Potter 2009: Chap. 11), and his conception of the form of a proposition as "the symbolizing relationship between the names in the proposition which makes it the case that the proposition says what it does" (Potter 2009: 114). Indeed, "[t]he point of this conception is in a sense the mirror image of the point of conceiving of the world as made up of facts, not complexes. In order to make judgments about the world, what we must perceive are facts, not complexes; and the symbols that express those judgments, likewise, must be facts, not complexes" (Potter 2009: 114). This leads in turn to the view that a proposition is "bipolar": it is related to its meaning—the fact(s) in question—truly or falsely (Potter 2009: 136), while its sense, what I grasp when I understand the proposition, consists of the conditions under which the proposition is true and the conditions under which it is false (Potter 2009: 151). Wittgenstein had also arrived at an early version of the Tractarian view that "the only explanation for logic's special character" was that it has no content, contradicting Frege's and Russell's conviction that "the fruitfulness of logic was proof that it has content" (Potter 2009: 157).

### *2.3 1914–1917: MSS 101–104, Bodleianus, and* Prototractatus

We can then see the path that leads to *Tractatus* in terms of a series of steps away from the point of departure summarized in the previous section, and set out in the October 1913 "Notes on Logic". Among the first of these steps is the more developed account of the nature of logical truth, and the emergence of the a-b notation, a topic already touched on in several letters to Russell written towards the end of 1913 (Potter 2009: 157). Shortly after, in February 1914, Wittgenstein broke off communications with Russell (Wittgenstein 2012: 68). As already noted, we find the initial articulation of the show/say distinction in the opening of the "Notes" that Wittgenstein dictated to Moore in Norway in April 1914 (Wittgenstein 1979: 108–119). *Tractatus* interpreters have made good use of the 1914–1916 *Notebooks*, an edition of Wittgenstein's three surviving wartime notebooks, MSS 101–103, approaching it as a dated chronological record of many further themes, beginning with the gnomic thesis that "Logic must take care of itself" in August 1914 (Wittgenstein 1979: 2), and the emergence of the Tractarian picture theory, as a general account of empirical propositions, and of the role of showing in their meaning, in September 1914 (Wittgenstein 1979: 7 ff.). In that first notebook, Wittgenstein is still clearly engaging in a dialogue with Russell.

Thanks to the work of Brian McGuinness (1989, 2002a, b), we know that MS 104, the source manuscript for *Prototractatus* (Wittgenstein 1971), provides a chronologically ordered log of the polished paragraphs that would later be rearranged and revised in the production of *Tractatus*. While a facsimile of MS 104 is included in the first and second editions of *Prototractatus*, the published text does not include an edited text of the manuscript in the order it was written. Instead, the remarks were rearranged by the editors in numerical order. Indeed, in the critical German-language edition of *Tractatus*, which includes the full published text of *Prototractatus*, together with detailed information about how each remark was revised (Wittgenstein 1989), there is no information about the original order of the remarks, and no tables or other apparatus that would aid the reader in studying MS 104.

When Wittgenstein began to assemble the material that would ultimately be rearranged and reorganized in the familiar numerical order from 1 to 7, he had not yet finished writing it, and had not yet worked out how to arrange the parts that he had written. Consequently, the manuscript of the remarks that we now know as *Prototractatus* could not be written up in the sequential, numerical order in which the book was published. However, sometime during the First World War, Wittgenstein worked out the ingenious numbering system that enabled him to organize, review, and repeatedly reorganize his work in progress, despite the very limited resources available to him while serving as a soldier. As a result, the manuscript containing the first known draft of his book (MS 104 in von Wright's numbering system, sometimes known as "Bodleianus", because it is owned by the Bodleian Library in Oxford), began with the first six whole-numbered remarks on the first page of the main text (Pilch 2015). That series is repeated, together with almost all the remarks with a single decimal through 4.4, on the next page.

After that, remarks were written down as Wittgenstein decided to make use of them, and each remark prefaced by a decimal number indicating its ultimate location in the sequence. The next page contains double decimal remarks appended to the whole number and single decimal remarks that formed the initial backbone for the growing book draft. Progressively higher-numbered remarks soon make an appearance, but throughout the process of construction recorded in MS 104, remarks are added to the tree-structure, not to a numerical sequence.

In October 1915, Wittgenstein wrote to Russell that he had recently done a great deal of work, and that he was "in the process of summarizing it all and writing it down in the form of a treatise (Abhandlung). … If I don't survive [the war], get my people to send you all my manuscripts: among them you'll find the final summary written in pencil on loose sheets of paper" (Wittgenstein 2012: 84–85). That loose-leaf "final summary" has not survived, but it is likely that it consisted of some kind of a tree-structure arrangement of his book in progress, as a sequentially ordered arrangement would have involved constant and extensive additions to what had already been composed, while inserting material into sheets containing remarks arranged in a tree-structure would have been simple. Certainly, it would have been impracticable to take in either the hypertextual structure or the sequential arrangement of the projected treatise by reviewing MS 104, the bound ledger containing a chronological ordered record of his additions to the book draft.

Thus, while the published *Prototractatus* looks very similar to the final *Tractatus*, the source manuscript on which that book was based was put together in a very different way. From each of the first six whole-numbered remarks, numerical sequences branch, starting with one-decimal series such as 1.1, 1.2; from these nodes, further branches stem.

When MS 104 was first discovered by von Wright in 1965, who took charge of preparing the text for publication over the next few years, the full significance of the order in which the remarks were written down was not yet appreciated. As a result, the focus of that book and of von Wright's introductory essay is on the path to the *Tractatus*, not the composition of MS 104. This is already made clear in the wording of the book's subtitle: "an early version of *Tractatus Logico-Philosophicus*". Consequently, the text of the first 103 pages was rearranged in the familiar numerical order, while last 15 pages of "corrections" were left out, as they belonged to a later stage of revision that could not be fully reconstructed from the available evidence. The immediate result of this enormous amount of careful and conscientious scholarly work was very disappointing: it was hard for the first generation of readers of *Prototractatus* to see what, if anything, there was to be gained or learned from this edition. The edited text looked too much like the familiar text of *Tractatus* to be instructively different, while the facsimile of the original seemed quite opaque. Indeed, while von Wright did not himself provide any further discussion of the "the most interesting differences between the two works" (Wittgenstein 1971: 4), his work made those materials available in a form which provoked others to identify those differences. This may well have been one of his most important contributions to our understanding of the complex relationship between MS 104, *Prototractatus*, and *Tractatus*.

However, until very recently only the most determined scholars have been in a position to study even the principal earlier stages, usually known as the "core" *Prototractatus*, which ends at a dividing line on page 28 of the manuscript, and "Proto-Prototractatus", which ends at a similar dividing line near the bottom of page 70.[4]

Recently, I have been working with Landon D. C. Elkind and Phillip Ricks on the University of Iowa *Tractatus* map, an online tool available at http://tractatus.lib.uiowa.edu/.[5] In addition to providing a subway-style map of the complete text of both the *Tractatus* and the *Prototractatus* (or the first 103 pages of MS 104), it also provides parallel access to the earlier stages, or "strata" of composition, contained within the source manuscript for *Prototractatus*. By choosing different start and end pages at the top of that map, one can look at different stages in the construction of *Prototractatus*: the chosen pages are in colour, the others are greyed out. In this way, one can look at the text of different stages in the construction of *Prototractatus*, and map the changing arrangement of the project as it was gradually assembled. This approach makes it much easier for the reader to appreciate the force of the suggestion, first proposed by Brian McGuinness, that Wittgenstein's Russell-inspired book on logic, received its most well worked out form in "Proto-Prototractatus", the draft of the book that he put together before the summer of 2016. Shortly afterwards, he not only added the current conclusion, remark number 7, but also many of the most striking and memorable remarks in the 6s, and significantly rearranged and reorganized much of the previous material in his book draft.

## 3 Conclusion

In the first part of this chapter, I proposed that most previous work on the relationship between Russell's and Wittgenstein's logical atomisms has had an overly narrow focus, because those researchers took it for granted that our principal focus should be on *The Philosophy of Logical Atomism* and *Tractatus*. Those two works are the most polished and detailed expositions of their respective logical atomisms at the end of the 1910s. However, any appraisal of the relationship of their philosophical views during this period has to take into account the fact that the two of them were only in close contact for a much shorter period during 1912 and 1913. Despite the fact that the bare outline of this chronology has long been well known, its implications for our understanding of the relationship between their conceptions of logical atomism have not yet been fully appreciated.

The second part of this chapter provides a brief outline of the four main stages of Wittgenstein's philosophical development during these years. The first stage is, his work on philosophical questions about the nature of logic while he was Russell's student. The second stage is, his most detailed and direct response to Russell's work on this topic at the end of that period in the "Notes on Logic". The third stage is, his further independent development of this project over the next two and a half years, leading up to the composition of "Proto-Prototractatus" in the first half of 1916. The final stage is, the rather different direction that his work took subsequently with the changes and additions that were made in turning that work in progress into the finished text of *Tractatus.* Consequently, if we want to compare Russell's logical atomism with Wittgenstein's, we should concentrate our attention on the "Notes on Logic" and "Proto-Prototractatus", not on *Tractatus.*

## Notes

1. In *Ludwig Wittgenstein: The Duty of Genius* Ray Monk gives an account of the Russell–Wittgenstein relationship during the pre-war years on which Wittgenstein's ideas were already at a very mature stage during his later conversations with Russell. As Monk sums up his interpretation in his chapter titles, Wittgenstein was first Russell's protégé, in 1912, and then his master, in 1913 (Monk 1990: Chaps. 3–4).
2. Wittgenstein (2016: 4.0312). Cf. Wittgenstein (1979: 37, dated 25 Dec. 1914).
3. Cf. Wittgenstein (2016: 5.5422): "The correct explanation of the form of the proposition 'A judges *p*' must show that it is impossible to judge a nonsense. (Russell's theory does not satisfy this condition.)"
4. Researchers can consult Schmidt (2016) and Pilch (2016) for facsimiles and transcriptions of many of the key documents, and there is a wealth of information about the structure of MS 104 and its relationship to both *Tractatus* and *Notebooks 1914–1916* in Geschkowski (2001). However, all this material is only available in German, and its overall structure is far from easy to take in.
5. For more information, see the welcome page on that site, and for much more information, see Stern (2016, 2018). Parts of the final section of this chapter are based on Part 4 of Stern (2016).

## References

### Works by Other Authors

Flowers III, F. A. and Ian Ground (eds.) (2016). *Portraits of Wittgenstein.* 2 vols., second edition. London: Bloomsbury.

Geschkowski, Andreas (2001). *Die Entstehung von Wittgensteins Prototractatus.* Bern: Books on Demand. Bern Studies in the History and Philosophy of Science.

Kang, Jinho (2017). "Wittgenstein against Russell's Theory of Judgment." Presented at SSHAP, Calgary, May 2017.

Landini, Gregory (2007). *Wittgenstein's Apprenticeship with Russell.* Cambridge: Cambridge University Press.

McGuinness, Brian (1989). "Wittgenstein's Pre-*Tractatus* Manuscripts." *Grazer Philosophische Studien* 33: 35–47. Reprinted with revisions in McGuinness 2002a.

McGuinness, Brian (2002a). "Wittgenstein's 1916 'Abhandlung'." In *Wittgenstein and the Future of Philosophy: A Reassessment After 50 Years: Proceedings of the 24th International Wittgenstein-Symposium*, eds. Rudolf Haller and Klaus Puhl. Vienna: Hölder-Pichler-Tempsky. 2002: 272–282.

McGuinness, Brian (2002b). *Approaches to Wittgenstein: Collected Papers.* London: Routledge.

Monk, Ray (1990). *Ludwig Wittgenstein: The Duty of Genius*, New York: Macmillan.

Pilch, Martin (2015). "A Missing Folio at the Beginning of Wittgenstein's MS 104." *Nordic Wittgenstein Review*, Vol. 4, No. 2: 65–97.

Pilch, Martin (ed.) (2016). Wittgenstein Source *Prototractatus* Tools. http://www.wittgensteinsource.org

Potter, Michael (2009). *Wittgenstein's Notes on Logic*, Oxford: Oxford University Press.

Schmidt, Alfred (ed.) (2016). Wittgenstein Source Facsimile Edition of *Tractatus* Publication Materials. http://www.wittgensteinsource.org

Stern, David G. (2016). "The University of Iowa Tractatus Map." *Nordic Wittgenstein Review* 5:2: 203–220. http://www.nordicwittgensteinreview.com/article/view/3437

Stern, David G. (2018). "Mapping the *Tractatus.*" In *Proceedings of the 40th International Ludwig Wittgenstein-Symposium*, eds. Christoph Limbeck and Friedrich Stadler. Göttingen: de Gruyter.

Wittgenstein, Ludwig (1913). "Notes on Logic." In L. Wittgenstein, *Notebooks 1914–16.* Ed. G. H. von Wright and G. E. M. Anscombe. Oxford: Basil Blackwell: 93–106.

Wittgenstein, Ludwig *TLP*$_3$ (2016). *Tractatus Logico-Philosophicus*, Side-by-side-by-side edition, version 0.43. http://people.umass.edu/klement/tlp/

Wittgenstein, Ludwig (1971). *Prototractatus: an Early Version of Tractatus Logico-Philosophicus*, eds. B. F. McGuinness, T. Nyberg, and G. H. Von Wright, trs. D. F. Pears and B. F. McGuinness. Ithaca, NY: Cornell University Press.

Wittgenstein, Ludwig (1979). *Notebooks 1914–1916*, second edition, G. H. von Wright and G. E. M. Anscombe (eds.), tr. G. E. M. Anscombe. Chicago: The University of Chicago Press.

Wittgenstein, Ludwig (1989). *Logische-philosophische Abhandlung: kritische Edition.* eds. B. F. McGuinness and J. Schulte. Frankfurt am Main: Suhrkamp.

Wittgenstein, Ludwig (2012). *Wittgenstein in Cambridge: Letters and Documents 1911–1951.* Fourth edition. Ed. Brian McGuinness. Oxford: Blackwell.

CHAPTER 6

# Russell in Transition 1914–1918: From *Theory of Knowledge* to "The Philosophy of Logical Atomism"

*Russell Wahl*

## 1 Introduction

Wittgenstein's *Tractatus* gives a clear presentation of a metaphysics of logical atomism, with its account of logic, its independent atomic facts, and its clear assertion that the only necessity is logical necessity (*TLP*$_2$: 6.37; 6.375). Russell's view, from his earlier papers to "The Philosophy of Logical Atomism" and his 1924 paper "Logical Atomism," was not quite so clear. Many have seen the difference between the two as a function of Russell's concerns with epistemology as opposed to Wittgenstein's purely logical concerns.[1] But much of Russell's earlier atomism does not seem to be so wedded to the principle of acquaintance, and in Russell's later papers on logical atomism he made reference to Wittgenstein who Russell suggested greatly influenced him, his views and approach remained quite distinct from Wittgenstein's.

R. Wahl (✉)
Idaho State University, Pocatello, ID, USA
e-mail: wahlruss@isu.edu

 
L. D. C. Elkind, G. Landini (eds.), *The Philosophy of Logical Atomism*, History of Analytic Philosophy,
https://doi.org/10.1007/978-3-319-94364-0_6

When Russell first used the term "logical atomism," in 1911, it is safe to say that he had not worked out the detailed picture we now think of as logical atomism. He used "atomism" initially almost as a synonym for "pluralism." He used it as a contrast to the monism of the British idealists which he had rejected. In 1911, when he dubbed his philosophy "analytic realism," he said this philosophy was analytic, "because it claims that the existence of the complex depends on the existence of the simple, and not vice versa, and that the constituent of a complex, taken as a constituent, is absolutely identical with itself as it is when we do not consider its relations" (*AR*: 133). He then went on to say that it was, therefore, an "atomic philosophy." And that the atoms are of two kinds: "universals ... and particulars ..." (ibid.). The remark about the complex depending on the simple was the key to understanding this as an atomist philosophy. This position was closely related to the doctrine of external relations Russell had adopted at the turn of the century. In 1911 Russell emphasized the point that these atoms are logical in the sense that they need not persist in time nor occupy space, and thus are not physical atoms (*AR*: 135).[2] There was not yet an atomism of atomic propositions. And in fact much of what is contained in Russell's atomism of the 1918 lectures is developed from but not present in 1911.

In this chapter, I look at the development of Russell's logical atomism from the time of his 1911 realism through his 1918 lectures. I am particularly interested in the changes that he made between the 1911 work done before his encounter with Wittgenstein and his position after this encounter. I hope thereby to clarify some of the differences between Russell's and Wittgenstein's atomism. I conclude with some of the points that were lasting in Russell's philosophy even after Wittgenstein's influence had receded.

There are several points that strike one looking first at the 1911 papers and then at the later lectures. The next six sections will focus on these points:

(1) There is a statement already in 1911 but emphasized more in 1918 that the particulars are logically independent of each other.
(2) There is the position that anything complex is a fact and facts are not to be confused with things. An important point connected to this claim is the Wittgensteinian doctrine that facts cannot be named. Nothing of this sort is present in 1911–1912.
(3) In 1911 Russell still held to the view present in *The Principles of Mathematics* that everything can be a logical subject of a proposition. On this view universals, relations, and so on, can all be logical subjects. By 1918, Russell had changed his view on this point.

(4) In 1911 and 1912, Russell did not really have a clear picture of what logic was. There was an elaborate logical system in *Principia Mathematica*, but when he came to ask what logic is, he did not have a satisfactory answer as is evident from his sketch in "What is Logic," from 1912. By 1918, he appears to have accepted Wittgenstein's view that logic consists of tautologies, although again he is still unsure about what that amounts to.

(5) Russell held the multiple-relation theory of judgment in 1911, and by 1918 he had abandoned it.

(6) Russell had used his method of analysis in his philosophy of mathematics. In 1911 he had not yet used the same techniques in his general philosophy of nature, but this changed with the difficult musings of "On Matter." By 1914 Russell was using the techniques we associate with logical atomism for his analysis of matter.

In "The Philosophy of Logical Atomism," Russell credited several of these changes, specifically points 2, 3, 4, and 5, to his conversations with Wittgenstein. One might think that despite Russell's first use of the term, Russell really became an atomist because of his interaction with Wittgenstein. However, I think that picture is too simple. Many of these changes were due to a change of mind inspired by Russell's interaction with Wittgenstein. Those changes, I argue, were all in place much earlier than 1918 and in fact were already adopted by Russell in 1914. These changes certainly played an important role in Russell's logical atomism after 1918, yet it was not the Wittgenstein of the *Tractatus* that influenced Russell. In the end what is enduring in Russell's atomism is the method of analysis mentioned in point 6 above. Russell's method of analysis antedated his encounter with Wittgenstein, although his application of it beyond the foundations of mathematics to the analysis of mind and matter came later. In the end his method of analysis was actually at odds with the view of logic embraced by Wittgenstein.

## 2 Particulars Are Logically Independent of Each Other

This is a key point in Russell's logical atomism. In "The Philosophy of Logical Atomism," we have a clear statement of it: "[E]ach particular that there is in the world does not in any way logically depend upon any other particular. Each might happen to be the whole universe; it is a merely empirical fact that this is not the case" (*PLA*: 179).

In "Analytic Realism," Russell held that, "in every complex there are two kinds of constituents: there are terms and the relation which relates them" (*AR*: 135). Terms are divided into particulars and universals, and Russell said of particulars (in order to distinguish them from traditional substances) that they "need not persist" and that they do not need to be causally independent of other existences. An important point he wished to emphasize in "Analytic Realism" was that particular sense-data were causally dependent on physiological conditions. But after this remark, he added in parentheses that from the logical point of view "any simple existence is independent of any other, and the only dependence is that of the complex on the simple" (ibid.). Russell used these remarks to distinguish his view from the more traditional view of substances. This is a long way from the atomism of Wittgenstein's *Tractatus*, with its rejection of causal connections. In addition, it is clear Russell allowed for complex particulars, something he rejected in 1918.[3] These particulars will have a logical dependence on their parts. Nonetheless, there is the germ of the independence thesis of the later atomism.

## 3 Anything Complex Is a Fact and Facts Cannot Be Named

In "Analytic Realism," Russell did not rule out complexes as particulars. In *Theory of Knowledge*, though, we have remarks that may suggest movement in the direction of the 1918 position. Russell defined a "complex" as anything analyzable, and described a fact as "what there is when a judgment is true, but not when it is false" (*TK*: 79). He did not want to say definitely that a complex is always the same as a fact, but said that there is at least a one-to-one correspondence between them, and so in *Theory of Knowledge*, they are treated as the same. In his discussion of the constituents of complexes, Russell divided them into particulars and universals, and held that there must be at least one universal (*TK*: 81), strongly suggesting that he is thinking of a complex as the ontological corollary of a true proposition.

In *The Principles of Mathematics*, Russell held that everything whatsoever, simple or complex, could be a logical subject and had to be able to be one on pain of contradiction. This is the doctrine of the unrestricted variable. It is clear that not only particulars, complexes, and universals were substituends of the variable, but that propositions were as well.[4] That propositions were also terms played an important role in Russell's substitution theory of classes and relations, which was developed during 1906–1908. The difficulties with that view brought on the rejection of propositions as entities and the advent of the multiple-relation theory of judgment. Of course, that theory continued to hold that complexes were entities that could be logical subjects, even if the old propositions were now not considered entities.

In *Theory of Knowledge*, Russell modified this position when working out the complex details of the development of the multiple-relation theory of understanding. In Part II Chaps. 3 and 5 of *Theory of Knowledge*, Russell was working on an account of belief or understanding complexes and the complexes which make given beliefs true or false. Russell was wrestling with the problem of distinguishing the complex *A-before-B* from the complex *B-before-A*. These complexes have the same constituents. The beliefs associated with such complexes he called "permutative," as permuting the terms results in a different belief and a different complex will be required to make the belief true. The belief *that A is similar to B*, on the other hand, is an example of a non-permutative belief, since there is no difference between the complexes *A-similar to-B* and *B-similar to-A*. At the time, Russell did not think there was a problem with the multiple-relation theory as applied to non-permutative beliefs, which is why the diagrams of belief facts in *Theory of Knowledge* are all of a belief in the similarity of two things. To come up with belief complexes which will discriminate the permutative beliefs, Russell introduced a position relation of a term to a complex, $xC_1\gamma$, where $C_1$ is the relation of being in the first position in the complex, and $x$ is a particular and $\gamma$ is the complex. What is interesting for our purposes is that Russell did not think that this relation is permutative because it is what he called heterogeneous (*TK*: 147).

We should pause to see that thinking of this relation as heterogeneous involves seeing a difference of ontological type between a particular and a complex. This type distinction is quite distinct from the type theory developed in *Principia Mathematica*. That type theory, as laid out in the Introduction to the first edition, talks about judgments of various orders and defines truth of judgments of the lowest level after the multiple-relation theory of judgment ($PM_2$, Vol. 1: 43). From these different levels, a system of orders of propositions and propositional functions is developed ($PM_2$, Vol. 1: 48–55). Gregory Landini has argued that it is best to understand this type theory as a theory of structured variables.[5] On the other hand, the type differences we have with these heterogeneous relations are ontological type differences. These are what Landini calls types*. Landini argues that there is a type* difference between universals and particulars, even though universals can be substituends for the unrestricted variable. He does not think the ramified theory of types and orders should be understood as a theory about reality.

In *Theory of Knowledge*, Russell perceived a restriction on what kinds of things can enter into certain places in a relation. Russell was instituting type restrictions as to ways things can be combined. There can be no complexes of the sort $xC_1y$ where $x$ is itself a complex or $y$ a simple particular. The heterogeneousness of the relation explains why these complexes are non-permutative. Thus there are type* differences between complexes and particulars. In a footnote (*TK*: 146), he indicated that there is also such a type* difference between relating relations and the other terms.[6] Seeing that Russell restricted these variables here, should we conclude that Russell held that there never was a relation which could hold sometimes between a particular and another particular and also between a particular and a complex? While Russell used the term "acquaintance" for our awareness of different types* of objects, it seems that even in *Theory of Knowledge* Russell recognized a "looseness in using the one word 'acquaintance' for immediate experience of different kinds of objects" (*TK*: 100). Russell actually did not use the word "type" for the difference between particulars, universals, and logical objects, but rather said they differed in "logical character" (ibid.).

Russell's account of position relations in *Theory of Knowledge* supposed that there can be singular reference to facts or complexes and that these can enter into relations with other entities, either facts or particulars. Perhaps being sensitive to the type* differences, Russell used Greek letters for variables restricted to complexes.

By 1918, though, Russell had adopted the position that facts cannot be named and so cannot be logical subjects. The discussion of this point began with a distinction between the different relations symbols have to what they mean. Names are proper symbols for particulars (including people), predicates are proper symbols for universals, and "a sentence (or a proposition) is the proper symbol for a fact" (*PLA*: 167). These relations are quite distinct, and it is a particularly grievous error, Russell held, to think of the relation of a proposition to a fact as anything like the naming relation. To argue for this he emphasized the bipolarity of a proposition (*PLA*: 167):

> It is perfectly evident as soon as you think of it, that a proposition is not a name for a fact, from the mere circumstance that there are two propositions corresponding to each fact.

Now in Russell's published work, there was nothing between the 1914 works and "The Philosophy of Logical Atomism" to indicate when this transition occurred. But when we examine some of Russell's letters and the notes Victor Lenzen took while attending his Harvard course in the spring of 1914, we can see that this change was already pretty firmly in place quite early. Here, for example, is an excerpt from a letter Russell sent to C. D. Broad in February 1914 before he embarked on his visit to America:

> The fact is that "acquaintance" cannot be applied straight off except to particulars, i.e. the only two-term cognitive relation of the form $S \rightarrow O$ has its converse domain confined to particulars. Knowledge by description is knowledge of a general proposition of the form $(\exists x)\varphi x$. In such cases we cannot properly speak of acquaintance with $\varphi\hat{x}$, "$\varphi\hat{x}$" must never be put in subject place, i.e. it must only occur in propositions where it is doing the proper work of a function. Universals, props., function, facts cannot be named, & cannot occur in subject-places; they are not "things." The symbols which are concerned with them are never simple, & do not name them. E.g. redness is introduced by "the meaning of '*x* is red' whatever *x* may be." This is Wittgenstein's theory & I am sure it is right.

Russell went on to say:

> There is, of course, in all such cases, immediate knowledge of the sort one calls "understanding the indefinable," which I formerly took in a lump as "acquaintance" but in fact it is necessary to distinguish. Immediate Knowledge of any indefinable other than a particular involves apparent variables and the propositional form.

Here it is not just facts that cannot be named, but also universals and functions. Something else takes the place of the old acquaintance relation with respect to these other types*. But these will not be simple two-place relations between a subject and an object.

In the notes for the class on March 17 Victor Lenzen noted that the acquaintance relation came in different sorts and that acquaintance with what Russell called "actual particulars" was a different relation from acquaintance with facts. In the case of verbs Lenzen noted "Never make a verb as subject only particulars subject."[7] Nonetheless, it would seem that the "verbs" were subjects in the new acquaintance relation. On the same day, he recorded Russell talking about the different acquaintance relation that we have with facts; this relation Russell called "perception." The relation had to be different from the acquaintance relation which holds between a subject and a particular, but it still seems to be a dual relation, this time between a subject and a fact, where the fact is in the place of a logical subject.

## 4 Words for Universals and Relations Cannot Occur as Logical Subjects; Universals and Relations Themselves Only Occur in Facts as Relating Relations or as Predicates, Never as Terms

As we can see from the letter to Broad, Russell adopted this view under the influence of Wittgenstein. In *The Principles of Mathematics*, Russell was quite clear that universals and relations had to be able to occur as terms, and that the contrary doctrine would lead to paradox. He thus rejected as paradoxical Frege's position that the concept horse was not a concept. In *The Principles of Mathematics*, he did hold that universals and relations, unlike what he later called particulars, could occur both as terms and as predicates in propositions. In his 1911 papers, he maintained that universals and relations can play both roles: they can occur as predicates (in the ontological sense) and as terms. The multiple-relation theory of judgment is formulated with them occurring as terms.

There are disputes about whether Russell's rejection of the1906–1907 substitution theory led him to think that there is an ontological distinction of types of entities. Landini's view is that Russell did hold a distinction in ontological types (the types* mentioned earlier) but these should not be identified with the ramified types of *Principia Mathematica*, which were never intended to apply to entities. Bernard Linsky argued that Russell did have an ontological theory of types, although he also does not identify universals with propositional functions (see Linsky 1999). Edwin Mares' construction of models for ramified type theory with the axiom of reducibility includes a simple theory of types for entities, but Mares also does not hold the ramified type theory applies to entities (see Mares 2007). Whatever your views about Russell's view of types of entities in *Principia Mathematica*, it should be clear that there is a change between *Principia Mathematica* and what we find in "The Philosophy of Logical Atomism," where only particulars, objects of the lowest type can be logical subjects.

We see the change in the letter to Broad quoted above, and we see it in the notes taken in Russell's class by Lenzen. There is as well a letter from Russell to Ralph Barton Perry written November 9, 1913, saying with regard to his theory of logical types that "all except the first are fictions." It is hard to think that he thought of the theory of types in ontological terms while maintaining that all higher types were "fictions." As he still seemed to think that in some sense there are universals and relations and facts, these type* differences should not be identified with the hierarchy of types in *Principia Mathematica.*

From the 1914 letter to Broad mentioned above, we see that the acquaintance relation that holds between a mind and a particular cannot hold between a mind and a universal. This letter gives some hint as to how Russell would handle such issues as *S is acquainted with redness.* It can't simply be that there is another relation A* which holds between a subject and a universal as in S-A*- $\hat{x}$-is-red, but rather some relation between S and "the meaning of '$x$ is red' whatever $x$ may be," where the predicate occurs as a predicate and not as a logical subject. The suggestion is that our acquaintance is with a complex where the universal or relation occurs as predicate. Of course, this analysis still seems to suppose that facts or complexes can occur as terms. Russell's letter to Broad seems to deny this is possible. Lenzen's notes, however, make reference to Russell's use of perception of facts, which would suggest that there is a two-place relation between a perceiving subject and a complex. In any case, what appears to be a radical change in doctrine in 1918 is already present in 1914.

## 5 Logic in 1912 and Again in 1918

In the opening sentence of *The Principles of Mathematics*, Russell gave a clear statement of his logicism by characterizing pure mathematics as,

> [T]he class of all propositions of the form p implies q where p and q are propositions containing one or more variables, the same in the two proposition, and neither p nor q contains any constants except logical constants. (*PoM*: 3)

He included in his list of the basic logical constants implication, the relation of a term to a class of which it is a member, and the notion of *such that.* Logic for Russell was in some sense the science of everything—there are no restrictions on its content. The propositions of logic are completely general applying to everything. The logicist's task was to show that mathematics did not employ any special intuition. Russell made this point especially clear in his chapter on Kant's Theory of Space (*PoM*: §434).[8] Russell had a metaphysical logic in *The Principles of Mathematics* including a robust realism about classes, logical relations, and various forms of propositions. Recent work on Russell's logic helps us distinguish the ontology of *The Principles of Mathematics* from that of *Principia Mathematica*. Russell had abandoned classes in favor of propositions in his substitution theory of classes and relations after 1906, and this no-classes theory is evident in the 1908 "Mathematical Logic as Based on the Theory of Types." By the time of *Principia Mathematica*, he had rejected propositions as entities in favor of his multiple-relation theory of judgment. He did not think that the logical connectives stood for special logical relations between propositions as entities. What exactly his views toward propositional functions were remains controversial. He did think there are universals and relations in intension. He saw logic as the study of the forms of propositions.

In his "The Regressive Method of Discovering the Premises of Mathematics" (1907), Russell argued that we discover the basic premises of logic by working backward from those items we think of as fairly obvious and in the case of mathematics not merely empirical generalizations. We seek more basic principles which may not initially be as well known, but from which we can deduce the propositions in question (*CPBR 5*: 572–575).[9] What is of special interest is that Russell did not see logic as completely different from any other science, and the search for the basic axioms appears to be an abductive one. Both of these positions are in sharp contrast to Wittgenstein's views on logic. Russell reported that Wittgenstein had an enormous influence on him (*MPD*: 83) and he later felt that he went too far in agreeing with him. Nevertheless, Russell's continued interest in the regressive method (see Sect. 7 below) suggests that Russell never quite accepted the full picture that Wittgenstein was painting.

Wittgenstein and Russell spent much of 1912 and 1913 talking about the nature of logic. One of the earliest records we have of Wittgenstein's views, other than snippets from letters to Russell, is Wittgenstein's *Notes on Logic*, dictated in October 1913. Russell read these with great care and used them in his 1914 lectures at Harvard. These notes, I believe, had a very strong effect on Russell. Russell himself stated this in his book *My Philosophical Development* (*MPD*: 83):

> At the beginning of 1914, Wittgenstein gave me a short typescript consisting of notes on various logical points. This, together with a large number of conversations, affected my thinking during the war years while he was in the Austrian army and I was, therefore, cut off from all contact with him.

A few pages later Russell said, "Wittgenstein maintains that logic consists wholly of tautologies. I think he is right in this although I did not think so until I read what he had to say on this subject" (*MPD*: 88).

If there is a governing idea in Wittgenstein's *Notes on Logic*, it is the bipolarity of propositions and the related view that the logic of propositions and all logical relations can somehow be represented by what he calls *a* – *b* diagrams. On Wittgenstein's view, these diagrams show that logic is tautologous and explain completely the function of logical constants. The logical constants are used to express these truth combinations and do not in any way represent logical objects. Now it is true that by this time Russell also did not think the logical constants represented special logical objects. He did think that there were certain more basic premises of logic which it was his business to discover. The account of logic in the *Notes on Logic* rejected this last part.

Several points from Wittgenstein's *Notes on Logic* were incorporated into the classes Russell taught at Harvard in the spring of 1914. These are documented in the notes Victor Lenzen took. We see especially the position that there are no real variables, only apparent ones (notes from March 14), and that what is complex must have a complex symbol—the structure of the symbol must be identical with the structure of the symbolized (notes from April 7) and so a proposition itself is a fact (notes from April 9).[10] However, while we find in them examples of truth tables and a discussion of the Sheffer stroke, there is no statement to the effect that the mark of logic is tautologyhood. In the 1918 *Introduction to Mathematical Philosophy*, Russell said that what makes purely symbolic or formal propositions logical is that they are analytic or that their contradictories are self-contradictions. He recognized that he hadn't fully specified what this amounts to, but said that whatever this characteristic is he will call "tautology" (*IMP*: 203). In "The Philosophy of Logical Atomism," we have the following (*PLA*: 211):

> Everything that is a proposition of logic has got to be in some sense or other like a tautology. It has got to be something that has some peculiar quality which I do not know how to define, that belongs to logical propositions and not to others ... They have a certain peculiar quality which marks them out from other propositions and enables us to know them a priori. But what exactly that characteristic is, I am not able to tell you.

Perhaps the hesitancy here is what kept Russell from mentioning this point in 1914, or perhaps Russell had changed his position between 1914 and 1918 and now felt more comfortable with Wittgenstein's position.

Wittgenstein himself had not come much further when he was still in communication with Russell. Here is a letter from Wittgenstein to Russell of November 1913, written after the *Notes on Logic* (Wittgenstein 1979: 128):

> As to what tautologies really are, however, I myself am not yet able to say quite clearly, but I will try to give a rough explanation. It is the peculiar (and most important) mark of non-logical propositions that one is not able to recognize their truth from the propositional sign alone ... But the propositions of logic—and they only—have the property that their truth or falsity, as the case may be, finds its expression in the very sign of the propositions.

Wittgenstein saw his *a* – *b* diagrams as replacing the need for any primitive propositions. There are no special premises for logic or mathematics. This would suggest among other things that Russell's regressive method for discovering such premises is ill-conceived. Yet as we shall see in Sect. 7, not only did Russell continue to accept the regressive method, it plays an important role in his later philosophy.

## 6 The Multiple-Relation Theory of Judgment Versus the New Beast in the Zoo

The story of Russell's development of the multiple-relation theory of judgment, together with Wittgenstein's criticisms, has been well told, even if in different narratives. I am not going to rehearse all of that here. What is clear is that he abandoned *Theory of Knowledge* without coming to a resolution on the theory of judgment and he had abandoned the multiple-relation theory by the time of "The Philosophy of Logical Atomism," where he not only rejects the multiple-relation theory of judgment, but any attempt to, as he put it, draw a map of a belief fact (*PLA*: 198).

Russell no longer defended the multiple-relation theory of judgment after 1913. This point has been hidden because in both his Harvard lectures and in *Our Knowledge of the External World* he continued to affirm that judgment was not a dual relation. However, while he still rejected a dual relation between a subject and a proposition, there is no continuation of the multiple-relation theory as it was conceived from 1910 until 1913. The rejection of universals or relations occurring as logical subjects, as we saw in section "Anything Complex Is a Fact and Facts Cannot Be Named," is inconsistent with the multiple-relation theory as it was developed. Russell mentions this in "The Philosophy of Logical Atomism," where he said (*PLA*: 198):

> You can make a map of an atomic fact, but not of a belief, for the simple reason that space-relations always are of an atomic sort or complications of the atomic sort ... The point is connection with there being two verbs in the judgment and with the fact that both verbs have got to occur as verbs, because if a thing is a verb it cannot occur otherwise than as a verb.

Russell then gave a "map" of a two-term relation between a subject and a proposition and argued that that wouldn't do. He concluded "I have got on here to a new sort of thing, a new beast for our Zoo, not another member of our former species but a new species. The discovery of this fact is due to Mr. Wittgenstein" (*PLA*: 199).

From the *Tractatus* we have the following remark on the theory of judgment: "[It] is clear that '*A* believes that *p*', '*A* thinks *p*', '*A* says *p*' are of the form "'*p*' says *p*": and here we have no co-ordination of a fact and an object, but a co-ordination of facts by means of a co-ordination of their objects" (*TLP*$_2$: 5.542). Whatever new beast in the zoo Russell had in mind, this does not seem to be it. Of course, Russell had no access to the *Tractatus* when he wrote this. He did, though, have access to the notes Wittgenstein had dictated to Moore in 1914. There Wittgenstein said (Wittgenstein 1979: 119):

> The relation of "I believe *p*" to "*p*" can be compared to the relation of '"*p*" says (*bebesagt*) *p*' to *p*: it is just as impossible that *I* should be a simple as that "*p*" should be.

(See also *TLP*$_2$: 5.542.) However, it is not clear that Russell really paid much attention to Moore's notes. Russell didn't mention anything of this position in "The Philosophy of Logical Atomism," even though he did mention that propositions were themselves facts.[11]

What there is in "The Philosophy of Logical Atomism" is for the most part already in the lecture notes from Russell's classes. During April 1914 Russell sketched the view that in such propositions as "I believe Smith hates Jones" there are "two verbs which occur as verbs." How exactly the two-verb theory is supposed to work is not well explained either in the 1914 notes or in "The Philosophy of Logical Atomism." At this stage, Russell wants to treat it as a new propositional form entirely.

At the end of Wittgenstein's *Notes on Logic*, which we know Russell used for his lectures, Wittgenstein made the following remarks after emphasizing that we may erroneously think that in *A* believes *p* we could substitute a proper name for "*p*." Wittgenstein said that to make the direction of *p* even more explicit we might say "A believes that '*p*' is true and 'not-*p*' is false." He then went on (Wittgenstein 1979: 106):

> Here the bipolarity of *p* is expressed, and it seems that we shall only be able to express the proposition "*A* believes *p*" correctly by the ab-notation; say by making "*A*" have a relation to the poles "*a*" and "*b*" of *a*-*p*-*b*. The epistemological questions concerning the nature of judgment and belief cannot be solved without a correct apprehension of the form of the proposition.

It isn't clear how Russell's position might be related to this one. Wittgenstein has a relation going between the subject *A* and the *a* – *b* poles of the proposition believed.[12] This doesn't quite fit with the two-verb view, unless we see the relation between *A* and the poles as one verb and the verb in the proposition *p* as the other. Russell didn't relate his two-verb beast to poles of a proposition in this way, but we do see that he already had this view in 1914.

It is often thought that Russell rejected the judging subject as an entity on the basis of Wittgenstein's criticism of his theory of judgment. But there is nothing in the *Notes on Logic* to suggest that Wittgenstein thought the account of judgment required the rejection of a simple subject. That suggestion only appears in the notes dictated to Moore. Russell himself came to reject the judging subject as a single entity in "On Propositions" written in the first part of 1919. There is a superficial similarity between Russell's view presented there and Wittgenstein's view in that the analysis holds that one fact, namely what Russell there calls an image-proposition, represents another fact. My own view is that Russell rejected the mental subject for reasons having to do with his adoption of neutral monism rather than Wittgenstein's theory of judgment, which he did not espouse in either the 1914 lectures nor in "The Philosophy of Logical Atomism." So it is unlikely that Russell's analysis in "On Propositions" is a result of Wittgenstein's work. He wrote "On Propositions" between February 23 and March 4, 1919 (*CPBR* 8: 276), and at that time Russell had not yet received the

*Tractatus* (*Logisch-Philosophische Abhandlung*) which Wittgenstein sent from prison in Cassino in June of that year.

## 7 Russellian Analysis

While Russell accepted much of Wittgenstein's account of logic in "The Philosophy of Logical Atomism," it seems that he did not abandon his position from "The Regressive Method of Discovering the Premises of Mathematics." For example, in the first section of "The Philosophy of Logical Atomism" he stated (*PLA*: 162):

> The sort of premiss that a logician will take for a science will not be the sort of thing which is first known or easiest known: it will be a proposition having great deductive power, great cogency and exactitude, quite a different thing from the actual premiss that your knowledge started from.

Now one could take Russell to be speaking of the special sciences and not logic here, as he was in particular addressing the vagueness of ordinary language about such things as Piccadilly. But if we look at his 1924 "Logical Atomism" we see that this method is also applied to the premises of logic (*LA*: 164):

> Exactly the same thing happens in the pure realm of logic; the logically first principles of logic—at least some of them—are to be believed, not on their own account, but on account of their consequences.

The real legacy of Russell's logical atomism is the method of analysis which he used in *The Principles of Mathematics* to define the entities of mathematics and which Whitehead had used to eliminate points and instants as entities (ibid.):

> When some set of supposed entities had neat logical properties, it turns out, in a great many instances, that the supposed entities can be replaced by purely logical structures composed of entities which have not such neat properties. In that case, in interpreting a body of propositions hitherto believed to be about the supposed entities, we can substitute the logical structures without altering any of the detail of the body of propositions in question.

Russell concluded this paragraph with the maxim, "Wherever possible, substitute construction out of known entities for inferences to unknown entities."

Russell saw his analysis of mathematics from *The Principles of Mathematics* as fitting this model, and in "Logical Atomism," he gave the examples of the reduction of electrodynamics to Maxwell's equations and of the analysis of matter (*LA*: 166–167). The discussion of the analysis of matter is interesting, since here we see Russell using the term "logical necessity," which we do not find in his earlier work. "One cannot help feeling that impenetrability is not an empirical fact, derived from the observation of billiard balls, but is something logically necessary" (*LA*: 167). Russell went on to say that this could only be the case if matter were a logical construction. Both Russell and Wittgenstein of the *Tractatus* held that the only necessity is logical necessity. But, unlike Wittgenstein, Russell saw this as setting a task for us to try to develop the logical constructions that will show how some purported necessity is just a logical necessity. It may not always be the case that we can develop construction that will show this for a particular case. In such a case, Russell would have to say that impenetrability, for example, perhaps was in the end an empirical fact carrying no logical necessity.

Wittgenstein has a different attitude in the *Tractatus*, where he asserted (*$TLP_2$*: 6.3751) that for two colors to be at one place in the visual field is logically impossible. He then gave a sketch of an analysis that might account for this involving the claim that a particle cannot be in two places at the same time and pointed out that the claim that a point in the visual field has a given color cannot be an elementary proposition. It seems that here Wittgenstein had adopted Russell's contribution to logical atomism, but he did not talk of logical constructions, nor did he see the need to give such an analysis of the claim that a particle cannot be in two places at the same time. This was the very point Russell thought required analysis, and the analysis required seeing matter as a logical construction.

## 8 Concluding Remarks About Russell and Ramsey

As far as I know, Russell didn't explicitly address the issue of the incompatibility of color predicates in his atomism papers. Given that particular sense-data having particular colors feature as examples of his atomic facts, it would seem that he would not have an account that treated them as logical constructions. Thus, he might have been more inclined to agree with what Ramsey said a few years later on this point. In 1927 Ramsey wrote "Facts and Propositions," which is primarily on the topic of judgment, but contains Ramsey's own version of logical atomism, although he did not use this term. Ramsey was concerned whether a speaker's own names might be of hidden complex objects so that his own apparent atomic propositions could turn out to be contradictory. In this case, what appeared to be atomic propositions would not be such, and further analysis would yield the true logically proper names which corresponded to the simples. Ramsey responds to this difficulty as follows (Ramsey 1927: 48):

> Whatever may be thought of this hypothesis [the hypothesis that further analysis will reveal such complexity], it seems to me that formal logic is not concerned with it, but presupposes that all the truth-possibilities of atomic sentences are really possible or at least treats them as being so. No one could say that the inference from "This is red" to "This is not blue" was formally guaranteed like the syllogism.

He went on to suggest that it might be best to look at this incompatibility as something that arises from the "physical constitution of men." Ramsey thought that this should have no more to do with logic than an (accidental) magnetic field prohibiting the movement of certain chess pieces would have to do with an account of which chess moves are possible. Ramsey's analogy makes it look as though he had decided that color incompatibility really isn't a question of logical necessity, any more than the accidental presence of a magnetic field would be a question of the rules of chess.

Ramsey thought that the issue wasn't relevant to logic because logic treats all the truth possibilities of atomic sentences as if they were really possible. In contrast, Russell's view seems to be that it is an open question whether there is an analysis that will reveal such an underlying logical complexity. But if it does so reveal it, then what we took to be an empirical generalization was actually a matter of logic. Ramsey instead wanted to say that what is an atomic sentence is relative to a language, and the language will just presuppose what is or is not an atomic sentence.

In some sense, Russell's position captures some of what Ramsey wanted. For in one sense it really isn't purely a matter of logic whether impenetrability turns out to be contingent. Ramsey would presumably allow there to be a language in which "A is in position *x*, *y*, *z*" and "B is in position *x*, *y*, *z*" are atomic sentences and thus in that language both would be possible. In another language, these could be treated as complex propositions which would exclude each other. Our choice of language would be a pragmatic one. Russell, on the other hand, would have us using the regressive method to try to find an analysis which shows the logical exclusions. If we succeed, we have revealed that it was all along a matter of logic. But this is set as a task which may not succeed. The task of analysis is to push back the unexplained necessities. Presumably those we have been unsuccessful at explaining should either no longer be treated as necessities or set as a task for further investigation.

Russell saw this method as a key to his logical atomism, but we can see that it makes Russell's project far more tentative than Wittgenstein's project in the *Tractatus.* Logic for Wittgenstein is something wholly distinct from any other science. The question of what is behind color incompatibility, for example, Wittgenstein would see as a scientific question and not a philosophical one.[13]

In the end, while Russell seemed to embrace Wittgenstein's view of logic, his atomism is not as pure as Wittgenstein's. The atomism appears to be set as a project rather than something laid down all at once as it seems it is with Wittgenstein. Thus Russell did not abandon the regressive method and he could state in response to a question that analysis could continue ***ad infinitum*** without reaching an ontological simple (*PLA*: 180).[14]

## Notes

1. Pears is perhaps the clearest on this claim, in Pears (1967, 1987: 63), where he draws the contrast between Russell and Wittgenstein. Many others have held this as well. See Sainsbury (1979: 13–14) where he sees Russell's atomism as a theory of meaning connected to the principle of acquaintance. These views of Russell's atomism have been criticized by, among others (Landini 2011: 163–164).
2. This contrast with physical atoms is also mentioned in "The Philosophy of Logical Atomism."
3. As James Levine pointed out to me, Russell explicitly includes complexes as particulars in *Papers* 6, 150.
4. As James Levine has pointed out to me, Russell at this time held that all complex unities (as opposed to aggregates) are propositions, although some are asserted and some are not (*POM*: 139). In Appendix B of *PoM* Russell introduced his first theory of types, which restricted aggregates which were classes as many from the level of individuals. Classes as one remained at the same type as individuals, provided its members were also individuals (*PoM*: 523).
5. See Landini (2007: 57).
6. This point can easily take care of Wittgenstein's criticism of the theory of judgment that it fails to prevent judging that the table penholders the book. (See Wittgenstein [1979: 103] for the criticism.)

7. Both James Levine and Bernard Linsky have said that they think that this note was actually taken in Russell's logic class. Levine suggests that on this matter Russell was being somewhat more careful in presenting Wittgenstein's views in the logic class. In any case we can see that Russell was rejecting the position that universals could occur in subject place even though he did not make that position as clear in the class on theory of knowledge.
8. This is also the section where Russell affirmed that both logic and mathematics are synthetic a priori.
9. For an important discussion of this paper, see Irvine (1989). In his 1906 paper "Les Paradoxes de la logique," Russell also outlined this same view. See (*CPBR 5*: 746–747; 280–281 for the English version). "The Regressive Method of Discovering the Premises of Mathematics" was read to the Cambridge Mathematics club and not published until Douglas Lackey's 1973 collection *Essays in Analysis*. This was also the first publication of the English version of "Les Paradoxes de la logique" which was initially entitled "On 'Insolubilia' and their solution by Symbolic Logic." It is perhaps because of this relatively late publication date that this aspect of Russell's work was not very well known. However, as I mention below, it is discussed in Russell's 1924 paper "Logical Atomism."
10. The Lenzen notes are in a notebook in the Bertrand Russell Archives at McMaster University.
11. Wittgenstein complained about Russell not digesting the notes to Moore in a letter to Russell written in December 1914. Russell apparently showed the letter to Moore who was angry that Russell had not asked him to explain the Notes. Apparently Russell did read the notes in early 1915, but he did not comment on them. See Wittgenstein (1997: 91–92). Here I agree with Michael Potter (2009) who also sees Russell as influenced by the *Notes on Logic* rather than the notes dictated to Moore.
12. It isn't clear what exactly the relation was that Wittgenstein thought held between the subject and these two poles. In fact, he was not clear as to what the two poles were.
13. It seems Wittgenstein had second thoughts about this when he wrote "Some Remarks about Logical Form," but he was clearly dissatisfied with what he did there as he decided not to read that paper, although it ended up being published. It is clear that in his later philosophy Wittgenstein gave up the atomism of the *Tractatus*.
14. I would like to thank the members of the 2017 Oberman Summer Seminar on the "Philosophy of Logical Atomism: A Centenary Celebration" for helpful comments. I especially wish to thank James Levine for detailed comments on an earlier draft. Thanks to the Bertrand Russell Archives in the William Ready Division of Research Collections, McMaster University Library, for permission to use unpublished materials.

## References

### Works by Other Authors

Irvine, Andrew (1989). "Epistemic Logicism and Russell's Regressive Method". *Philosophical Studies* 55: 303–327.

Landini, Gregory (2007). *Wittgenstein's Apprenticeship with Russell.* Cambridge: Cambridge University Press.

Landini, Gregory (2011). *Russell.* London and New York: Routledge.

Linsky, Bernard (1999). *Russell's Metaphysical Logic.* Palo Alto: CSLI.

Mares, Edwin (2007). "The Fact Semantics for Ramified Type Theory and the Axiom of Reducibility." *Notre Dame Journal of Formal Logic* 48: 237–251.

Pears, D. F. (1967). *Bertrand Russell and the British Tradition in Philosophy.* London: Collins.

Pears, D. F. (1987). *The False Prison Vol. I.* Oxford: Oxford University Press.

Potter, Michael (2009). *Wittgenstein's Notes on Logic*, Oxford: Oxford University Press.

Ramsey, Frank P (1927). "Facts and Propositions." *Proceedings of the Aristotelian Society*, Vol. 7: 153–170.

Sainsbury, R. M. (1979). *Russell.* London: Routledge & Kegan Paul.

Wittgenstein, Ludwig (1979). *Notebooks 1914–1916*, second edition, G. H. von Wright and G. E. M. Anscombe (eds.), tr. G. E. M. Anscombe. Chicago: The University of Chicago Press.

Wittgenstein, Ludwig (1997). *Cambridge Letters: Correspondence with Russell, Keynes, Moore, Ramsey, and Sraffa*, ed. B. McGuinness and G. H. von Wright. Oxford: Blackwell Publishing.

PART III

# Metaphysics: Fundamentality and Negative Facts

CHAPTER 7

# Russell on Ontological Fundamentality and Existence

*Kevin C. Klement*

Until recently, many have perhaps assumed that metaphysics, or at least that branch of it called *ontology*, is concerned with issues of existence, and that one's metaphysical position is more or less exhausted by one's position on what entities exist. In his "On What There Is", Quine argued that the *ontological commitment* of a theory or set of views is determined by what things its quantifiers range over: "To be is to be the value of a variable", as he succinctly put it (Quine 1948: 15). Quine's views were never universal, but the weaker assumption that one's ontological commitments are at the center of one's metaphysics is very widespread. Recently there has been some pushback against this broad Quinean framework. Kit Fine has suggested that "we give up on the account of ontological claims in terms of existential quantification" (Fine 2009: 167). Jonathan Schaffer claims that the Quinean approach has created a "tension in contemporary metaphysics" (Schaffer 2009: 354), one that can only be resolved by returning to a more "Aristotelian" conception of metaphysics. The positive proposals of

K. C. Klement (✉)
University of Massachusetts Amherst, Amherst, MA, USA
e-mail: klement@philos.umass.edu



L. D. C. Elkind, G. Landini (eds.), *The Philosophy of Logical Atomism*, History of Analytic Philosophy,
https://doi.org/10.1007/978-3-319-94364-0_7

such figures vary, but often they suggest we focus our ontological investigations on what is *fundamental*, or "what grounds what" instead.

Quine was the first to point out the ways in which Russell was both an inspiration and a forerunner of his position. Notably, there was Russell's analysis of existence claims using the existential quantifier, and his well-known arguments that one can resist positing Meinongian unreal objects by accepting his theory of descriptions. However, it would be a mistake to read Russell as nothing more than a proto-Quinean. This will perhaps already be conceded for the periods when Russell still thought there were notions of "existence" not explicable by means of the quantifier, or embraced a distinction between existence and mere being or subsistence (e.g., *PoM*: §427; *EIP*: 486–489; *PoP*: 156). However, I shall argue that this is true for *mature* Russell, even when (starting roughly 1913) he officially held the position that all existence claims are to be understood quantificationally. In particular, while mature Russell understood "*F*s exist" as expressing $\ulcorner(\exists v)Fv\urcorner$, he would not have taken this necessarily to settle the metaphysical or ontological status of *F*s. Russell had, running alongside his account of existence, a conception of belonging to what is, as he variously put it, "ultimate", "fundamental", the "bricks of the universe", the "furniture of the world", something "really there". This contrasts with that which has only a "linguistic existence", which he also described as "logical fictions" or "linguistic conveniences". This hints at something like an Aristotelian conception of metaphysics in Russell, though he would prefer to speak of "analysis" rather than "grounding" for the relationship between the derivative and the fundamental. The overall position is explicit in his late 1957 paper, "Logic and Ontology", but is evident earlier, including in the 1918 *Philosophy of Logical Atomism* lectures. His Aristotelian conception of metaphysics is not entirely divorced from his quantificational analysis of existence, though the relationship is somewhat complicated. It does not help that Russell's way of speaking on these issues is often unclear, and seemingly inconsistent. I attempt to sort things out below.

## 1 "Logic and Ontology"

His position is presented most clearly in one of Russell's last philosophical writings, "Logic and Ontology" (1957). This piece was a response to G. F. Warnock's "Metaphysics and Logic", and represents Russell's reaction to later developments in analytic philosophy concerning the relationship between logic and metaphysics, including some of Quine's work. At the center of Russell's position is the claim that the connection between language and the world requires a *meaning* or *naming* relationship between words or symbols and things in the world. However, only *some* words or symbols need have this relationship, depending on the kinds of words or symbols they are:

> The relation of logic to ontology, is, in fact, very complex. We can in some degree separate linguistic aspects of this problem from those that have a bearing on ontology. ... Sentences are composed of words, and if they are to be able to assert facts, some, at least, of the words must have the kind of relation to something else which is called "meaning". If a waiter in a restaurant tells me, "We have some very nice fresh asparagus", I shall be justly incensed if he explains that his remark was purely linguistic and bore no reference to any actual asparagus. This degree of ontological commitment is involved in all ordinary speech. But the relation of words to objects other than words varies according to the kind of word concerned ... A large part of the bearing of mathematical logic upon ontology consists in diminishing the number of objects required in order to make sense of statements which we feel to be intelligible. ... (*LO*: 628)

In ordinary speech most words bear "ontological commitment": the asparagus must really be there. However, mathematical logic has a deflationary effect on ontological commitment. Later in the essay he writes:

> What mathematical logic does is not to establish ontological status where it might be doubted, but rather to diminish the number of words which have the straight-forward meaning of pointing to an object. (*LO*: 629)

He interprets his own work as having shown that terms "for" classes, numbers, and perhaps other "abstract" or "logical" symbols needn't have "reference"; such discourse apparently *can be* "purely linguistic". Surely he does not mean to *equate* numbers, classes, and so on, with linguistic expressions, so how is this to be understood?

In the essay, he reiterates his well-known view that existence claims are to be interpreted by means of the existential quantifier.

> I come now to the particular question of "existence". ... I maintain that the only legitimate concept involved is that of ∃. This concept may be defined as follows: given an expression *fx* containing a variable, *x*, and becoming a proposition when a value is assigned to the variable, we say that the expression $(\exists x).fx$ is to mean that there is at least one value of *x* for which *fx* is true. I should prefer, myself, to regard this as a definition of "there is", but, if I did, I could not make myself understood. (*LO*: 627)

He writes here that this is the "only legitimate concept" of existence, so he is not returning to an existence/being or existence/subsistence distinction. However, he *immediately* goes on to deny that the truth of an existentially quantified statement always suffices to bring about ontological commitment or establish the reality of the apparent "things" quantified over:

> When we say "there is" or "there are", it does not follow from the truth of our statement that what we say there is or there are is part of the furniture of the world, to use a deliberately vague phrase. Mathematical logic admits the statement "there are numbers" and metalogic admits the statement "numbers are logical fictions or symbolic conveniences". Numbers are classes of classes, and classes are symbolic conveniences. An attempt to translate ∃ into ordinary language is bound to land one in trouble, because the notion to be conveyed is one which has been unknown to those who have framed ordinary speech. … we find that if we substitute for *n* what we have defined as "1", we have a true statement. This is the sort of thing that is meant by saying there is at least one number, but it is very difficult, in common language, to make clear that we are not making a platonic assertion of the reality of numbers. (*LO*: 627–628)

Russell defines a cardinal number as a certain kind of class, that is, a class of classes including all and only those classes cardinally similar to a given class. He might write "there are numbers" in *PM*'s notation as follows:

$$(\exists\beta)(\exists\alpha)(\beta = \mathrm{Nc}`\alpha)$$

This claim follows almost immediately, as Russell suggests above, from something such as:

$$1 = \mathrm{Nc}`0$$

To see that this formula does not "ontologically commit" us to numbers, recall that class-terms in Russell's logic are "incomplete symbols" defined using higher-order quantification. The quantifiers used in existential claims about classes are eliminable in virtue of higher-order quantifiers as well (for details, see *PM* *20). These claims only ontologically commit us to whatever such higher-order quantifications commit us to, and nothing further. Nonetheless, it is true to say "numbers exist" if we mean (∃N).

Russell seems to admit that when we read this in ordinary language as "numbers exist" it can mislead and suggest a platonic reality of numbers that (∃N), when properly understood, doesn't require.

Russell insists that *some* symbols must have reference to external reality in order for language to express facts. Still, given his account of "incomplete symbols", he thinks it is possible for languages to include certain *apparently* unified symbols which are not meaningful in this way. They may have *parts* that make contact with reality without doing so themselves. They do not, as wholes, *name* anything. Nonetheless, as the no classes theory shows, he thinks one can introduce variables that take the place of such expressions, and use them to make true existence claims. He somewhat sloppily words this by saying that "numbers are symbolic conveniences", but it is apparent what he means. It is perfectly intelligible to speak of numbers, use symbols that *seem* like names of them, and even make existence claims about them, but once we understand how the symbols are being used, it becomes apparent that there is no need to posit entities that the symbols name or variables range over.

In the case of numbers and other classes, it might seem that Russell escapes commitment to them only by committing himself instead to special entities as the values of higher-order "propositional function" variables. But here too, Russell is poised to deny that any such entities really are "there" as part of the "furniture of the world". Some existential quantifications using these variables will come out as true, but again, this is not enough to guarantee genuine ontological status. In the same essay Russell presses the point, distancing himself from Quine:

> Quine finds a special difficulty when predicate or relation-words appear as apparent [bound] variables. Take, for example, the statement "Napoleon had all the qualities of a great general". This will have to be interpreted as follows: "whatever $f$ may be, if '$x$ was a great general' implies $fx$, whatever $x$ may be, then $f$(Napoleon)". This seems to imply giving a substantiality to $f$ which we should like to avoid if we could. … We certainly cannot do without variables that represent predicates or relation-words, but my feeling is that a technical device should be possible which would preserve the difference in ontological status between what is meant by names, on the one hand, and predicate and relation-words, on the other. (*LO*: 629)

I shall try to clarify the position Russell is taking here in what follows. I hold that, despite some minor changes, this 1957 position was already in place during the core "logical atomist" period of the 1910s. I start by discussing Russell's views on quantification—the heart of his account of existence—to make it clearer why existence claims do not always guarantee metaphysical status for what exists.

## 2 Russell's Views on Quantification

In previous works, I have argued for interpreting Russell as endorsing a "substitutional" semantics for quantification, as opposed to an "objectual" semantics (Klement 2004, 2010, 2013). I have been surprised by the pushback on this (e.g., Soames 2008, 2014), because the textual evidence strikes me as conclusive. However, there are legitimate worries about what this commits Russell to in terms of the requirements of any adequate language, and whether or not it undermines any alleged advantages of the theory of descriptions. Let us first sort out the interpretive issue, and leave discussion of the alleged problems for the next section. It is perhaps a tad anachronistic to attribute to Russell a clear understanding of the difference between objectual and substitutional semantics. It would be decades before the difference was described in the literature. Nonetheless, I think there is enough evidence to make it clear that Russell's views were *extremely close* to what we would now call substitutional semantics, on which the truth of a formula of the form $\ulcorner(\exists v).\varphi v\urcorner$ is to be understood in terms of the truth of at least one substitution instance $\ulcorner\varphi c\urcorner$, and the truth of $\ulcorner(v).\varphi v\urcorner$ understood in terms of the truth of *all* such instances. What is even clearer, however, is that Russell had a *truth-based*, rather than a *satisfaction*-based, understanding of quantification.

On the modern "objectual" understanding of quantification, the truth of $\ulcorner(v).\varphi v\urcorner$ is specified not in terms of the *truth* of anything else, but rather in terms of a distinct notion of *satisfaction*. Whereas truth is a property, a sentence, proposition or other truth-bearer, either has or lacks, satisfaction is a *relation* between an object (or *n*-tuple or sequence of objects) and something else (either an open sentence, or the semantic value thereof). The objects entering into this relation are the objects being "quantified over", and hence, there must *be* such objects to make sense of the semantics. Russell himself used the word "satisfy" or "satisfaction" in an analogous way (*PoM*: §24; *IMP*: 164), but unlike later thinkers he defined satisfaction in terms of truth rather than vice versa. He always understood quantification as involving an open sentence, or what an open sentence represents—a "propositional function"—the role of which was to represent *all propositions* of a certain form. The quantified proposition is understood as true if all these propositions are true, which explains in part his occasional tendency to prefer the wording "$f(x)$ *always*" over "$f(x)$ for every $x$", though he used both (*ML* 5: 593; $PM_1$: 127; *IMP*: 158). This basic description of a quantified statement as involving the truth of *all instances* of a class of propositions alike in form is consistently found throughout his writings (*PoM*: §42; $PM_2$: xx; *PLA*: 203; *IMP*: 158; *IMT*: 164; $LP_2$: 164). This truth-based account is incompatible with the kind of objectual semantics that makes the satisfaction relation prior to truth.

This is not yet enough to show that Russell held a substitutional theory of quantification in the modern sense. I have been speaking of quantified *propositions* and their relationship to a class of *propositions* all sharing a form, deliberately sidestepping the complications arising from Russell's changing views on the nature of "propositions". On his early view of propositions as language- and mind-independent complex objects, to say that the proposition $(x).\varphi x$ requires the truth of the propositions $\varphi a$, $\varphi b$, $\varphi c$, and so on, is not to say that the truth of the linguistic formula "$(x).\varphi x$" is to be understood as involving the truth of the linguistic formulas "$\varphi a$", "$\varphi b$", and so on, which is what one would expect on a modern substitutional semantics. It probably would be a mistake to interpret very early Russell as understanding quantification substitutionally. However, sometime around 1907 Russell abandoned "Russellian propositions". Thereafter, he used "proposition" in a variety of ways, sometimes tying it to his ever-changing theories of judgment (*TK*: 114–115; *PLA*: 196; *OP*: 296), sometimes defining a proposition as an assertoric sentence (*TK*: 80 footnote 1; *PLA*: 166; *OP*: 281), unfortunately sometimes both in the same work.

Russell focuses nearly all his work on theories of judgment and belief on those whose content would be expressed by *elementary* or *atomic* sentences. We never get a clear account of how the infamous multiple-relations theory of judgment would be applied to general or existential judgments.[1] This lacuna in his theories of judgment is perhaps best explained by his assumption that it is only the words occurring in atomic or elementary judgments that "refer" or "mean" things in objective reality, and hence an account of the kind of truth involving the relationship between the mind and the world need only tackle atomic or elementary judgments.[2] More complex logical forms—quantified forms, molecular forms, and so on—presuppose atomic forms, and their truth and falsity is derivative upon that of the simpler forms. Russell is explicit about this dependence in *Principia*:

> Whatever may be the instances of propositions not containing apparent [bound] variables, it is obvious that propositional functions whose values do not contain apparent variables are the source of propositions containing apparent variables, in the sense in which the function $\varphi\hat{x}$ is the source of the proposition $(x).\varphi x$. For the values for $\varphi\hat{x}$ do not contain the apparent variable $x$, which appears in $(x).\varphi x$ … this process must come to an end … ($PM_1$: 50)

> … it follows that "$\varphi x$" only has a well-defined meaning … if the objects $\varphi a$, $\varphi b$, $\varphi c$, etc., are well-defined. ($PM_1$: 39)

Quantified propositions depend for their significance on propositional functions not containing quantifiers, which depend in turn on the significance of their non-quantified values. This dependence is reiterated many times in Russell's later writings:

> … propositions containing non-logical words are the substructure on which logical propositions are built … (*V*: 151)

> Let us begin with purely linguistic matters. There are certain words which are called "logical words"; such as "not", "or", "and", "if", "all", "some". These words are characterized by the fact that sentences in which they occur all presuppose the existence of simpler sentences in which they do not occur. (*PoU*: 267)

Notice that the dependency mentioned here is explicitly one between *sentences.*

This dependence is arguably a cornerstone of logical atomism itself. I find it difficult to understand what sort of dependence is involved her except a *semantic* one: the truth or falsity of non-atomic (or non-elementary) statements depends recursively on the truth or falsity of atomic/elementary statements. In *Principia* itself one even gets the impression that the dependence is, ultimately, *only* on them. *Principia* speaks of "complexes" or facts corresponding only to elementary judgments, and explicitly denies that quantified statements point to single complexes ($PM_1$: 46). Only elementary propositions connect to the world. Later on, Russell does introduce *general facts*, as in (*PLA*: Lecture V), but he provides little insight into their nature, as he admits himself (*PLA*: 207–208). They seem to be "meta-facts" about what atomic facts there are, not involving any new "things" or "entities" beyond those in the atomic facts. The official position in 1914s *Our Knowledge of the External World* is that knowledge of all atomic facts, along with the knowledge *that they are all the atomic facts*, fixes the truth or falsity of all propositions (*OKEW*: 50). The same is suggested in the 1925 second edition of *Principia* (*$PM_2$*: xv). Perhaps this one meta-fact about atomic facts is the *only* general fact we need countenance. If so, then it seems that Russell's metaphysics should admit no more entities than those involved in making atomic statements true, and the general "totality" fact that the ones there are are all there are. Of course, Russell accepts many "existence" claims regarding things not involved in atomic or elementary judgments (classes, numbers, etc.). These employ higher-order quantifiers. Assuming the restrictions of ramified type-theory are obeyed, the truth-conditions of a statement involving quantifiers of order $n+1$ can be defined in terms of the truth or falsity of their values, which can only involve further quantifiers of order $n$; these are defined in terms of the truth or falsity of those of order $n-1$, and so on, until one gets to elementary, non-quantified propositions. It is pretty clear that if Russell *had* accepted an objectual understanding of higher-order quantification, he would be committed to many entities besides simple individuals and their properties and relations, entities entering into satisfaction relations unanalyzable into facts about simple individuals and their simple properties. But, in fact, Russell's picture of the world during his logical atomist period seems only to countenance simple individuals, their properties and relations, the atomic facts made therefrom, and meta-facts thereabout (e.g., *OKEW*: 47).

Since he accepts existentially quantified higher-order claims, in some sense, "propositional functions" (as he calls their values in informal discussion) "exist". Nonetheless, this does not mean that they are part of the "furniture of reality". They too may have a mere "linguistic existence", like classes and numbers. Russell is fairly clear about this in a number of places: he says a propositional function is "an incomplete symbol" (*T*: 498), "not a definite object" (*$PM_1$*: 48), "nothing but an expression" (*MPD*: 53), "a mere schema, a mere shell" (*IMP*: 157), "nothing" (*PLA*: 202). There is some sloppiness about use and mention here, but the point is that although we can speak about open sentences as making existentially quantified higher-order formulas true, they are not meaningful by *naming* entities. An open formula which is a substituend of a higher-order variable may contain names as parts, and these names hook onto the world, even if the open sentence as a whole does not. If the open sentence does not contain such names, it may also contain further quantifiers, with variables whose substituends will contain names (or their instances will, and so on). Eventually such higher-order quantified statements will make reference to the world, but not simply by using a *name*.

In *Inquiry into Meaning and Truth*, Russell makes his substitutional understanding of such quantifiers explicit when he writes, "In the language of the second order, variables denote symbols, not what is symbolized" (*IMT*: 202). This way of putting it is somewhat misleading; as I have argued elsewhere, substitutional quantification is not the same as objectual quantification over expressions (Klement 2010: 648–653), but Russell was writing for an audience that likely would not pick at this nit. In the same context (*IMT*: Chap. 13), he sometimes rewords a quantified sentence back into English as "all sentences of the form … are true" or claims that they may be interchanged with the infinite conjunctions (if universally quantified) or infinite disjunctions (if existentially quantified) of their values. Throughout, Russell speaks of *sentences*, not propositions. This is clearly an endorsement of the view that the truth-conditions, at least, for a formula of the form $\ulcorner(v)\varphi v\urcorner$ consists in the truth of all the instances $\ulcorner\varphi c\urcorner$ where $c$ is any closed symbol of the appropriate logical type. Russell here limits his remark to "the language of the second-order", though presumably the same would hold for higher orders. This suggests that something is different about higher-order variables as opposed to first-order variables. Another indication that he sees a difference comes where he speaks of different meanings of "there is" or "there are" as early as *The Philosophy of Logical Atomism*. He claims that of the different meanings of "there are", "[t]he first only is fundamental" (*PLA*: 233), by which he means the first-order quantifier $(\exists x) \ldots x \ldots$. Moving only one type up, to classes of individuals, Russell says "you have travelled already just as much away from what there is" as if you have gone up any number of types (*PLA*: 233), since "[t]he particulars are there, but not classes". Clearly, Russell thinks that first-order quantification is ontologically committing in a way that higher-order quantification is not. It is perhaps this difference that has led Gregory Landini to argue that Russell accepts a "nominalistic" or substitutional semantics for variables of most higher-types, but not for individual variables.[3]

However, I believe the evidence suggests that Russell accepts a substitutional account for all types. When discussing the hierarchy of different senses of truth in *Principia*, he writes:

> Let us call the sort of truth which is applicable to $\varphi a$ "*first* truth." (This is not to assume that this would be first truth in another context: it is merely to indicate that it is the first sort of truth in our context.) Consider now the proposition $(x).\varphi x$. If this has truth of the sort appropriate to it, it will mean

> that every value of $\varphi x$ has "first truth." Thus if we call the sort of truth that is appropriate to $(x).\varphi x$ "*second* truth," we may define "$\{(x).\varphi x\}$ has second truth" as meaning "every value for $\varphi x$ has first truth," … ($PM_1$: 42)

Russell means this example to illustrate how to think about the truth or falsity of quantified formulas of *any* given order in terms of the truth or falsity of formulas in the order just below it. Hence, his remark is not *specifically* targeted at first-order quantification. However, the use of the variable "*x*" and constant "*a*" strongly suggests that first-order quantified formulas are included in his remarks. If the remark meant to apply *only* at higher levels, he likely would have used "*f*" or "$\varphi$", rather than the conventionally first-order "*x*" and "*a*". Russell has already abandoned Russellian propositions by this point, so this passage suggests that we should understand the truth of the sentence "$(x)\varphi x$", where *x* is an individual variable, as meaning that every sentence "$\varphi n$", for every "logically proper name" *n*, has (elementary) truth.

The position is even clearer in later works, such as *An Inquiry into Meaning and Truth*, where he writes:

> The next operation is *generalization*. Given any sentence containing … a name "*a*", we may say that all sentences which result from the substitution of another name in place of "*a*" are true, or we may say that at least one such sentence is true. … For example, from "Socrates is a man" we derive, by this operation, the two sentences "everything is a man" and "something is a man", or, as it may be phrased, "'*x* is a man' is always true" and "'*x* is a man' is sometimes true". The variable "*x*" here is to be allowed to take all values for which the sentence "*x* is a man" is significant, i.e., in this case, all values that are proper names. (*IMT*: 196)

"Everything is a man" *means* that every *sentence* differing from "Socrates is a man" by the substitution of a proper *name* for the *name* "Socrates" is true. Russell's wording is clearly substitutional at the linguistic level, and he clearly has in mind a first-order variable.

This is not to say that there is no important difference between the first-order quantifier and others. The first-order quantifier carries existential import with it, because unlike other quantifiers, the substituends for its variable are proper names, and proper names *must* refer to something outside language in order to have meaning. It is in the name/name-bearer relationship that Russell thinks "the rubber meets the road", or language confronts reality.

A Quinean might argue that Russell's own theory of descriptions makes genuine proper names unnecessary: one can use a description such as "the *x* such that *x* Socratizes" instead of "Socrates".[4] For this to work, Socrates himself must be a value of a variable. Accepting an objectual semantics, the Quinean thinks that ***quantification*** can make a connection between language and the world. Russell himself, employing a substitutional semantics, explicitly denies that his theory of descriptions makes proper names unnecessary. Famously, Russell analyzes "an *F* exists" as stating that "*Fx*" is true for at least one *x*, and "the *F* exists" as stating that there is at least one and at most one such *x*. Given his understanding of first-order quantification, this means that there must be a ***name*** that can be substituted for this "*x*". He says so explicitly:

> An object ambiguously described will "exist" when at least one such proposition is true, i.e. when there is at least one true proposition of the form "*x* is a so-and-so," where "*x*" is a name. ... With definite descriptions, on the other hand, the corresponding form of proposition, namely, "*x* is the so-and-so" (where "*x*" is a name), can only be true for one value of *x* at most. (*IMP*: 172)

Russell himself argues that his theory of descriptions cannot make the study of names superfluous, because the truth of quantified statements, ***including*** those using descriptions, presuppose instances of the quantified formulas with names in place of the variables:

> In connection with certain problems it may be important to know whether our terms can be analysed, but in connection with names this is not important. The only way in which any analogous question enters into the discussion of names is in connection with descriptions, which often masquerade as names. But whenever we have a sentence of the form,
>
> "*The x* satisfying $\varphi x$ satisfies $\psi x$"
>
> we presuppose the existence of sentences of the forms "$\varphi a$" and "$\psi a$", where "*a*" is a name. Thus the question whether a given phrase is a name or a description may be ignored in a fundamental discussion of the place of names in syntax. (*IMT*: 96)

Russell thinks even first-order quantification cannot be made sense of without presupposing names as the values of the first-order variables, which of course would only be true if he understood them substitutionally as well. It also underscores how fundamental he thinks names are to how language connects to the world. To re-invoke "Logic and Ontology", names are those symbols that *do* point to something outside words, that make it so our asparagus must really be there. Russell finds it possible to imagine languages in which names do not stand for particulars, but only for universals (*HK*: 84; *IMT*: 95), but professes himself "totally incapable" of imagining a language without names (*IMT*: 94).[5]

In conclusion, (1) Russell's substitutional semantics for variables also applies to first-order variables, and (2) despite this, there is something special about these variables compared to others, in that the substituends for them must be the kinds of symbols that are meaningful by pointing to extra-linguistic entities. This is why Russell at times speaks of them as more "fundamental" than others, and doesn't speak of their values as if they were "nothing but an expression" or as having a mere "linguistic existence", as he does with higher-order variables.

## 3 Objections to a Substitutional Semantics for Russell

I think it is fair to say that a substitutional semantics for quantification is relatively unpopular, and indeed, prior to Kripke (1976), many thought it too problematic to be taken seriously. In Russell's case, it is natural to worry about whether the approach is compatible with other views he held. I here focus on two worries, one dealing with the application of Russell's theory of descriptions in his epistemology, another dealing with the requirement that there be infinitely many simple proper names and the coherence of a language with so many names. Both these issues are pressed by Scott Soames in his recent book.[6]

These worries involve a presupposition to the effect that it would be impossible for someone to understand a quantified statement, interpreted substitutionally, unless that someone understood *all* the expressions that were substituends for the variable. This is not a presupposition Russell shared. Soames writes:

> A remark in Russell [*IMP*] shows that he did not think of the quantification employed in his logical system as substitutional. On pp. 200–201 he says, "It is one of the marks of a proposition of logic [which contains no nonlogical vocabulary] that, given a suitable language, such a proposition [sentence] can be asserted by a person who knows the syntax without knowing a single word of the [nonlogical] vocabulary." Although the remark is true on an objectual understanding of quantification, it is incompatible with treating quantifiers in a "proposition of logic" substitutionally. (Soames 2014: 528–529 footnote)

If it were true that one could not understand a quantified statement without understanding all the vocabulary involved in its instances, this would surely pose a problem for Russell. Russell employed his theory of descriptions in his epistemology to make a distinction between "knowledge by acquaintance" and "knowledge by description" (*KAKD*: 147–161). If "the *F* is *G*", means, as the theory of descriptions says it does, "$(\exists x)((y)(Fy \equiv y = x) \,.\, Gx)$", and the quantifier here is understood substitutionally, then if it is true, one of the proper names of the language, "*c*" say, must be a name of the thing that is uniquely *F*. Russell is clear that a proper name can only be understood by direct acquaintance with its meaning. If understanding "the *F* is *G*" meant that I needed to understand the name "*c*", knowledge of something by description would be impossible without also having knowledge by acquaintance of the same thing. This, clearly, would be disastrous for Russell's epistemology.

To solve this, one must either drop the assumption that the quantifiers in the analyzed descriptive statement are substitutional, or reject the supposition that understanding such quantifiers even when substitutionally interpreted requires understanding all the names that are their substituends. Soames cites the following remark from Hodes in favor of the latter supposition:

> If a quantifier prefix in the sentence ... is to be interpreted substitutionally, and a relevant substituend contained an un-understood word, the speaker would not understand a relevant substituend and so would not understand that quantifier prefix and so would not understand that sentence! (Hodes 2015: 397)

I must confess, however, that this assumption seems to me to be wholly without merit. Understanding the truth-conditions of $\ulcorner(x).\varphi x\urcorner$—substitutionally understood—means that I must know that it is true just in case $\ulcorner\varphi n\urcorner$ is true for all proper names, *n*. This does not require that I have examined or understand each such instance $\ulcorner\varphi n\urcorner$, or name *n*. It requires at most that I understand the *difference* between a symbol that is a proper name and a symbol that is not, a difference in logical form. As Russell makes clear in the passage from *Introduction to Mathematical Philosophy* Soames mistakenly quotes in favor of his view, there's no reason to think I need to understand any specific proper names in order to understand the *form*, that is, the *syntax*, of a proper name. (Compare: if someone tells me that every sentence in so-and-so's article on quantum gravity is true, I can understand well enough what is required for *that* to be true, even if I don't understand half the words, and hence, half the sentences, in the article. If it's in another language, I might understand none. At most I need to understand the difference between what are sentences in the article, and what aren't.)

In the following passage Russell comes close to addressing the issue head on:

> There remains one question concerning generalization, and that is the relation of the range of the variable to our knowledge. Suppose we consider some proposition "*f*(*x*) is true for every *x*", e.g., "for all possible values of *x*, if *x* is human, *x* is mortal". We say that if "*a*" is a name, "*f*(*x*) is true for every *x*" implies "*f*(*a*)". We cannot actually make the inference to "*f*(*a*)" unless "*a*" is a name in our actual vocabulary. But we do not *intend* this limitation. We want to say that everything has the property "*f*", not only the things that we have named. There is thus a hypothetical element in any general proposition; "*f*(*x*) is true of every *x*" does not merely assert the conjunction
>
> $$f(a).f(b).f(c)\ldots$$
>
> where *a*,*b*,*c*... are the names (necessarily finite in number) that constitute our actual vocabulary. We mean to include whatever will be named, and even whatever *could* be named. This shows that an extensional account of general propositions is impossible except for a Being that has a name for everything; and even He would need the general proposition: "everything is mentioned in the following list: *a*,*b*,*c*...", which is not a purely extensional proposition. (*IMT*: 203)

This comes only a few pages after the passage quoted earlier in which Russell gives an explicitly substitutional account of "generalization". Here, however, he is clear that the substitution instances that are involved in the general truth go beyond those names that are in my present *personal* vocabulary. Instead, the generalization includes all names used by others, names only used in the future, and even merely possible names. We need not have an "extensional" list of such names; it is enough if we understand "intensionally" the difference between a name and something else.

This passage brings up the other alleged problem with Russell's adoption of a substitutional semantics. In order for every individual to be captured in the range of the quantifiers, every individual would have to have a name. If there are infinitely many individuals (as would be required by the so-called axiom of infinity, which Russell at least does not reject), there would need to be infinitely many names (cf. Soames 2014: 528). No one person's vocabulary is infinitely large, as we have seen Russell admits in the previous quotation. It does not immediately follow from this that a language must contain only finitely many names, as even a fluent person need not understand every word in the language. Of course, if there are finitely many speakers, as there are for any actual languages, each of whom uses a finite vocabulary, the sum total of those vocabularies would still be finite. Russell intends that the names involved in the truth-conditions of quantified statements go beyond even the sum total of everyone's actual vocabulary. He writes:

> This principle of assigning names may be used to define various possible philosophies. Let our list of names consist of all those that I can assign throughout the course of my life. If, then, from the fact that "$P(a)$", "$P(b)$", … "$P(z)$" are all true, I do not allow myself to infer that "$P(x)$" is true for all values of $x$, that is a denial of solipsism. If my list of names consists of all those that sentient beings can assign, the denial of the inference is an assertion that there are, or may be, things that are not experienced at all. (*RC*: 29)

Russell is neither a solipsist, nor someone who thinks existence is limited to what is experienced. If we are to interpret his views of quantification substitutionally, whether or not there are infinitely many, we must acknowledge that in some sense there are, or can be, names no one does or ever will understand. This is puzzling.

The puzzle is lessened somewhat by the consideration that Russell usually had in mind a "logically ideal language". He was of course aware that this language had not been fully developed, and hence that no one *actually* used such a language. However, he actively and knowingly assumed about such a language that it would have a name for every simple thing. This comes across both in his later reminiscences about his early work, as well as in that work itself. In *My Philosophical Development*, he wrote:

> I thought, originally, that, if we were omniscient, we should have a proper name for each simple, and no proper names for complexes, since these could be defined by mentioning their simple constituents and their structure. (*MPD*: 166)

In *PLA*, he is explicit that each of us would understand only a small subset of the logically perfect language's total vocabulary, but that nonetheless, every simple object would have a name therein. He also bemoans the fact that actual languages don't have names for true simples, true particulars, a complaint he also makes elsewhere (*AMi*: 193). He writes:

> In a logically perfect language, there will be one word and no more for every simple object, and everything that is not simple will be expressed by a combination of words, or a combination derived, of course, from the words for the simple things that enter in, one word for each simple component ....The language which is set forth in *Principia Mathematica* is intended to be a language of that sort. It is a language which has only syntax and no vocabulary whatsoever. ... It aims at being the sort of language that, if you add a vocabulary, would be a logically perfect language. Actual languages are not perfect in this sense, and they cannot possibly be, if they are to serve the purposes of daily life. A logically perfect language, if it could be constructed, would not only be intolerably prolix, but, as regards its vocabulary, would be very largely private to one speaker. ... I shall, however, assume that we have constructed a logically perfect language, and that we are going on state occasions to use it ... (*PLA*: 176)

Although Russell endorses a substitutional semantics even for first-order variables, he does so in the context of a theoretical language that in fact has a name for every simple object. He realizes that such a language not only isn't in use (even on "state occasions"), but could not practically be in use. One might worry whether or not Russell's intended semantics is intelligible if it requires making reference to a language of this sort. Must languages actually be in use to exist? Some might allege that languages are abstract objects as argued in (Katz 1980), or nothing more than pairings of possible expressions with semantic values à la (Lewis 1975), but such views do not seem very Russellian.

Clearly, however, Russell's acceptance of a substitutional theory of quantification involves not simply supposing that $\ulcorner(x).\varphi x\urcorner$ is true when $\ulcorner\varphi n\urcorner$ is true for every name *n* which *is or was* actually in use, or even every name *n* that ever will be in use: it must mean that it is true for every name *n* that *could* be in use, or *would* be in use *if* we had a logically perfect language. The modal terminology here could allow Russell to deflect certain worries some might have about his substitutional semantics. But it might create other worries. The only account of modality Russell himself provides is itself spelled out in terms of quantification, and so it could only circularly be applied here (*PLA*: 203). One common, and very compelling, interpretation of his logical atomism would exclude his countenancing any modal notions except *logical* possibility and necessity,

(Landini 2011: Chap. 4), and it is also unclear that these could be spelled out non-quantificationally. Can the modal or theoretical notions be dropped from the statement of the semantics? He claims more than once that "omniscience" might help, but as Russell is no theist, this does not quite help enough. Perhaps it is enough to suggest that understanding quantified statements with his intended semantics depends only on an understanding that it requires the truth of all statements that *would take a given form* if properly expressed or analyzed, which does not require being able to list, or even understand, all such sentences. This puts knowledge of logical form at the center of his account, which seems appropriate.

There are puzzles in this view remaining, and legitimate questions one may raise. But I think that some kind of substitutional view is clearly what Russell had in mind, even if he did not make it fully clear. Moreover, unless we attribute to Russell something like a substitutional account, not only do certain aspects of his logical atomism not make sense (e.g., the dependence of other propositions on the atomic ones), but Russell's entire metaphysical outlook, explicitly outlined in works like "Logic and Ontology", where he separates existence questions from those of genuine metaphysical status or ontological commitment, would fall apart.

## 4 Russell's Metaphysics: Why There Is What There Isn't

The title of the final lecture of *The Philosophy of Logical Atomism* is "Excursus into Metaphysics". Clearly, he thinks the subject was not exhausted by his discussion of existence in lectures V and VI. What's puzzling is that the subtitle is "What There Is", and assuming "there is" is a kind of quantifier, this suggests that quantification can be of some use in understanding Russell's metaphysics. Hopefully, we have seen enough of Russell's views to explain away this puzzle. Quantification is understood substitutionally. Some quantifiers use variables whose substituends are symbols that are not meaningful by *naming* or *representing* extra-linguistic entities. First-order quantifiers, ranging over particulars, use variables whose substituends *are names* of things. These variables carry metaphysical commitment; the other quantifiers don't.

Russell is an ideal language philosopher, and thinks that our ordinary language expressions of existential statements, for example, "there are numbers", as we have seen, are "bound to land one in trouble". Ordinary language is ill-suited to represent properly the difference in form between expressions of differing types. The infinitely many meanings of "there is" or "there are" (*PLA*: 232) are all pronounced or appear the same in ordinary language. Upon hearing "there are" in ordinary language, we are apt to interpret it as standing for the *ultimate* meaning of "there are"—the first-order meaning. When Russell is presenting his philosophical views in ordinary language, he is apt to claim that "there are" no such things as numbers, or classes, or to claim that propositional functions are "nothing". In those contexts, he means that there are no such things in the ranges of the ontologically committing quantifiers. At other times, however, he expects his reader to understand that his ordinary language quantification talk is to be adjusted in interpretation to something that would be more perspicuously represented with a different-type quantifier. The "no" in the title of Russell's "no classes" theory is a kind of quantifier, but that theory does not say there are no classes in the sense in which it best makes sense to quantify over classes: it is only that *no individuals*, no genuine things in the extra-linguistic world, are classes. Russell only *apparently* contradicts himself when, in one paragraph of *The Philosophy of Logical Atomism*, he says that "there are classes" and "there are particulars" can both be interpreted as true so long as one understands that these are two different meanings of "there are" (*PLA*: 230), but then in the next paragraph goes on to say his theory allows one to do without "supposing for a moment that there are such things as classes" (*PLA*: 231–232). Ordinary language renditions of his views cannot do them justice.

One might worry that Russell's ordinary language presentation of his metaphysical views is in "too much" trouble. By his own lights, the "there are" which is used in first-order quantification cannot even be meaningfully applied to classes, so the "no" of the title of the "no classes theory" is meaningless. Most likely, Russell would claim that what is meant is that there are no individuals which have the kinds of *formal properties* (cf. (*PLA*: 236)) which would make them appropriate to play the role classes play in logic. Russell is committed to a class for every propositional function:

$$(\varphi)(\exists\alpha)(x)(x\varepsilon\alpha \equiv \varphi!x)$$

Part of what he means when he says, in ordinary language, that "there are no classes" is presumably that there are no individuals suitably like classes for which there is a relation structurally *analogous* to ε which all and only satisfiers of certain functions bear to them, that is:

$$\sim(\exists R)(\varphi)(\exists y)(x)(xRy \equiv \varphi!x)$$

There are no individuals that can play the role classes play.

There are many places where Russell speaks as if "there are" no such things as physical bodies (tables, chairs, Piccadilly street)—and after his conversion to neutral monism, no such things as minds either (*PLA*: 170; *PaM*: 273–274). All of these he calls "logical fictions", and thinks that all there "really" are are simple particulars arranged in certain ways, and bearing certain relations to each other, such that we group them together in the same class. But these classes still *exist* in the sense in which classes exist; Russell would not *deny* that there are over a million people living in Britain, or that there are exactly three chairs in this room. He means that the symbols for these so-called things are not names; the truth or falsity of claims about them is reducible to the facts regarding ultimate, simple things. We need not presuppose there are things having their sort of formal properties at the *fundamental* level. For Russell, this is the true meaning of Ockham's razor, the sense in which, as he put it in "Logic and Ontology", his mathematical philosophy diminishes the number of objects in our ontology. It is not that a well-shaved philosophy will accept fewer existence claims, where those claims are interpreted in a derivative way. Rather, a well-shaved philosophy will posit fewer things at the "ultimate" or "fundamental" level: the level of those things involved in making true the real facts that, in a much more indirect fashion, ultimately make discourse about non-fundamental things possible.

In the metaphysics of *The Philosophy of Logical Atomism*, Russell considers the "simple" things that make up reality to be such things as sense-data, and their properties and relations. These are what are involved in atomic facts, which make atomic propositions true or false. These, he says, "have a kind of reality not belonging to anything else" (*PLA*: 234). Constructs out of them do not have the same kind of reality: there is some derivative sense in which they exist, but all this means is that we can use certain complicated symbols, and also regard these symbols as substituends of variables. The reality of constructs is thus reduced to "linguistic convenience". We thereby reduce our "metaphysical baggage", the apparatus our view of the world has to "deal with". He makes it clear that real metaphysical commitment involves regarding certain symbols as *names* of things:

> If you think that 1, 2, 3 and 4, and the rest of the numbers, are in any sense entities, if you think that there are objects, having those names, in the realm of being, you have at once a very considerable apparatus for your metaphysics to deal with … (*PLA*: 234)

Russell himself is happy to make claims about numbers, quantify over them and assert, for example, that for every number, *there exists* a higher one. He denies that doing so commits him (directly at least) to any kind of metaphysical outlook on what there is "ultimately". In a 1958 review of a work on mathematical infinity by E. R. Emmet, Russell writes:

> He [Emmet] comes to an astonishing conclusion (page 679): "An indefinite [infinite] number is not a positive 'thing' that is there, but a negative absence of definiteness." Does Mr. Emmet consider that the natural numbers are positive "things" that are "there"? If so, he is astonishingly Platonic; but if not, I am at a loss to see in what way the number of inductive numbers differs from any other number in respect of being "there". (*MI*: 364)

Russell's views had not changed much between 1918 and 1958. Russell is happy to admit that infinite numbers are not "positive things" that are "really there", but does not think this is any reason to ignore or downplay their mathematical properties, or treat them as any different from finite numbers.

For Russell, then, metaphysics addresses the question as to what the "ultimate constituents" of the world are, what is "fundamentally real". What sort of logical "fictions" or derived "objects" can be constructed from them is mainly of negative interest: if we can show that things we might take to be fundamentally real are logical constructions instead, we remove the need to take them as part of our metaphysics. Russell's "supreme maxim in scientific philosophizing", to "substitute constructions out of known entities for inferences to unknown entities" (*RSDP*: 11; *LA*: 164) is the directive to reduce one's conception of what there really is to as few things as possible, things easily known or experienced, and treat things with "smooth logical properties" (*PoM*: xi) as logical constructions. One is then left with the task of identifying the "smallest apparatus" (*PLA*: 235) or "minimum vocabulary" (*HK*: 242ff.) with which one can fully describe what is "really out there", or give a complete catalog of the world. Given Russell's general understanding of the logical forms of language, extra-logical vocabulary only occurs within atomic statements: so Russell's metaphysics is mainly the attempt to identify what vocabulary is needed to account for the simplest of truths—atomic propositions—upon which the truth of all others ultimately rests. Russell's exact understanding of atomic propositions changes, but he consistently holds that in an analyzed language, the symbols making them up are those that represent some part of extra-linguistic reality.

Early on, when he held that "individual" or "term" was the "widest word in the philosophical vocabulary" (*PoM*: §47), he held that all words expressing an atomic proposition stand for individuals and these are all included in the range of the first-order quantifier (*AIT*: 261; *PoL*: 290). On this view, all metaphysically real things would be individuals. Hence, during this period, Russell writes that individuals are "[s]uch objects as constitute the real world as opposed to the world of logic" (*STCR*: 529), "being[s] in the actual world" (*AIT*: 44), entities which "exist on their own account" (*PM*$_1$: 162) and "do not disappear on analysis" (*PM*$_1$: 51).[7]

Later, under Wittgenstein's influence, he came to think that particulars and universals had different logical types (*PLA*: 182; for discussion, see (Klement 2004)), and hence that there would be no one logical type, and thus no one style of variable, encompassing both. It is for this reason, presumably, that in the 2nd edition to *Principia* (*PM*$_2$: xxxii), he discusses adding a new style of variable for the universals in atomic propositions (though ends up deciding it is not necessary)—a clear indication that he does not consider the "propositional function" variables of *Principia* already as objectual variables over universals. Presumably if he had added such a variable, it too would be ontologically committing. Still later, he came to doubt particulars altogether and to think that *all* the "names" in atomic propositions might be taken to stand for universals. Then the only ontologically committing variables would be those whose substituends would be names of universals rather than "proper names" in the usual sense (*HK*: 84; *IMT*: 95). Naturally, as his overall metaphysics changed, so did his account of the kinds of symbols entering into atomic propositions, as well as the kinds of variables that might replace those symbols.

One might object that this is "too linguistic" a conception of metaphysics. What is metaphysically real is one thing; what is involved in our unanalyzed sentences is another. The way we set up our languages is to some extent a matter of convention: what counts as primitive vocabulary in one language might not in another. Is metaphysics itself language-relative? Again, one must bear in mind that Russell has in mind primarily a *logically ideal* language where the logical forms of its expressions closely mirror the logical forms of the reality they depict. Even after Russell backed away from the view that a logically "perfect" language was anything like a realistic aim to search for, he seems to have been confident that the minimum vocabularies of *adequate* languages for scientific research would not differ much concerning what counts as fundamentally real. He writes:

> The theory of incomplete symbols shows that it is possible to construct a minimum vocabulary for logic which does not contain the word "class" or the word "the". I incline to think—though as to this I have some hesitation—that the contradictions prove, further, the impossibility of constructing a minimum vocabulary containing the word "class" or the word "the", unless highly complicated and artificial rules of syntax are imposed upon our language. For similar reasons, no acceptable minimum vocabulary will contain words for numbers, i.e. every acceptable minimum vocabulary will be such that numbers are defined by means of it. (*RC*: 23)

This commits Russell to a fairly narrow conception of "acceptability" that he doesn't spell out, at least not here, and it shows that he does not think such issues are completely "conventional" or "relative to language choice" in a broadly Carnapian vein.

So we can see the many ways in which, in spite of his proto-Quinean views on the relationship between existence and quantification, Russell's metaphysics can be understood as broadly Aristotelian, in Schaffer's sense. He is interested in what is *fundamental*. But his metaphysics also has certain features that differentiate it from contemporary forms of neo-Aristotelian metaphysics. First, as we have seen, existence questions are not entirely divorced from questions about what is metaphysically real or "ultimate": some, but not all, quantifiers, are ontologically committing, and sometimes metaphysical theses are best expressed using those quantifiers. Related to this is the even more important point that Russell is very deflationist about the non-fundamental: he is willing to say that in at least *some* sense, non-fundamental things are "nothing", not "there", mere "fictions" and so on. Fine's, Schaffer's, and other contemporary "Aristotelian" approaches to metaphysics focus largely on the *relation* of grounding: but the mere fact that grounding is a relation presupposes that there are, really are, relata of this relation. Some understand grounding as a relationship between objects, some as a relationship between facts, but generally, they accept that both the grounders and the groundees are fully "there" to enter into this relation. Russell would of course prefer to speak of "analysis" rather than "grounding", and the things that are "analyzed" are, in a sense, analyzed "away". Their existence is *merely* linguistic, and so are the relations into which they enter: all truths about them ultimately resolve into truths about the ultimate things. Only the ultimate things can enter into genuine relations. The rest is just, as Russell often says, a *façon de parler*, or way of speaking.

## Notes

1. Of course, such accounts exist in the secondary literature. At ($PM_1$: 45), there is an obscure passage suggesting that a general judgment "collects together" a number of elementary judgments, but he clearly does not mean that someone who makes a general judgment makes *each* of the specific elementary judgments collected together individually. Soames (2014: 526) cites this passage as something that doesn't "sit well" with the interpretation of Russell as having a substitutional theory of quantification, but also doesn't explain how it sits any better with any other interpretation.
2. Among elementary judgments, Russell did not make a distinction between atomic and molecular in *PM* itself, but did soon thereafter. For a proposed explanation for this, see Klement (2015: 213–214).
3. See Landini (1998: Chap. 10); Landini (2011: Chap. 3). There are no formulas of *PM* expressible only using individual variables. To get the hierarchy of senses of "truth" up and running, Landini must also allow predicative second-order variables to be interpreted objectually, which seems to undermine Landini's own conclusion that Russell's understanding of higher-types is purely "nominalistic".
4. Quine's own attitude about this strategy is more complicated than common lore would suggest; see Fara (2011).
5. This remark sits a bit uneasily with his claim that the logical language of *PM* represents the core of a logically ideal language, but only including its syntax, not its vocabulary. *PM* does not use any specific names in it: can he not imagine it? This tension is relieved by the fact that Russell seems to think that although *PM* does not use any particular names, the intended semantics of its formula presuppose that names *should be added* to round it out, and that without them we do not have a full "logically ideal language"; see (*PLA*: 176; *IMP*: 201).
6. Soames presses other worries in his earlier (Soames 2008), which I have responded to in Klement (2010). It is sometimes not altogether clear whether Soames objects to *interpreting* Russell as having a substitutional view of quantification, or objects to Russell's *having* such a view, but these are separate issues.
7. *Principia*'s theory of types, even in the first edition, is often wrongly read as implying that universals would not be values of *Principia*'s individual variables; I correct this misunderstanding in Klement (2004).

## References

### Works by Other Authors

Fara, Delia Graff (2011). "Socratizing." *American Philosophical Quarterly* 48: 229–238.

Fine, Kit (2009). "The Question of Ontology." In *Metametaphysics*, eds. David Chalmers, David Manley, and Ryan Wasserman. Oxford: Oxford University Press. 2009: 157–177.

Hodes, Harold T. (2015). "Why Ramify?" *Notre Dame Journal of Formal Logic*, Vol 56: 379–415.

Katz, J. J. (1980). *Language and Other Abstract Objects*. Totowa, NJ: Rowman & Littlefield Publishers.

Klement, Kevin C. (2004). "Putting Form Before Function: Logical Grammar in Frege, Russell and Wittgenstein," *Philosopher's Imprint* 4 no. 2: 1–47.

Klement, Kevin C. (2010). "The Functions of Russell's No-Class Theory." *Review of Symbolic Logic* 3: 633–664.

Klement, Kevin C. (2013). "*PM*'s Circumflex, Syntax and Philosophy of Types." In *The Palgrave Centenary Companion to Principia Mathematica*, eds. Nicholas Griffin and Bernard Linsky. Basingstoke: Palgrave Macmillan. 2013: 218–247.

Klement, Kevin C. (2015). "The Constituents of the Propositions of Logic." In *Acquaintance, Knowledge and Logic: New Essays on Bertrand Russell's* The Problems of Philosophy, eds. Donovan Wishon and Bernard Linsky, Stanford: CSLI Publications. 2015: 189–229.

Kripke, Saul (1976). "Is There a Problem About Substitutional Quantification?" In *Truth and Meaning*, eds. G. Evans and J. McDowell. Oxford: Clarendon Press: 325–419.

Landini, Gregory (1998). *Russell's Hidden Substitutional Theory*. Oxford: Oxford University Press.

Landini, Gregory (2011). *Russell*. London and New York: Routledge.

Lewis, David (1975). "Languages and Language." In *Language, Mind, and Knowledge*, ed. Keith Gunderson. Minneapolis: University of Minnesota Press. 1975: 3–35.

Quine, W. V. (1948). "On What There Is." In *From a Logical Point of View*. Cambridge, Massachusetts: Harvard University Press. 1953: 1–19.

Schaffer, Jonathan (2009). "On What Grounds What." In *Metametaphysics*, eds. David Chalmers, David Manley, and Ryan Wasserman. Oxford: Oxford University Press. 2009: 347–383.

Soames, Scott (2008). "No Class: Russell on Contextual Definition and the Elimination of Sets." *Philosophical Studies* 139: 213–218.

Soames, Scott (2014). *The Analytic Tradition in Philosophy*, vol. 1, *The Founding Giants*. Princeton: Princeton University Press.

CHAPTER 8

# The Near Riot Over Negative Facts

*Bernard Linsky*

## 1 Introduction

On February 5, 1918, near the end of the third of the *Philosophy of Logical Atomism* series, Russell said:

> When I was lecturing on this subject at Harvard I argued that there are negative facts, and it nearly produced a riot: the class would not hear of there being negative facts at all. I am inclined to think that there are. However, one of the men to whom I was lecturing at Harvard, Mr. Demos, subsequently wrote an article in Mind to explain why there are not negative facts. It is in *Mind* for April 1917. (*PLA*: 187)

There are two sets of handwritten notes on Russell's lectures in "Advanced Logic" (Philosophy 21) at Harvard in 1914. One set, which is well known to professors of Modern English Literature, if not philosophers, consists of notes by T.S. Eliot, who was then a graduate student writing his dissertation. Another set of notes, which is to be found in a small notebook in the University Archives in the Watkinson Library at Trinity College, Connecticut, are by Harry T. Costello, Russell's teaching

B. Linsky (✉)
Department of Philosophy, University of Alberta, Edmonton, AB, Canada
e-mail: bernard.linsky@ualberta.ca



L. D. C. Elkind, G. Landini (eds.), *The Philosophy of Logical Atomism*, History of Analytic Philosophy,
https://doi.org/10.1007/978-3-319-94364-0_8

assistant for Philosophy 21. Russell had brought a copy of the first typescript of Wittgenstein's "Notes on Logic" with him to Harvard in March of 1914, and it was Costello who published them in 1957 in the *Journal of Philosophy*.[1] Costello's notes on the lectures, and a letter from Russell to Ottoline Morrell, suggest that the lecture that "nearly produced a riot" was given in Philosophy 21 on April 11, 1914. Reproduced below are transcriptions of notes from that day by Costello which provide circumstantial evidence for this claim. The evidence they provide is admittedly indirect, for the words "negative facts" do not appear in them. (Nor do they appear anywhere in either Costello's or Eliot's lecture notes.) That lecture does present ideas from Wittgenstein's "Notes on Logic", and Wittgenstein discusses negative facts in them. There is also a draft of a letter that Costello sent to Russell in the Bertrand Russell Archives which is attached to a short essay. Costello's essay mentions Wittgenstein by name. Another letter in the Archives, this from Raphael Demos to Bertrand Russell, suggests (see Sect. 6 below) that Demos attended the third lecture of *The Philosophy of Logical Atomism*, so he may have led Russell to overstate the reaction in order to single out Demos in the audience. It seems clear, to me at least, that the commotion over negative facts came in the midst of a discussion of Wittgenstein's "Notes on Logic". I will conclude with a discussion of Russell's views on Demos' paper, and, more generally, logical reasons why Russell should have taken negative facts seriously, unlike his dismissal of conjunctive and disjunctive facts. This is more speculative, in that it isn't clear that any of these thoughts went through Russell's mind, but they could influence us in thinking about the topic anew.

## 2 Wittgenstein's "Notes on Logic" and Negative Facts

Although Costello's notes do not indicate that Russell presented the "Notes on Logic" as such to the students as a reading, the record of his lectures contain statements of doctrines that are expressed in Wittgenstein's notes. Wittgenstein's name appears twice in the Costello material, once, spelled "Wittenstein" in the notes for March 23, and a second time, spelled correctly, in a handwritten letter that Costello sent to Russell in May.

A letter from Russell to Ottoline Morrell dated April 15, 1914 indicates that it was most likely the lecture of April 11 in the logic course that provoked the response. Russell reported that:

> Today my whole morning has been occupied with pupils. I put forward a theory in my logic lectures which roused the vehement opposition of the whole class; now, one by one, they are coming round. The great thing is to find something about which people feel strongly. (RA 18168)

It is proposed here that they were responding to the lecture of April 11, and that the material in the lecture about the "sense" of a proposition produced the response that Russell later described as nearly a "riot". The written responses of Costello, and of course Demos' later paper in *Mind* show that at least these two students responded strongly. Here are the notes from April 11 by Costello:

**Costello's April 11, 1914 Notes**

If $x^{\text{Fact}} \rightarrow_q x$ false. You must (to distinguish $p \vee q$ from $p \cdot q$) consider $(p, q)$ what $p \vee q$ means. A proposition not only denotes a fact but is a fact.

Not all facts are atomic facts. How about beliefs etc. (several verbs). (Relations or Predicate = Universals)

General indefinables (not quite same but more general including e.g. $p \vee \sim q$ = simplest type of function.) Apparent variable facts must also be added corresponding to $(x)\ \phi x$ and $(\exists x)\ \phi x$, not a set of simpler atomic facts. All that be expressed by verbs (adjectives) are *universals*. The sense (what happens when true and when false) of the propositions $(x).\ \phi x$ includes $(\exists x)\ \phi x$.

$(x).\ \phi x$ = (sense = direction of meaning)

$t\ (x)\ (t\ \phi x\ f)\ (\exists x)\ f$

$(\exists x)\ \phi x$

$t \cdot (\exists x)\ (t\ \phi x\ f)\ (x) \cdot f^2$

We want to say not-not- $p$ is identical with $p$. So *not* does not occur in not $p$. "not" simply reverses the sense.

Theory of types. $\kappa = \hat{\alpha}\ (\alpha \sim\epsilon\ \alpha)\ .\supset.\ \kappa \epsilon \kappa\ .\equiv.\ \kappa \sim\epsilon\ \kappa$

$\phi(\phi\hat{x})$ meaningless.

Functions and propositions are the only important entities not things—they require different symbols and really types are chored with eliminating errors of symbolism. Redness = $\hat{x}$ being red. $\phi x$ subject predicate propositions are such as contain only two components = $x_{\varpi}$.

Dual relations $xRa$ triple $R(x,a,b)$ Each is a type in *narrow* sense. (Wider sense: adding these together) For any one of the symbols above $x_R$ (a wider type).

For molecular derivatives of $x_R$ by $p|q$ (Sheffer)

If you have an operation sometimes it will begin to duplicate (full group). Denote these symbols by $\phi x$. We can go no further without *generality* by apparent variables.

Eliot's notes are very similar to Costello's for this day. In particular, the parts close to Wittgenstein's discussion of negative facts are almost identical. Neither Costello nor Eliot uses the words "negative facts'" in their notes from April 11 (or anywhere else in their notes on the course). Russell's discussion of negation in the lecture, however, presents Wittgenstein's notion of propositions having two poles, in that they point either toward a fact, in the true direction, or away from a fact in the false direction. A negated proposition ~*p* will point away from the "fact" that *p*, indicating that it is not true. If ~*p* itself is true, however, what fact makes it true? It would be the *negative fact* that *p* is not the case. That may have been the way that the notion of negative fact entered into the discussion of the April 11 lecture.

It is certainly odd, given Russell's later report of the near riot that he provoked, that the words "negative fact" don't show up in either set of notes. There is no direct evidence of any sort that Russell talked about negative facts at all in the course. The expression "negative facts" does, however, appear in Wittgenstein's 1913 "Notes on Logic" (Wittgenstein 1913: 97):

> There are *positive* and *negative facts*: if the proposition "This rose is not red'" is true, then what it signifies is negative.

So, it is plausible that in a discussion of Wittgenstein's account of the truth and falsity of propositions that the notion of "negative fact" would have been in the air, at least.

The timing of the letter to Ottoline four days after April 11, 1914 makes it seem most likely that this is the class where he could have "put forward a theory in my logic class which roused the vehement opposition of the whole class…". In the letter Russell says that "now, one by one, they are coming round", can be read as saying that they were coming "round" to his office for personal conversations, not "round" to his view in the sense of agreeing with him. In any case he finds that it is a "great thing" to find something about which people "feel strongly'". So the letter to Ottoline is not conclusive either, but the evidence converges on April 11 as the most likely date of what Russell years later described as nearly a "riot".

## 3 Costello's Letter About Negative Facts

There is a handwritten letter about negative facts from Costello to Russell that Costello likely handed to Russell in May of 1914, while Russell was still in Massachusetts. It was written on two connected pages of a notebook (6 × 8½″), torn out of their stitching in the middle. Russell kept the letter and it survives in the Bertrand Russell Archives. It seems to be a reply to Russell's comments on a paper by Costello on negative facts, as Costello quotes comments by Russell and then responds to them. The references in the letter to Wittgenstein and Meinong should be noted. The notion that a description of a fact, "the fact that *p*", refers to a possible fact and that this is comparable to one of Meinong's assumptions shows that a familiarity with Meinong was assumed in the class. The reference to Wittgenstein accompanies a different point from the misspelled occurrence in the notes for March 23. This suggests that Wittgenstein was discussed at some length, perhaps in discussions outside of the formal lectures. The "accompanying paper" which Costello describes as having been presented to Josiah Royce's seminar must have been another piece which does not survive among either the Costello or Russell papers. Russell may have returned it to Costello before he left Harvard.

First, some background for Costello's letter. These are Costello's notes for Royce's seminar on May 12, 1914, reporting the following. Since Costello's letter refers to this session we may conclude that the letter to Russell was sent after May 12. We find:

> Most of the meeting was taken up by the comments Professor Royce made on the preliminary doctorate examinations in logic, the present general low level of logical study and teaching at Harvard, and the hope that the coming year would see a pronounced improvement in these regards. Particularly were students who took the examination ignorant of the nature of implications, of the fact that a false proposition implies all propositions. Hence they had not seen that, to the question discussed by St. Thomas Aquinas, "Could God sin if he would" the answer was in the affirmative. (Comment: I examined Mr. Russell on these questions. He "flunked" completely on the St. Thomas one!) He then called attention to Dr. Lewis' discussion of the question whether a false proposition implies all propositions. Professor Royce held Dr. Lewis had wrongly introduced questions of our knowledge and psychological powers, since he had wished to define *implies* as "can be deduced from." These formal questions are worth discussing, for a student should acquire the power to study the implications of a doctrine—abstraction made from whether or no he believes it. (Costello 1963: 179)

The references to "Dr. Lewis" are to C.I. Lewis and his views on material implication. Royce said that Russell "flunked" the question on St. Thomas, but most likely not because he held that this question should be interpreted as a material conditional with a false antecedent. He may have had some irreligious witticism in his reply that Royce did not relay to his students.

The letter to Russell occupies the first page of the sheets from Costello's notebook, and is signed by Costello. The letter is as follows:

> Mr. Russell:
>
> The accompanying paper I beg you to read, if you have time, and give me some constructive criticism. It is only a first draft (written on both sides of the paper (!), numbers refer to notes at the end), and breaks off in the middle because I intended to read this much in Prof. Royce's seminary before continuing. The latter hope was frustrated because Prof. Royce himself took my time in expounding more important things (namely "that a false proposition implied all propositions"!)[3]
>
> The paper I had intended to publish … was, intended to be in three parts, of which one and a half are here. The third part … some points suggested by Husserl's Vol II Part VI.[4] The rest of part II was to treat of the distinction between "indicative descriptions" and "iconic descriptions", the ambiguities of the word "not", etc. It is getting so long I shall probably have to split it in separate points, for it threatens to grow into a book!…
>
> H. T. Costello

Costello's letter to Russell is followed by a three page note on negative facts. I quote at length from the Costello note:

> Note on Mr. Russell's criticism of my opinions about negative facts.
>
> Mr. Russell seems to admit my contention that it is formally possible to construct logic with only apparent variable negative facts, which was the constructive part of what I contended for. But apparently, he fails to see why I think this desirable. I think it desirable because I believe that in the apparent variable negatives there is a notion of negation which is distinct from the negation which would be present in the particular negative facts if there were any. It belongs instead with the negation which we encounter in passing from one type to another. If I say that what is asserted of terms cannot be asserted of relations, this is distinct from saying it can be asserted "in a negative fashion'" of relations. My contention is that we can get along with only two sorts of negatives: (1) the symbolic "not", which is a symbolic device for inverting the polarity of propositions, and (2) the negative which

> says of the *x*'s, "there are, besides these, no more *x*'s. I pointed out though you experienced all the *x*'s there were, you would not thereby perceive that these were all, so that this negative in a way escapes experience. Hence I am willing Mr. Russell should write my formulae as $\sim(\exists x).\ \phi x$ only I beg thereupon to enter on a further enquiry, to consider what this symbol really means. And at present I am disposed to maintain that it cannot be equated with the sum of a set of particular negative facts.
>
> ... in experience positive particular facts are given and "negative facts" are products of mental synthesis of elements given separately.
>
> Also I have tried to imagine a world in which there were only "negative facts", and I still must confess the conception of a "negative fact" baffles me; somehow I can't "see" it.
>
> One more point by the way. Mr. Russell says of me: "I think you have not realized sufficiently that facts cannot be subjects". I am not sure of the relevance of this remark, but as to the remark itself I agree I may be guilty, for I am not sure they can't. For example we seem to say of a proposition *p* that it is true. This seems to be an assertion about *p* and not about its parts. Now *p* is a symbol for a fact. But a symbol for a fact is itself a fact (Wittgenstein).[5] Therefore we seem to have said of a certain sort of a fact it is true, that is to have said something about a fact as such. Of course the assertion, with facts for subjects, if such there be, would not contain arbitrary verbs (relations); they would be comparable to verbs like "believe" which seem to take whole facts as objects. I am not sure we don't need also a fuller theory of descriptions of facts, *possible* facts as well as real (of the form "that so and so", etc. cf. Meinong's assumptions), which descriptions shall be irreducible to descriptions of particular things ("the so and so").

It is hard to make much out of Costello's views, and there is no record of a response from Russell. The letter is valuable, however, as evidence that Demos was not the only student in the class who responded to the discussion of negative facts.[6] The letter also contains the other (correctly spelled) occurrence of the name "Wittgenstein" in the material on Russell's visit to Harvard that is to be found in all this material.[7]

## 4 Costello's Points of Disagreement with Russell

Costello reused the notes for his lectures in Philosophy 21 the next year (1914–15) when he had a one year appointment at Yale. The body of the notes include markings to indicate how far he got in each lecture at Yale. The first three pages, dealing with the details of Russell's visit, were not reused and are followed in the box of notes in the Trinity College library by four pages of notes handwritten in pencil torn out of a notebook. They are titled "Points of disagreement with Russell". After these pages follow the notes that were used again at Yale. These "Points" include references to the published version of *Our Knowledge of the External World*, and so would not have been written before late in 1914. They may have been made when Costello first read that book, and so could be from his year at Yale. What is relevant for my purpose is that they give a very tidy and clear summary of Costello's objection to Russell's view on negative facts:

> I. Negative Facts. What corresponds to "There is not an elephant in this room" and makes "There is an elephant in this room" false? Russell holds there are simple negative facts of this sort directly observed. I hold that "negative facts" are always inferred, or rather, are constructed of parts drawn from different sources, or are, in a word, more complex than positive facts. That "this paper is not green" cannot be observed. I can observe it is white, but it might also be green too, unless I know the *general truth* that it cannot be both. This suggests that "negative facts" have a kinship with "general facts" rather than with particulars. Ontologically the elimination of the myriads of "negative facts" would be a great step towards simplification.

In the discussion of atomic and molecular propositions in *Our Knowledge of the External World* (in the second lecture, 54–55), there is a discussion of *atomic facts* and then a discussion of *atomic* and *molecular* propositions, but no discussion of negation, whether of negated propositions or of *negative facts*. Instead, there is a view that seems close to what Costello presents as his own (quoted earlier). Russell says:

> It is easy to see that general propositions, such as "all men are mortal", cannot be known by inference from atomic facts alone. If we could know each individual man, and know that he was mortal, that would not enable us to know that all men are mortal, unless we *knew* that those were all the men there are, which is a general proposition... Thus general truths cannot be inferred from particular truths alone, but must, if they are to be known, be either self-evident, or inferred from premisses of which at least one is a general truth. But all *empirical* evidence is of *particular* truths. (*OKEW*: 56)

If all truths known through perception are of particular atomic facts, it is not clear how we would know any true negated atomic empirical proposition, unless there are negative facts of perception. We might know a general empirical proposition, as Costello suggests, and thereby infer that the atomic proposition is false. It looks as though Russell avoided any talk of negative facts in his Lowell lectures and saved that for his more advanced students at Harvard.

## 5 Demos on Negative Facts

Raphael Demos' 1917 paper on negative facts is "A discussion of a certain type of negative proposition". The "certain type" of proposition to which he refers is negated atomic propositions, *~p*, for *p* an atomic proposition. After arguing against defining negation in terms of a propositional attitude such as "rejection" or "denial", he proposes instead to use Russell's theory of descriptions and propose that a negated proposition is a definite description which refers indefinitely to some positive fact. Using the Sheffer Stroke " | " as a sign of "incompatibility" (since "$p|q$" can be read as "not both *p* and *q*" or "*p* and *q* are incompatible", one can view Demos as defining $\sim p$ as $(\exists q)\ (q\ \&\ (p|q))$. That is, "*p* is false" means that there is a true proposition *q* that is incompatible with *p*. We find:

> Corresponding to the exclamation of the simple phrase we have the assertion of the negative proposition, and just a "Rain!" is really "Rain exists" ("There is rain"), so "not-*p*" is really "not-*p* is true", or "an opposite of *p* is true", or "some proposition is (no line breaks after "so" above, or "is") true which is a contrary of *p*". (Demos 1917: 193)

## 6 The Reply to Demos in Lecture III

The last pages of the third lecture of *The Philosophy of Logical Atomism* lectures present Russell's response to Demos' view:

> I find it very difficult to believe that theory of falsehood. … If I say 'There is not a hippopotamus in this room', it is quite clear there is some way of interpreting that statement according to which there is a corresponding fact, and that fact cannot be merely that every part of this room is filled up with something that is not a hippopotamus.[8] … If I say "*p* is incompatible with *q*", one at least of *p* and *q* has got to be false. It is clear that no two facts are incompatible. … If you were making an inventory of the world, propositions would not come in. Facts would, beliefs, wishes, wills would, but propositions would not. … This theory of Mr. Demos's that I have been setting forth here is a development of the one hits upon at once when one tries to

> get round negative facts, but for the reasons that I have given, I do not think it really answers to take things that way, and I think you will find that it is better to take negative facts as ultimate. (*PLA*: 189–190)

Russell sees the issue as finding a "truthmaker" for true negative propositions such as "There is not a hippopotamus in the room", and a "falsemaker" for false propositions, such as "Socrates is alive". If one treats the Sheffer stroke "$p|q$" as asserting that the propositions $p$ and $q$ are incompatible (not both true) then the issue is what sort of fact makes that so. What might be the truthmaker for the proposition $p|q$? He does not want to include facts about propositions in the world, and facts are all "compatible" with each other in that one obtaining does not force other facts to obtain or not. One thinks immediately of the problem of color exclusion, that *a*'s being red might be incompatible with *a*'s being green. Russell, however, does not allude to this phenomenon here. Rather, he is thinking of the Sheffer stroke as a truth functional connective. There is no need for a "disjunctive fact that $p$ or $q$" because the truth of the fact that makes $p$ true, or the fact that makes $q$ true, or both, is enough to make the disjunction true. Similarly, there is no need for an "incompatibility fact" to make $p|q$ true, because all one needs is that not both are true, or, equivalently that "either not $p$ or not $q$". But the truthmaker for this will be whatever makes the one (or both) of $\sim p$ or $\sim q$ true. So, there are no "incompatibility facts". Thus, while the Sheffer stroke does have the effect of simplifying the number of primitive connectives in logic, simple propositions using the stroke, such as $p|q$ do not have corresponding truth makers, as negative propositions $\sim p$ will have, in negative facts.

Raphael Demos spent some of the war years in Britain, including the early months of 1918 when Russell delivered the logical atomism lectures. A letter in the Russell Archives seems to show that Raphael Demos attended the third lecture. On February 11, 1918 Demos wrote the following to Russell:

> … I was very much interested in your discussion last Tuesday of my article on Negation. Your account is quite accurate and fair and even adds to the elucidation of the theory, as when you suggest various motives which might lead me to make a negative assertion in my sense. I appreciate the difficulty to which you call attention, namely that the theory entails regarding propositions as entities which enter into relations, and it looks as though one is doing away with negative facts by introducing another type of entity, such as

> propositions. If I were to write the article anew, I would not put the theory as a disproof of negative facts, but as a possible way of dealing with negative propositions, and allow the view as to negative facts as another possibility,—leaving the matter at that after the fashion of the logician who presents various sets of postulates for a given system. (*RA* 76494)

This letter almost certainly proves that Demos was in the audience when Russell was speaking. He might have been responding to a report on the lecture from a friend who also attended the lectures such as Eric Neville, who is identified in the discussion of the preceding lecture, or Dorothy Wrinch, who likely also was there. This letter is dated as "Feb. 11" from the address "7 Victoria St. Emmanuel Rd, Cambridge". It seems most likely that Demos had taken the short train ride from Cambridge to Dr. Williams' Library in Bloomsbury on the Thursday before. The recorder of the logical atomism lectures does not list the identity of the questioners which followed Lecture III. It is hard to see Demos as asking any of the questions that were transcribed, so it seems that Demos sat quietly after having the attention of the audience drawn to him in the lecture, and sent the letter in part to express views that he had not stated in the discussion period.

## 7 An Argument in Support of Negative Facts

Russell did not write about negative facts after these lectures, and even the attention to them in Lecture III may have been prompted the presence of Demos in the audience. Wittgenstein himself says little about negative facts after the "Notes on Logic" beyond one remark in the *Tractatus* (2.06). It might seem, then, that the "riot" over negative facts was in fact only a minor episode incidental to Russell's philosophy of logical atomism. To conclude I will try to explain how the logic of negation is unlike that of conjunction and disjunction, and so how it was plausible that Russell was concerned about the existence of negative facts for good reason.

Russell's interest in negative facts may seem to treat connectives inconsistently, given that he is not in the least interested in *conjunctive* or *disjunctive* facts. It would seem that there is no call for a fact to make a conjunctive proposition $p$ & $q$ true. It is obvious that this is the case if $p$ is true and $q$ is true. There is no such easy way to avoid postulating *negative* facts. Following the "semantic account" of negation in the clause in the Tarskian definition of truth for sentences: "'not s' is true if and only if 's' is not true", one would say for propositions: "$\sim p$ is true if and only if $p$ is not true." This seems to allow one to avoid any special category of fact with which such a proposition will correspond. Why should one think that ~*p* is made true by a "negative fact" when all one needs to say is that ~*p* is true just in case *p* is not true? It seems that nothing is needed to make the negative proposition true.

The problem is if one thinks that despite the metalinguistic sounding talk about "truth" of propositions in all of Russell's logical works, that he would still be interested in capturing the behavior of negative propositions in axioms or inferential rules. Although this has been vigorously disputed, it does seem that there are traces of what van Heijenoort called the "universality of logic" in Russell's views, by which it would appear redundant or difficult somehow to try to talk about the negation of a proposition ~*p* being true in terms of *p* not being true.[9] If one considers a natural deduction system, the rules of *conjunction elimination* and *conjunction introduction* seem to capture all there is to conjunction and so all there is to conjunctive propositions. That one can infer a disjunction from either one of the disjuncts captures the idea behind rejecting disjunctive facts. If *p* is true or *q* is true, then we know there is a truthmaker for *p* ∨ *q*. Similarly, that *p* & *q* can be derived only if *p* and *q* are already derived shows that there is no need for a truthmaker for the conjunction beyond the individual truthmakers for *p* and for *q*.

The difficulty with negation is that there is no way to capture the nature of negation with such simple introduction and elimination rules. Here I follow Bergmann, Moore and Nelson's *The Logic Book* (2013) and will rely on it to make this point about negation in classical logic. Their system "SD" or sentential derivations contains familiar looking introduction and elimination rules for negation. Their introduction rule (~I) says that if a sentence *q* and its negation ~*q* can be deduced from a subsidiary derivation from *p*, then ~*p* follows from the undischarged assumptions of the derivation other than *p*. Likewise the elimination rule tells us that if from an assumption of a negation ~*p*, we can derive an explicit contradiction (*q* and ~*q*), then one has a derivation of *p* based only on the undischarged assumptions of the sub-derivation.

Upon careful examination one sees that these rules are slightly odd. For one thing, the connective being "defined" by these rules actually occurs in the rules in a way unlike that of, say the conjunction rules. These negation rules are not "pure", that is, they do not use only the connective that is being introduced or eliminated, as do the conjunction rules, for example. Secondly, the elimination rule ~*E*, also known as "indirect proof" is very strong. Intuitionists reject indirect proof as unjustified in going far beyond its premises. More surprisingly is that in the context of the rest of a classical natural deduction system, one can prove the ~I rule from this rule already! It could be "derived rule" of a system with fewer rules, SD minus ~I.[10]

Natural deduction systems sometimes include the *falsum* symbol "⊥", a propositional *constant* in the formulation of propositional logic.[11] The elimination rule for ⊥ is that anything follows from ⊥. The latter rule is known as "*explosion*" by paraconsistent logicians, who reject it. But classical logicians are happy with the rule. Negation, or "~" is then defined, $\sim p =_{df} p \supset \bot$. (Even intuitionists like this definition, they just disagree about when an implication can be proved). Symmetrically, the "⊥ introduction rule" would seem to be that ⊥ follows from any contradictory sentences, $q$ and $\sim q$. The latter, however, uses the defined symbol ~, and when analyzed the premises of the suggested rule are just $q$ and $q \supset \bot$. But then the rule for introducing ⊥ after inferring a contradiction is already just an instance of *modus ponens.* There is still only one rule for negation, just an elimination rule. But following the examples of "and" and "or", an introduction rule is needed to show what the truthmaker for negations would be.

How much of this could Russell have had in mind? He did have the notion of *explosion*, for he held that every proposition follows from a logically false proposition. In *The Principles of Mathematics* used $(r \supset r) \supset_r (p \supset r)$ as the definition of negation (*PoM*: §18). In "The Theory of Implication" (1906) Russell decided to use ~ as a primitive notion on the grounds that proving that a proposition $p$ implies all other propositions $q$ wouldn't concern someone who was willing to accept that everything is true, and so is not a good "method" for defining negation. Russell writes:

> My reason for not adopting this method is not its artificiality or its difficulty, but the fact that it never enables us to know that anything whatever is *false*. It enables us to prove the *truth* of whatever can be proved by the method adopted above, and it does not enable us to prove the truth of anything that is in fact false. … but if any man is so credulous as to believe that everything is true, then the method in question is powerless to refute him. For example, we get the law of contradiction in the form $\vdash p \,.\, \sim p \,.\, \supset .\, (s) \,.\, s$; but this does not show that $p \,.\, \sim p$ is false, unless we assume that $(s).\ s$ is false. (*CPBR* 5: 60–61)

Russell thus came to see that negation must be a primitive connective unless there were further assumptions about the proposed definition, such as about the truth of all propositions, or about the incompatibility of propositions. Would he have worried about the axioms that capture negation in the same way that the *Principia Mathematica* axioms capture disjunction, the other primitive? If negation must be treated as a primitive as it is with respect to introduction rules, then there is a need for a special *truthmaker* for negative propositions, a "negative fact" which will account for *falsehood*. It is clear, at least, that Russell had thought about negation as special among the propositional connectives.

Recall Russell's objections to Demos' arguments. Demos suggests that we define negation in terms of incompatibility, the semantic interpretation of the Sheffer stroke. If Russell were to base all of propositional logic on the Sheffer stroke, as he was to do in the second edition of *Principia Mathematica* in 1925, then he would be treating it as the one primitive of logic.[12] Consequently, it would seem, the notion that it expresses, incompatibility, would be the primitive notion of propositional logic. One would then define negation, conjunction, and so on, in terms of the Sheffer stroke, and so certainly not have any need for negative facts. Russell does not agree. He still thinks that the notion of "incompatibility" presupposes negation in some way. Russell doesn't think that the new logically primitive connective, the Sheffer Stroke, is philosophically primitive. He thinks that there is no such thing as an incompatibility between facts that could make the primitive propositions of logic such as $p|q$ true. From studying the development of Russell's propositional logic from *The Principles of Mathematics* from 1903, through "The Theory of Implication" in 1906 and then to *Principia Mathematica* in 1910, it is clear that the choice of primitive connectives changes, and is not even reflected in the primitive propositions of each system.[13] *Principia Mathematica* has all its axioms expressed in terms of disjunction, "$\vee$", and material implication, "$\supset$", which is itself not a primitive connective, but instead *defined* in terms of the real primitives, "$\sim$" and "$\vee$", namely "not" and "or". Similarly then a logically primitive notion expressed by the Sheffer Stroke does not require a philosophically primitive kind of truthmaker. The arguments for negative facts are independent of whether the Sheffer Stroke is a logically primitive connective.

All this suggests that negation must be treated differently from conjunction and even disjunction in terms of rules that give their meaning, and so, perhaps some reason to think that they will differ in relation to facts. They seem to suggest that there must be *negative facts* even if *conjunctive* and *disjunctive* facts can be avoided. How much of this might have gone through Russell's mind is unclear. It is clear that he thought about the notion of negation and also that he allowed for a gap between the choice of primitive connectives for the purpose of constructing an axiomatic system, and the discovery of primitive logical notions through logical analysis. He just might have thought that negation is prior to the Sheffer Stroke, and that a theory of facts to explain the truth of propositions might therefore require a primitive notion of negative fact.

## Notes

1. Costello (1957a). See Potter (2009) for a study of the composition of Wittgenstein's Notes.
2. Wittgenstein uses "*a*" and "*b*" instead of "*t*" and "*f* " with this notation both in the "Notes on Logic" and *Tractatus.*
3. This is likely the session of May 12, cited above. The notes for that day continue: "Dr. Costello then read a note on two topics: one on the nature of the so-called logical entities, insisting on their abstractness; the other on the contrast between statistical probability and the kind of probability you get when you verify an hypothesis, which may be called systematic probability."
4. Presumably, this refers to Husserl's *Logical Investigations.*
5. The *Notes on Logic* has: "Propositions, which are symbols having reference to facts, are themselves facts" (Costello 1957a: 236).
6. Grover Smith (Costello 1963: 5) reports that the enrolled students were A.P. Brogan, N.N. Sen Gupta, Raphael Demos, E.W. Friend, C.E. Kellogg, Victor F. Lenzen, and Robert L.M. Underhill. T.S. Eliot is described as an "auditor". Lenzen's notes from Philosophy 21 do not survive, although his notes for Russell's other course, Philosophy 9, "Theory of Knowledge", are preserved in the Bertrand Russell Archives.
7. One is also struck by the familiarity with Meinong at Harvard which is indicated here, and frequently in Costello's notes on Royce's seminar. T.S. Eliot mentions Meinong in his presentation to Royce's seminar, and also several times in his thesis (Eliot 1964), which was completed in 1916.
8. In a letter to Ottoline Morrell on November 1, 1911, Russell says of Wittgenstein, who he had just met and did not know well enough to know that he was Austrian: "My German engineer very argumentative and tiresome. He wouldn't admit that it was certain that there was not a rhinoceros in the room." Wittgenstein followed Russell to his rooms after the lecture and continued the argument. On the next day (November 2) the argument continued: "He thinks nothing empirical is knowable—I asked him to admit that there was not a rhinoceros in the room, but he wouldn't." (Monk 1990: 39). This issue is presented as a question of epistemology, but still seems to be a debate over how a negative proposition (or fact) is known.

9. See Van Heijenoort (1967) and Korhonen (2013) for discussions of the notion of the "universality of logic" for Russell.
10. *Proof*: (1) First observe that Double Negation Elimination (DNE) can be derived using ~*E*. Assume that ~~*A*. Then further if we suppose ~*A*, that subordinate assumption leads to a contradiction, ~*A* and ~ ~*A*, so by ~E we can derive *A*, based only on the first assumption, ~ ~*A*. Thus DNE is shown valid. (2) Now we show that ~ I can be derived using ~E. Suppose that a contradiction follows from some sentence *B*. We know that the same contradiction can be derived from ~ ~*B*, by one use of *DNE* followed by the derivation of the contradiction. But then we have a contradiction derived from ~ ~*B*, and so a proof of ~*B* by ~E. But now we have derived ~*B* on the supposition that *B* leads to a contradiction, and this is exactly what the rule ~ I allows.
11. ⊥ is also called "bottom" to recognize its place as the bottom value in a Boolean algebra, as well as "eet" to indicate that it is the inverse of "tee" the name of the letter "T".
12. See Linsky (2011) for an account of the Introduction to the Second Edition of *Principia Mathematica* where Russell shows how the body of the text can be based on the Sheffer stroke.
13. See Linsky (2016) for an account of the changes in Russell's choices of primitive connectives for propositional logic between *The Principles of Mathematics* and *Principia Mathematica*.

## References

### Works by Other Authors

Costello, Harry T. (1957a). "Notes on Logic." *Journal of Philosophy*, Vol. 54, No. 9: 230–245.

Costello, Harry T. (1957b). "Logic in 1914 and Now", *The Journal of Philosophy*, Vol. 54, No. 9: 245–265.

Costello, Harry T. (1963). *Josiah Royce's Seminar, 1913–14*, ed. Grover Smith. New Brunswick: Rutgers University Press.

Demos, R. (1917). "A Discussion of a Certain Type of Negative Proposition". *Mind*, Vol. 26 No. 102: 188–196.

Eliot, T. S. (1964). *Knowledge and Experience in the Philosophy of F. H. Bradley*. New York: Columbia University Press.

Korhonen, Anssi (2013). *Logic as Universal Science: Russell's Early Logicism and its Philosophical Context*, Basingstoke: Palgrave Macmillan.

Linsky, Bernard (2011). *The Evolution of Principia Mathematica: Bertrand Russell's Manuscripts and Notes for the Second Edition*, Cambridge: Cambridge University Press.

Linsky, Bernard (2016). "Propositional Logic from *The Principles of Mathematics* to *Principia Mathematica*." In *Early Analytic Philosophy: New Perspectives on the Tradition*, ed. Sorin Costreie. London: Springer. 213–229.

Monk, Ray (1990). *Ludwig Wittgenstein: The Duty of Genius*, New York: Macmillan.

Potter, Michael (2009). *Wittgenstein's Notes on Logic*, Oxford: Oxford University Press.

Van Heijenoort, Jan (1967). "Logic as Language and Logic as Calculus." *Synthese*, Vol. 17: 324–330.

Wittgenstein, Ludwig (1913). "Notes on Logic." In L. Wittgenstein, *Notebooks 1914–16*. Ed. G. H. von Wright and G. E. M. Anscombe. Oxford: Basil Blackwell: 93–106.

CHAPTER 9

# Can We Be Positive About Russell's Negative Facts?

*Katarina Perović*

## 1 Introduction: Negative Facts in *Lecture III* of *The Philosophy of Logical Atomism*

It is quite surprising that Russell—the philosopher who famously insisted on the importance of having "a robust sense of reality" in metaphysics—was at one time committed to entities such as *Socrates not being alive* and *a hippopotamus not being in this room*. In his 1918 *The Philosophy of Logical Atomism Lectures* (*PLA*), Russell briefly discusses negative facts and confesses that his open defense of such entities "nearly produced a riot" when he lectured at Harvard, in 1914; the class, apparently, "would not hear of there being negative facts at all" (*PLA*, 187).

Despite such a reception four years earlier, Russell in his *Lecture III* of PLA states that he is still inclined to believe that there are, or at least that there *may be*, negative facts. The motivation for admitting such entities stems from wishing to provide facts that make certain positive statements false, as well as facts that make certain negative statements true. An example

K. Perović (✉)
University of Iowa, Iowa City, IA, USA
e-mail: katarina-perovic@uiowa.edu





L. D. C. Elkind, G. Landini (eds.), *The Philosophy of Logical Atomism*, History of Analytic Philosophy,
https://doi.org/10.1007/978-3-319-94364-0_9

of the former is the fact of *Socrates not being alive* making *false* the statement "Socrates is alive" and an example of the latter is the fact of *a hippopotamus not being in this room* making *true* the statement "A hippopotamus is not in this room".

Interestingly, the case that Russell makes in favor of negative facts is itself a negative one. There isn't much discussion at all of how negative facts are to be characterized ontologically, or of why they may be a good sort of entity to have in one's metaphysics. Instead, Russell argues that an account that rejects negative facts and that postulates negative propositions in their place is unsatisfactory. Such an account was produced by one of Russell's students from Harvard—Raphael Demos.

In his 1917 paper "A Discussion of a Certain Type of Negative Proposition" in *Mind*, Demos argues that negative propositions are to be considered as mind-independent entities, whose negative character is entirely independent of a judging mind. He also makes a case that negative propositions should not be considered at face value, as containing a negative constituent, because such a treatment of negative propositions would end up committing one to negative facts as their truthmakers. And negative facts were simply unacceptable for Demos; he thought of them as not given in experience, and believed that any knowledge of apparent negative facts could actually be derived from perceptions of a positive kind. Thus, the key to avoiding negative facts, according to Demos, had to be found in a specific treatment of negative propositions as not formally different from positive propositions. Negative propositions were essentially negative modifications of the content of the rest of the proposition (and did not involve negations of predicates). The meaning of "not", according to Demos, was simply to be interpreted as "the opposite", and hence a negative proposition "non-p" was to be interpreted as "the opposite of p". In this way, a simple negative proposition for Demos amounted to nothing more than an ambiguous description of some true positive proposition.

Russell criticizes Demos most extensively on this last point. He notes that Demos's avoidance of negative facts comes at a high cost of making "incompatibility fundamental and an objective fact, which is not so very much simpler than allowing negative facts" (*PLA*, 189). The incompatibility that Russell is talking about in this context is incompatibility between propositions. And this, for Russell, is problematic because it commits Demos to *facts about incompatible propositions*. Thus, Demos is not just committed to propositions, a commitment that Russell does not share, but he is also committed to there always being positive interpretations of negative propositions, and, finally, to fundamental facts about incompatibility between propositions. For Russell, all these commitments do not add up to an improvement on negative facts. He writes (*PLA*, 189):

> It is perfectly clear, whatever may be the interpretation of 'not', that there is *some* interpretation which will give you a fact. If I say 'There is not a hippopotamus in this room', it is quite clear there is some way of interpreting that statement according to which there is a corresponding fact, and the fact cannot merely be that every part of this room is filled up with something that is not a hippopotamus. You would come back to the necessity of some kind or other of fact of the sort that we have been trying to avoid. We have been trying to avoid both negative facts and molecular facts, and all that this succeeds in doing is to substitute molecular facts for negative facts, and I do not consider that that is very successful as a means of avoiding paradox, especially when you consider this, that even if incompatibility is to be taken as a sort of fundamental expression of fact, incompatibility is not between facts but between propositions.

This passage needs some unpacking. Russell says that the statement "There is not a hippopotamus in this room" cannot be made true by a "mere fact" *that every part of this room is filled up with something that is not a hippopotamus.* By this, he seems to suggest that even if one were to admit molecular conjunctive facts, such as *part* $x_1$ *of the room is filled with A, and part* $x_2$ *of the room is filled with B, and part* $x_3$ *of the room is filled with C, and so on*, until all parts of the room are exhausted, such facts would still not be sufficient truthmakers for the negative statement. What is needed is an additional fact about incompatibility. But what sort of incompatibility would this be—between facts, propositions, or properties?

It certainly can't be incompatibility between facts. Russell is adamant that "no two *facts* are incompatible" (*PLA*, 189). This is because for there to *be* two incompatible facts, they would both have to exist. There would have to be a fact that *part* $x_1$ *of the room is filled with something that is not a hippopotamus* (say, a desk) as well as a fact that is incompatible with it—the fact that *part* $x_1$ *of the room is filled with a hippopotamus.* Thus, incompatible facts would amount to a reifying of contradictions. In addition to this, one would have to admit a *relation* of incompatibility holding between such facts, and this would be contrary to Russell's commitment to logical independence of atomic facts.

Incompatibility could thus only hold between *propositions*, as argued by Demos. But Russell finds this to be unsatisfactory for two reasons: firstly, he does not believe that propositions can enjoy independent reality; and secondly, he thinks that facts of incompatibility between propositions should not be left brute, with no further explanation or ontological ground.

And yet, providing an ontological ground of incompatibility between propositions will boil down sooner or later, according to Russell, to the negative facts that Demos was trying so hard to avoid. Russell makes this point in his 1919 paper "On Propositions" (*OP*, 287–288):

> Usually it is said that, when we deny something, we are really asserting something else which is incompatible with what we deny. If we say 'roses are not blue', we mean 'roses are white or red or yellow'. But such a view will not bear a moment's scrutiny. It is only plausible when the positive quality by which our denial is supposed to be replaced is incapable of existing together with the quality denied. 'The table is square' may be denied by 'the table is round', but not by 'the table is wooden'. The only reason we can deny 'the table is square' by 'the table is round' is that what is round is *not* square. And this has to be a fact, though just as negative as the fact that this table is not square. Thus it is plain that incompatibility cannot exist without negative facts.

According to this passage, incompatibility between propositions should best be analyzed in terms of incompatibility between qualities, and incompatibility between qualities is in turn grounded in negative facts. Going back to Russell's examples in this passage, "the table is square" can be denied by "the table is round" in virtue of the fact that the two qualities *being square* and *being round* are incompatible. Incompatibility between these two qualities cannot be left as ontologically primitive and thus must be grounded in a further fact—a general negative fact *that what is round is not square*. It is this reasoning that makes Russell conclude that not only is the incompatibility account proposed by Demos unappealing, but it ultimately does not even avoid negative facts.[1]

In all of Russell's discussion of negative facts we get little indication as to what makes negative facts *negative* and how negative facts ought to be characterized ontologically. It might be tempting to deduce from the last passage that it is negative properties such as *not being square* that make facts negative. But we must remember that Russell never mentions negative properties; in fact, negative qualities would not sit well at all with the passage on page 188 of PLA in which Russell insists that "not" should never be taken to apply to the predicate but to an entire proposition. Strictly speaking, for Russell, there is no fact of *the table being not-square*; instead, there can only be a negative fact of it *not being the case that the table is square*.

One might think of the latter as an absence of some kind—what else is the fact of *it not being the case that the table is square* if not an absence of the fact of *the table being square*? But this won't do, according to Russell. He writes that "the absence of a fact is itself a negative fact; it is the fact that there is *not* such a fact as [*the table being square*]" (*OP*, 288). To this, an opponent of negative facts could object to Russell that he is reifying negative facts at will. For what is to stop him from saying "it is a fact that there is a fact that there is no such fact as *the table being square*" and then claim that this too is a further fact? Russell does not say.[2]

The students that attended Russell's lecture must have been puzzled about his conception of negative facts; at the end of the lecture, we see them pose questions that reveal a desire for a criterion, a definition, a "formal test" for detecting negative facts. Russell's replies are rather unhelpful. He says that there is no formal test, no clear indication (and certainly no indication at the level of language) of when it is that we are dealing with negative facts. He concludes the lecture by saying that: "you could not give a general definition [of a negative fact] if it is right that negativeness is an ultimate" (*PLA*, 191).

This is quite unsatisfying. Regardless of the prospects of providing a *definition* of negative facts, Russell should at least try to give us a sense of whether such entities are ontologically plausible. For if they are not, that is, if they are metaphysically incoherent entities, then one might just reply to Russell that the reason that he can't give a definition of a negative fact is not because it is a primitive, but because it is difficult to define something that is implausible and cannot exist.

In Sect. 3, I will flesh out in more detail different possible characterizations of negative facts available to Russell. Before I do so, however, I would like to situate Russell's interest in negative facts—when and in what context he came to entertain them, what scope such a commitment had for him, and at which point he gave them up.

## 2 Russell's Qualms About Negative Facts

We should not overstate Russell's commitment to negative facts. Though it is true that in *Lecture III* he argues in their favor, in other PLA lectures we find him saying things that do not sit well with a genuine commitment to negative facts.

Consider, for instance, Russell's discussion of facts and propositions in his first lecture. Here he makes sure to clarify that facts and propositions must be kept clearly distinguished and that a fact is "the kind of thing" that makes a proposition true or false (*PLA*, 163). His examples of facts are as follows (*PLA*, 163):

> If I say 'It is raining', what I say is true in a certain condition of weather and is false in other conditions of weather. The condition of weather that makes my statement true (or false as the case may be), is what I should call a 'fact'. If I say, 'Socrates is dead', my statement will be true owing to a certain physiological occurrence which happened in Athens long ago. If I say, 'Gravitation varies inversely as the square of the distance', my statement is rendered true by astronomical fact. If I say, 'Two and two are four', it is arithmetical fact that makes my statement true.

But at this point, Russell adds something that may be taken to conflict with his endorsement of negative facts. He writes (*PLA*, 163):

> On the other hand, if I say 'Socrates is alive', or 'Gravitation varies directly as the distance', or 'Two and two are five', *the very same facts which made my previous statements true show that these new statements are false.* (Italics mine.)

This passage can be interpreted in one of two ways: (1) Russell is suggesting that it is *the same "positive" facts* that can serve as truthmakers for truths and falsemakers for falsehoods, or (2) Russell is assuming that *the same facts simpliciter* (*positive or negative*) can serve as truthmakers for truths and falsemakers for falsehoods.

If Russell meant to say (1), he would be ruling out negative facts right off the bat in the first pages of *Lecture I*, and this would make his subsequent endorsement of negative facts in *Lecture III* very odd indeed. But given that Russell does not explicitly refer to positive or negative facts in these passages, it seems more likely that he meant to say (2) rather than (1). He is concerned with characterizing facts in general terms and takes them to be the sort of entity that can in certain circumstances serve as truthmakers, and in other circumstances serve as falsemakers. One unproblematic example is *2 + 2 = 4* acting as a truthmaker for "2 + 2 = 4" and a falsemaker for "2 + 2 = 5".

The only wrinkle in this interpretation is that Russell also cites *a certain physiological occurrence which happened in Athens long ago* acting as a truthmaker for "Socrates is dead" and a falsemaker for "Socrates is alive". His appeal here to the seemingly positive fact of *a certain physiological occurrence* rather than the negative fact of *Socrates not being alive* suggests that at least at this point of his lectures he might be preferring positive facts to negative ones.[3]

In *Lecture IV*, Russell does not discuss negative facts explicitly. However, while puzzling about the nature of belief, he writes (*PLA*, 198):

> You will notice that wherever one gets to really close quarters with the theory of error one has the puzzle of how to deal with error without assuming the existence of the non-existent. I mean that every theory of error sooner or later wrecks itself by assuming the existence of the non-existent. As when I say 'Desdemona loves Cassio', it seems as if you have a non-existent love between Desdemona and Cassio, but that is just as wrong as a non-existent unicorn. So you have to explain the whole theory of judgment in some other way.

I do not want to engage in any detail Russell's shifting position on the nature of beliefs and judgments during this period. But it must be noted that for some reason Russell here chooses *not* to make use of his commitment to negative facts, discussed in *Lecture III*, a mere ten pages prior. If indeed Russell had seriously considered admitting negative facts, wouldn't a negative fact of *Desdemona not loving Cassio* come in handy as a false-maker for "Desdemona loves Cassio"? Why dismiss such a possibility so strongly by equating it with non-existent unicorns?

I believe that the answer to this question is simply that Russell's commitment to negative facts was exploratory rather than thoroughgoing. He was toying with the idea for a few years, but not incorporating it fully into the rest of his philosophy even during those very years.

A quick look at a timeline of Russell's engagement with negative facts also confirms this interpretation. We see him begin mentioning negative facts as he was abandoning the *Theory of Knowledge* manuscript, in late May, 1913, after his notorious interactions with Wittgenstein. In "Props" (Appendix B.I, Folio 1 and 2), Russell talks of proposition "xRy" "pointing indifferently" to positive or negative facts. Understanding of a proposition involves, Russell writes, *neutral facts.* Judgment too involves a neutral fact, a fact that is neither positive nor negative, but stands in a relation to the positive or a negative fact. The following figure provided by Russell might help us understand better what he had in mind:

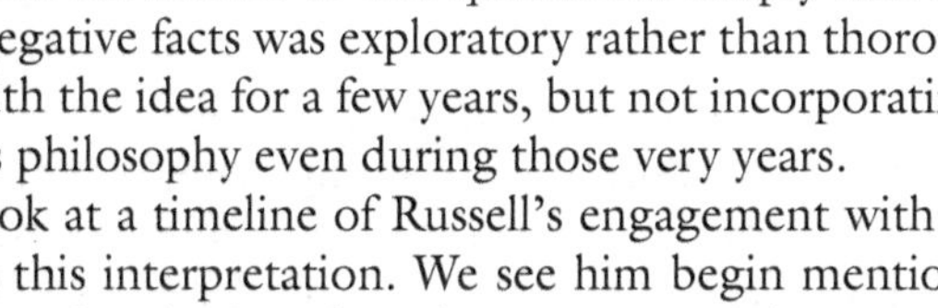

The sign "±" is meant to represent the neutral fact; "+" sign stands for a positive fact and the "−" sign stands for a negative fact. The arrow is meant to capture the type of correspondence that a neutral fact has to either of the two facts. Russell writes about it as follows (*CPBR 7*, 197):

> Judgment asserts *one* of these. It will still be a multiple relation, but its terms will not be the same as in my old theory. The neutral fact replaces the *form*.
>
> Call neutral fact "positively directed" when it corresponds to a positive fact, "negatively directed" when it corresponds to a negative fact.

So, the judgment will assert either the positive fact or the negative fact; however, neither the positive *nor* the negative fact can be a part of the judgment itself, for one might judge something that is simply not the case and Russell certainly did not wish to reify inexistent facts or make it so that one needs to know whether the judgment is true or false before being able to judge. This is why he is in these pages toying with the idea of making neutral facts become a part of the judging relation; this way one would still be judging something—a *fact*, but judging such a fact would not automatically commit one to the existence of a positive or a negative fact that is being judged. If the judgment is true, then the fact exists and the neutral fact will be pointing either positively or negatively at positive/negative fact respectively; if the judgment is false, then the fact simply does not exist. What *does* exist whenever there is judging going on is, rather problematically, a brand new category of *neutral facts*.

During his Harvard lectures in the spring semester of 1914, Russell returned to negative facts. This was the discussion that supposedly "produced a riot" and that prompted one of the students—Raphael Demos—to write the article in *Mind* in 1917 which denounced negative facts. Unfortunately, there isn't much to go on when reconstructing the content of the particular lecture in which Russell discussed negative facts. There are some lecture notes from students[4] but these do not give us enough to reconstruct what Russell's arguments in favor of negative facts were at the time. Since it is known that Russell brought with him to Harvard Wittgenstein's *Notes on Logic*,[5] it has been speculated that it was in the context of discussing Wittgenstein's views on facts that Russell might have wound up attempting to defend negative facts.

Curiously though, there is very little on negative facts in Wittgenstein's *Notes on Logic*. (There isn't that much more in *Tractatus* either.) Wittgenstein says that there are both positive and negative facts and he gives an example of a negative fact as what is "signified" by the proposition "this rose is not red". It doesn't seem as if Wittgenstein meant "signifying" here to stand for a correspondence to a negative fact. Just a few pages later Wittgenstein states that "the chief characteristic" of his theory is that "in it, *p* has the same meaning as not-*p*". This is also echoed in *Tractatus*'s 4.0621 which states that "The propositions '*p*' and '¬*p*' have opposite sense, but there corresponds to them one and the same reality". From all of this, it seems right to say that Wittgenstein's negative facts were negative on the level of sense, certainly not in any more substantive ontological way.

For Russell, on the other hand, in the years between 1918 and 1921, negative facts appeared to have more ontological weight than that. The most extensive discussion of negative facts we find in *Lecture III* of PLA and in his article "On Propositions" (1919). There is also a brief mention of negative facts in Russell's *Analysis of Mind* (1921), and after that Russell seems to give them up entirely.

Negative facts are conspicuously absent from "Logical Atomism" (1924), *Analysis of Matter* (1927), and *An Outline of Philosophy* (1927). And by the time he returns to the topic in his later works, he is adamantly opposed to negative facts.

In *Inquiry Into Meaning and Truth* (1940) and *Human Knowledge: Its Scope and Limits* (1948), Russell explicitly rejects negative facts as truthmakers. In *Inquiry* he writes (*IMT*, 99–100):

> The phrase "this is white" is in your mind, *this* is before your eyes, and "this is grey" is a sentence describing your experience. But "this is not white" is not a sentence describing what you see, and yet, on the basis of what you see, you are sure that it is true, in other words, that "this is white" is false. It might be argued that you know the general proposition "what is grey is not white", and that from this, together with "this is grey", you infer "this is not white". Or it might be said that you can confront the word "white" with what you see, and perceive an incompatibility. Either view has difficulties.

Here, Russell is considering different ways that we arrive at the truth of "this is not white". It may be that we compare the color white that we have seen before with the color grey that we are seeing right now and conclude that "this is not white" is true. Or, it may be that we know general propositions about color-incompatibilities and together with what we are seeing, we are deducing that "this is not white" is true. Either way, Russell in *Inquiry* is doing everything to avoid postulating a negative fact *this not being white* as the source of our perceptions and the truthmaker for "this is not white".

In *Human Knowledge* (1948), Russell is even more explicit. He denies that negative facts are needed as truthmakers for statements such as "The sun is not shining". In fact, he thinks that negation is altogether unnecessary for a complete description of the world (*HK*, 500):

> If the sun is shining, the statement "The sun is shining" describes a fact which takes place independently of the statement. *But if the sun is not shining, there is not a fact "sun-not-shining" which is affirmed by the true statement "The sun is not shining"*. Now, clearly I can believe, and believe truly, that the sun is not shining. But if "not" is unnecessary for a complete description of the world, it must be possible to describe what is happening when I believe that the sun is not shining without using the word "not". (Italics mine.)

Finally, in *My Philosophical Development* (1959), Russell treats "not" in the same way he treats quantifiers "some" and "all"; that is, he is very explicit that "not" has no ontological bearing: "If I say 'A precedes B', my sentence is a relation between *three* words, whereas what I wish to assert is a relation between *two* things. The complexity of the correspondence grows greater with the introduction of logical words such as 'or' and 'not' and 'all' and 'some'" (*MPD*, 188). Note that "not" is treated here as merely a "logical word" with no "thing" corresponding directly to it; Russell acknowledges, however, that statements containing such logical words correspond in complex ways to their truthmakers.

## 3 The Ontology of Russellian Negative Facts

I have thus far established the timeline of Russell's engagement with negative facts as well as the exploratory nature of his interest in such entities. It has emerged that even during the times that Russell was committed to negative facts, such commitment did not come to permeate other aspects of his philosophy. It is now time to investigate whether Russell's negative facts, during the limited time that he did entertain them, can at all be characterized in an ontologically coherent way. This task is by no means an easy one due to the constraints Russell imposes on negative facts in *PLA* and *OP*. The following three he is quite explicit about.

### *3.1 No Negative Constituent*

Russell certainly did not want to have a negative constituent corresponding to "not" within negative facts. In "On Propositions", he says clearly that a negative fact "contains no more constituents than a positive fact of the correlative positive form". For him, "the difference between the two forms is ultimate and irreducible"; he adds: "We will call this characteristic of a form its *quality*. Thus facts, and forms of facts, have two opposite qualities, positive and negative" (*OP*, 287).

It must be noted that there is nothing in this last comment of Russell's to suggest that he took the talk of positive quality and negative quality in an ontologically committal way. He does not say that it is positive qualities that constitute positive facts, or that it is negative qualities that constitute negative facts. All he seems to be saying here is that there are primitive qualities of being positive and being negative that apply to positive and negative facts and forms, respectively.

### *3.2 No Absences*

Russell was also quite explicit that he did not conceive of negative facts as absences. In *OP* he says that *A not loving B* cannot be analyzed as "a mere absence of a fact composed of *A* and loving and *B*". Why not? Because, as we have seen in Sect. 1, for Russell, "the absence of a fact is itself a negative fact; it is the fact that there is *not* such a fact as *A* loving *B*" (*OP*, 288). Thus, rather than explicating negative facts in terms of absences, for Russell, it is the latter that need to be explicated in terms of the former.[6]

### *3.3 No Negative Properties*

Russell was opposed to analyzing negative facts in terms of negative properties. However, since he did not distinguish different ways in which negative properties can be understood, he also did not end up explicitly rejecting all of the respective conceptions of negative facts. I will do some filling in on his behalf.

Consider the presumed negative fact of *this table not being round*. Such facts can be analyzed in terms of particulars, such as tables, instantiating negative property universals, such as *not round*, *not red*, and so on. Negative property universals, in their turn, can be understood in one of the following three ways:

1. By appealing to a negative constituent within the universal. Since Russell did not want to be committed to any kind of a negative constituent within facts, by extension he would not have looked favorably at having a negative constituent appear within a universal within a fact.
2. By appealing to absent universals. It is also very doubtful that Russell would have liked to analyze negative facts in terms of particulars instantiating absent universals. Indeed, it is not clear what it might be for a table to instantiate an absence of roundness, and an absence

of redness, if not an instantiation of some "positive", non-absent universal such as *rectangularity* and *brownness*. Instantiations of absent universals in any other more substantive sense would lead to a proliferation of shadowy existents; it would turn out that all particulars instantiate indefinite numbers of universals by instantiating absences of properties that do not characterize them.

3. As opposite universals. Recall that Russell also dismissed the idea that a negative property could be fruitfully replaced with an opposite property. His motivation for a denial of such substitutions had to do with his view that a replacement of *not round* with its opposite *square* can only work if there is a further fact that grounds such opposition, namely the general negative fact that *what is round is not square*.

But assume for a moment that Russell were willing to allow for primitive incompatibility between universals. In that case, he might have gone with ontologically characterizing negative facts with only "positive" opposite universals at hand. This might be a simple task with some properties: *being wise* has an opposite—*being foolish*; *having positive charge* has an opposite—*having negative charge*. The trouble that Russell would have soon encountered though is that a great deal of properties do not have straightforward candidates for opposites. For instance, is an opposite of *being red* just any other color universal under the determinable color—being yellow, blue, violet, and so on? Even if one could make this work for some universals, it wouldn't work well for the lowest determinates (such as *being crimson* and *having mass of 1 kg*) and for the highest determinables (such as *having mass, having color*, or even *having a quality*). There are no clear candidates for opposites of lowest determinates: is the opposite of *crimson* to be thought of as another specific shade of *red*, say, *vermillion*, or is it some other determinate shade under determinable *color*? Is the opposite of *having mass of 1 kg* just any other determinate mass property no matter how close or far it is quantitatively (*having mass of 2 kg*, *having mass of 3 tons*, and so on) or should the choice be restricted in some way?

Similar issues arise for the highest determinables; it would need to be decided whether instantiating the opposite of a determinable *being colored* is just to instantiate any other determinable physical quality, say, *having mass*, *having shape*, and so on. This way forward is certainly peculiar, for there is no clear sense in which *having mass* or *having shape* are opposite to *being colored*. With determinates, one can see how something cannot be *1 kg* and *2 kg* at the same time, or be *red* and *yellow* all over. But with respect to highest determinables there is nothing prima facie incompatible about an object having shape, mass, *and* being colored. Perhaps a way out of this difficulty might be just to stipulate different types of relationships between opposites for determinates and determinables of different levels. This would not be a particularly elegant strategy, and it would make the categorizations of determinables and determinates a matter of convention, but it might nonetheless be the best strategy available for proponents of opposite universals.

Be this as it may, Russell was certainly disinclined to go with the opposite universals account of negative facts. He thought that opposition between universals needed to be further grounded in negative facts of the type *what is round is not square*. Leaving the incompatibility between opposite universals primitive did not even seem to be an option that Russell considered. Perhaps he reasoned that what it really means to claim that *being round* and *being square* are opposites is that they cannot both be exemplified by the same particular; exemplifying one of these properties would exclude the possibility of exemplification of the other by the same particular. Since Russell did not have a worked out theory of possible facts or states of affairs, and since his logical atomism committed him to logical independence of atomic facts, we can see how he would have been at a loss about how to account for incompatibility between qualities while holding on to his other commitments.

The above three constraints on Russell's characterization of negative facts—no negative constituent, no absences, no negative properties—leave very little room for maneuver. It thus comes as no surprise that some have tried to interpret Russell's negative facts in some very creative ways—by invoking, for example, the relation of negative exemplification.

### *3.4 Negative Exemplification*

Drawing on Hochberg (1969) and his discussion of different conceptions of negative facts, Brownstein (1973) has hypothesized that Russell might have "taken negative facts to consist of objects standing in a relation like negative exemplification" (Brownstein 1973, 48). He discusses two ways in which a relation of exemplification can be construed as "negative": the first one involves a negative constituent as one of its relata, while the second one dispenses with the negative constituent and focuses on the way in which exemplification functions. It is this second way that Brownstein has in mind for Russell. Negative exemplification, in this view, joins differently its constituents from ordinary exemplification (cf. Brownstein 1973, 47). But what exactly this type of joining amounts to—whether it joins into an existing entity, into a possible entity, or fails to join at all—Brownstein does not specify.

Barker and Jago (2012) are similarly evasive. Although they call the tie slightly differently (they refer to it as "anti-instantiation"), they have a similar sort of thing in mind. For them, anti-instantiation is a primitive non-relational tie which ties differently from instantiation, and should not be thought of as a full-blown constituent of a negative fact.[7]

The problem with reading Brownstein's negative exemplification and Barker and Jago's anti-instantiation into Russell's account of negative facts is twofold. Firstly, we would be introducing into a Russellian account something rather obscure that these authors themselves fail to elucidate; and secondly, we would be forcing onto Russell a tie he makes no mention of and makes no real room for.

Let's take a closer look. Given Russell's opposition to internal relations, a relation of negative exemplification would need to be construed as an external relation. In keeping with his opposition to negative constituent analysis, *negative* exemplification would have to be understood either as a genuine relation with some sort of primitive negative quality, or as a relation that simply does not relate. Now, although Russell did in fact believe that relations can occur in two distinct ways in a fact—as relating relations or as terms being related, it would be difficult to construe this as an openness to relations that by their very nature do not relate. Hence, the second option is extremely unlikely. Construing it as a genuine relation, on the other hand, would have been more in the spirit of Russell, but then he would have had to explain the sense in which such relations are negative. Keeping in mind that Russell was unwilling to account for negative facts in terms of negative properties, it is hard to imagine him somehow being open to an account that relies upon negative relations.

### *3.5 Negative Particulars*

Finally, one might wish to construe Russellian negative facts as neither involving negative properties nor negative relations, but rather negative particulars. The thought might be that a negative fact of *this table not being red* is to be analyzed as there simply not being a table in one's vicinity that is red. And then rather than picking out an object and saying what quality it does not have, one might think that what is going on is a non-existent particular having a certain property. Obviously, this would bring Russell right back to Meinongian ontology of non-existent (but subsistent?) particulars such as unicorns, Centaurs, and so on, bearing properties and standing in relations. It is possibly the most outrageous conception of negative facts yet and it is hard to imagine Russell giving it any weight.

In light of all of the above, what can we conclude about the ontology of Russell's negative facts? It seems safe to say that Russell would not have wanted negative facts characterized in any of the five ways outlined above. He was explicit in his rejection of negative constituents, absences, and negative properties playing a role in his account of negative facts. But he failed to mention negative exemplification and negative particulars. And yet given his attitude toward relations more generally, and negative properties in particular, it seems highly unlikely that Russell would have been open to negative exemplification. The prospects are even dimmer for a negative particulars account of negative facts.

Perhaps the most promising account is the one that analyzes negative facts in terms of negative universals, and the latter in terms of opposites. This sort of account would require one to come up, hopefully in a principled and non-arbitrary way, with different sorts of opposites at different levels of the determinable/determinate scale. We mustn't forget, however, that this sort of account was not available to Russell who was against primitive incompatibility between universals. Indeed, it was in part the worries about primitive opposing qualities that motivated Russell's arguments against Demos's account in *PLA*.

## 4 Two Non-ontological Approaches to Negative Facts

The task of the previous section was to explore whether Russell's negative facts could be characterized in an ontologically coherent way. It turned out that by following Russell's constraints on negative facts, in letter and in spirit, no coherent account could be extrapolated.

One last attempt at saving Russell's account is to argue that he did not bother himself with the details of the ontological characterization of negative facts because either (1) he did not think of them as being on the same ontological plane as positive ones, or (2) he thought of them as being exactly the same as "positive" facts. Let's take a closer look at these two possibilities.

### *4.1 Quasi-linguistic Account*

Recall that the main motivation Russell gives for postulating negative facts has to do with their truthmaking role for sentences such as "A hippopotamus is not in the room" and the falsemaking role for statements such as "Socrates is alive". When considering the truthmaker for the sentence "A hippopotamus is not in the room", Russell indicates that even if we were to enumerate all the entities in the room, we would need a further negative fact; perhaps something like *the fact that* the chair that is in the room is not identical to a hippo, and *the fact that* the table that is in this room is not identical to a hippo, and so on. Similar considerations apply to the falsemaker of "Socrates is alive". What makes this statement false, for Russell, is not an absence of Socrates. Neither is it the presence of all the other entities, for even if we were to enumerate all the people in the world that *are* alive, this would not count, according to Russell, as a sufficient falsemaker for "Socrates is alive". What we would need to add is a further general negative fact—*the fact that* Socrates is not one of the people who are alive; or, perhaps we would need to add a multitude of negative facts which state that Socrates is not identical to Alice who is alive, and that Socrates is not identical to Michelle who is alive, and so on.

Russell's repeated use of the phrase "the fact about" and "the fact that" (italicized above), might be interpreted as showing that Russell had a linguistic or quasi-linguistic view of negative facts. But, if purely linguistic, then negative facts would be nothing but *statements* about incompatibility between certain particulars ("Socrates is not identical to Alice") and *statements* about incompatibility between universals ("*being square* is not identical to *being round*"). The challenge then would be to explain how such statements can act as truthmakers and falsemakers. The latter are usually thought of as mind-independent portions of the world that provide an ontological ground of truth and falsehood; statements, on the other hand, are the sorts of things that *we* make and thus, in this sense, are mind-dependent.

A possible reply to this worry is to invite us to revise our conception of truthmakers and falsemakers. One could argue that certain types of truths and falsehoods (such as negative ones) may very well require a mix of truthmakers—all of the relevant "positive" worldly facts, plus a statement-like negative fact(s) that are *about* such worldly facts (stating that there aren't any more of them, or that one fact is distinct or incompatible with another, etc.). Note, though, that these sorts of statement-facts cannot be treated like propositions, for Russell was adamantly opposed to propositions carrying any ontological weight, as we have seen in his criticism of Demos.

### *4.2 Complex Correspondence Account*

Another attempt to explain Russell's failure to properly ontologically characterize negative facts might go in the other direction entirely. On this account, negative facts are negative in name only. It is convenient to call them "negative" because they serve particular truthmaking and falsemaking roles, but in and of themselves they are exactly the same as "positive" facts. They are the "I know not what" that makes true a statement "The Hippopotamus is not in this room". Despite it not being clear what the nature of this truthmaker is, it still needs to be a something. And the reason why we cannot properly get at it, however, is because the correspondence between the statement and what makes it true is very complex, so much so that we cannot properly gauge the structure of such facts at all.

Russell says something to this effect when he compares propositions and facts in "On Propositions".[8] He writes (*OP*, p. 318):

> Thus the simpler kinds of parallelism between propositions and fact are only to be looked for in the case of positive facts and propositions. Where the fact is negative, the correspondence necessarily becomes more complicated. It is partly the failure to realize the lack of parallelism between negative facts and negative word-propositions that has made a correct theory of negative facts so difficult either to discover or to believe.

This diagnosis seems accurate. What remains surprising, however, is that despite the difficulty "to discover or to believe" a correct theory of negative facts, Russell—for a few years between 1914 and 1921—decided to try to believe in them. Agnosticism, or his later outright rejection of negative facts, might have been more wise.

## 5 Conclusion

In this chapter I have explored the type and the scope of Russell's commitment to negative facts. In Lecture III of his *PLA* lectures he thought that he needed negative facts to serve as truthmakers and falsemakers of certain statements. No appeal to opposite propositions *a la* Demos could do the trick, according to Russell, for there would still have to be further facts—facts about incompatibility between propositions—to supply sufficient truthmakers for, say, a statement "This table is not round". Such facts about incompatibility would themselves have to be negative facts for Russell.

Despite this, Russell's commitment to negative facts was not long-lived and was not pervasive. Russell fails to employ negative facts as truthmakers within the context of discussion of belief in Lecture IV, and he seems to prefer a positive characterization of Socrates's death in *Lecture I*. This, together with Russell's later abandonment of negative facts, seems to suggest that his interest in them was merely exploratory.

Even so, I have tried to understand how exactly Russell might have gone about characterizing negative facts. I have outlined five different ontological accounts of negative facts and shown that neither of them would have appealed to Russell. He would not have wanted negative facts characterized in terms of negative particulars, or negative exemplification relations, or negative constituents of any kind. Considering that Russell was not a fan of absences, it is likely that he would have also rejected an account of negative facts that invoked negative property universals understood as absences. I concluded that perhaps Russell would have had the best chance of characterizing negative facts in terms of opposite universals, though he made clear that he was unfavorable to such a view.

Finally, I considered whether Russell left the ontology of negative facts out of the story because he thought of such entities in quasi-linguistic terms. Not finding much support for this reading, I concluded with the discussion of a different alternative: namely, that Russell left negative facts in the dark because he thought that the correspondence involving negative facts was so complex to be inscrutable.

In conclusion, *can* we be positive about Russell's negative facts? If by "positive" we mean "confident" about what he had in mind, the answer is "no". Russell said a number of different things about negative facts and it is impossible to be confident about what sort of entity he was trying to describe. If by "positive" we mean "feel good", the answer again has to be "no". Russell has failed to show that only negative facts can fulfill a certain ontological role, and he has failed to describe them in a way that makes them seem ontologically coherent, within the constraints that he imposes on them.

## Notes

1. Rosenberg (1972, p. 37) spells out Russell's objection to Demos along the same lines.
2. See Oaklander and Miracchi (1980, pp. 452–453) for further criticism of Russell's move from absent facts to negative facts.

3. It is, of course, possible that Russell thought of the physiological occurrence mentioned here as involving at some point a negative fact (the fact of there *not being a heartbeat*, or there *not being breathing*, etc.). What is certain is that even if he thought of it in such a way, there is no written evidence of it in *Lecture I*.
4. See in particular Harry Costello's notes, which have recently been uncovered and are being edited by Bernard Linsky.
5. I owe this to Bernard Linsky and his paper "The Near Riot over Negative Facts" (in this volume).
6. For a recent attempt to provide truthmakers for negative truths in terms of absences see Kukso (2006). He argues that absences should not be identified with negative facts or negative states of affairs; the latter are usually thought of as entities, whereas for Kukso absences are not entities.
7. They write: "If *the lake's being frozen* is the state of affairs that results when the (thin particular) lake is tied to the property of *frozenness* in one way, then *the lake's not being frozen* is the state of affairs that results when the (thin particular) lake is tied to the property of *frozenness* in another way. If the first way is instantiation, then the second is anti-instantiation. The first way gives one kind of non-mereological whole, the second gives another. Neither kind of tie nor corresponding kind of whole is reducible to the other kind of tie or whole" (Barker and Jago 2012, p. 120).
8. In *Lecture I* of the logical atomism lectures he writes something similar but rather than talking about complex correspondence, he mentions a relation of "being true to the fact" and "being false to the fact" (PLA, p. 168). Similar language can also be found in *Lecture III*, two pages prior to his discussion of negative facts. Herein, Russell talks of corresponding truly and corresponding falsely: "The essence of a proposition is that it can correspond in two ways with a fact, in what one may call the true way or the false way. [...] Supposing you have the proposition 'Socrates is mortal', either there would be the fact that Socrates is mortal or there would be the fact that Socrates is not mortal. In the one case it corresponds in a way that makes the proposition true, in the other case in a way that makes the proposition false. That is one way in which a proposition differs from a name" (PLA, p. 185).

## References

### Works by Other Authors

Barker, Stephen and Mark Jago (2012). "Being Positive About Negative Facts." *Philosophy and Phenomenological Research*, vol. LXXXV/1, 117–138.

Brownstein, Donald (1973). "Negative Exemplification." *American Philosophical Quarterly*, Vol. 10, No. 1: 43–50.

Demos, R. (1917). "A Discussion of a Certain Type of Negative Proposition." *Mind*, Vol. 26, No. 102: 188–196.
Hochberg, Herbert (1969). "Negation and Generality." *Nous*, Vol. 3 No. 3: 325–343.
Kukso, Boris (2006). "The Reality of Absences." *Australasian Journal of Philosophy*, Vol. 84, No. 1: 21–37.
Oaklander, L. Nathan and Silvano Miracchi (1980). "Russell, Negative Facts, and Ontology." *Philosophy of Science*, Vol. 47, No. 3: 434–455.
Rosenberg, Jay F. (1972). "Russell on Negative Facts." *Nous*, Vol. 6 No. 1: 27–40.
Wittgenstein, L. (1961). *Tractatus Logico-Philosophicus*, edited and translated by D. F. Pears and B. F. Paris: Gallimard.

# PART IV

# Language: The Theory of Judgment and Descriptivism

CHAPTER 10

# Russell's Discussion of Judgment in *The Philosophy of Logical Atomism*: Did Russell Have a Theory of Judgment in 1918?

*Anssi Korhonen*

## 1 Introduction

Russell's discussion of the nature of judgment in *The Philosophy of Logical Atomism*, Lecture IV, is brief and, on the face of it, rather tentative. Russell himself acknowledges this, as when he apologizes to his audience for "pointing out difficulties rather than laying down quite clear solutions" (*PLA*: 199). Here, as in most of the lectures, Russell is concerned with the discovery of the logical form of a certain range of facts. In this case, these are the facts expressed by ***propositional verbs***, such as "believing", "doubting", and "wishing".[1] The analysis of such facts—facts containing two or more verbs ("Othello believes that Desdemona loves Cassio")—will introduce a new logical form, one that is distinct from the form of atomic facts, or facts containing just one verb.[2] Their analysis also presents fresh difficulties distinct from those involved in the analysis of molecular propositions.

A. Korhonen (✉)
Department of Philosophy, History and Art Studies,
University of Helsinki, Helsinki, Finland



L. D. C. Elkind, G. Landini (eds.), *The Philosophy of Logical Atomism*, History of Analytic Philosophy,
https://doi.org/10.1007/978-3-319-94364-0_10

Russell gave his lectures just before he set out to think hard about a complex set of issues relating to *neutral monism* and its acceptability: the nature of presentation (awareness) and the elimination of what Russell sometimes called the "pin-point Subject",[3] the nature of belief and the problem of error, and the nature of demonstrative reference or "emphatic particulars", as Russell called them. Two passages by later Russell himself comment on this development:

> It was in 1918 [...] that I first became interested in the definition of 'meaning' and in the relation of language to fact. Until then I had regarded language as 'transparent' and had never examined what makes its relation to the non-linguistic world. (*MPD*: 145)

> The problem of meaning is one which seems to me to have been unduly neglected by logicians; it was this problem which first led me, about twenty years ago, to abandon the anti-psychological opinion in which I had previously believed. (*RPsyL*: 362)

Russell had been occupied by neutral monism already before the Great War.[4] But as these two later comments indicate, he now set out to tackle his problems from a new perspective that introduced "meaning" and psychology.[5] The change of perspective is anticipated already in *PLA*. Consider, for example, the term "proposition", which points in two directions. On the one hand, Russell continues to hold that propositions are *nothing* (1918a: 196); and if you introduce propositions as objects of attitudes, you may keep that provided you remember that it is not the truth and that in reality "you have to analyze up the proposition and treat your belief differently" (1918a: 197). On the other hand, he also holds that for the purposes of the sort of logical discussion that is the concern of the lectures, "it is natural to concentrate upon the proposition as the thing that is going to be our typical vehicle on the duality of truth and falsehood" (1918a: 165). A proposition, understood in this way, is just a complex *symbol* (1918a: 166); it is a piece of language, a sentence in the indicative mood or maybe a sentence-nominalization. Russell holds, furthermore, that *facts* are truth- and falsehood-makers for propositions (1918a: 163), so that once we concentrate on the proposition as the truth-value bearer, we are thereby focusing, precisely, on the "relation of language to fact".

Still further, Russell has this to say about propositions *qua* symbols:

> When I speak of a symbol I simply mean something that "means" something else, and as to what I mean by "meaning" I am not prepared to tell you. [...] I think that the notion of meaning is always more or less psychological, and that it is not possible to get a pure logical theory of meaning, nor therefore of symbolism. (*PLA*: 167)

Here, as elsewhere in the lectures, Russell sets psychology aside, not because it is irrelevant to his concerns but because he is not yet prepared to take up issues that belong to psychology.[6] That development would indeed take place only slightly later. What is left are the sort of "more purely logical questions" (*PLA*: 196) with which he and Wittgenstein had been occupied before the war.

The discussion of belief in the logical atomism lectures presents Russell in transition on the problem of judgments. On the one hand, he continues to advocate the basic intuition behind the *multiple relation theory of judgment* (*mrtj*): "The belief does not really contain a proposition as a constituent but only contains the constituents of the proposition as constituents" (*PLA*: 197). On the other hand, it also seems, since we know what happened after the logical atomism lectures, that Russell was able to overcome difficulties in the analysis of belief only by re-psychologizing the proposition; this step is explicit in *On Propositions: What They Are and How They Mean* (1919) and subsequent work, as in *The Analysis of Mind* (1921).

Much of the scholarly commentary on Russell's discussion of judgment in the logical atomism lectures has been in line with this dualism. Thus, the interpretative claim has been advanced by many that the Russell of *PLA* no longer believed—or no longer really believed—in his *mrtj* but that he was not yet in a position to formulate an adequate alternative. Wittgenstein's influence, moreover, is usually seen as decisive here, so much so that the complex dialectic of Russell's changing views is seen exclusively through the prism of Wittgenstein's influence on his former tutor, or at any rate as so many reactions on Russell's side to points derived from Wittgenstein: first, it is held that it was the criticism that Wittgenstein directed at his *mrtj* that left Russell without a theory of judgment in 1913; second, this criticism is also seen as the decisive factor that made Russell abandon work on the *Theory of Knowledge* project that same year; third, it is further held that the reasons which were operative in 1913 were still decisive in 1918. Thus, although the Russell of the logical atomism lectures had come to accept the core of Wittgenstein's criticism, or what he took to be the core of that criticism, the lesson was still essentially negative in 1918: that is, Russell now knew, thanks to Wittgenstein's influence, what judgment could not be, but was not yet in a position to say what it was.[7]

There is no denying Wittgenstein's influence on Russell. The textual evidence is quite clear on this point, as when Russell explains in the short Preface to the published versions of the lectures that they "are very largely concerned with explaining certain ideas which I learnt from my friend and former pupil Ludwig Wittgenstein" (*PLA*: 160). But the evidence also suggests a picture of "Russell and judgment in 1918" that is more nuanced than the standard story. In fact, I'm inclined to go so far as to argue that in 1918, when he gave the lectures, Russell *did* have a theory of judgment. It was still a version of the multiple relation theory, although in a radically new form, which was necessitated by things that he got from Wittgenstein.[8]

## 2 The Multiple Relation Theory: Five Observations

There are no less than five observations suggesting that the standard story needs revision.

First, when Russell definitively repudiated the multiple relation theory 1919, his explanation of why the renunciation was necessary had, on the face of it, nothing to do with the criticism that Wittgenstein had directed at his *mrtj* in 1913 (*OP*: 295):

> The theory of belief which I formerly advocated, namely, that it consisted in a multiple relation of the subject to the objects constituting the "objective", i.e. the fact that makes the belief true or false, is rendered impossible by the rejection of the subject. The constituents of the belief cannot, when the subject is rejected, be the same as the constituents of its "objective".

Saying that Russell's stated reason has nothing to do with Wittgenstein's criticism of the *mrtj* presupposes a reasoned opinion on what that criticism was. But whatever exactly it may have been—and there is no scholarly consensus on that point—it is reasonably clear that Wittgenstein was not yet concerned with the "rejection of the subject" in 1913 and, consequently, that this could not have been the point of his criticism.

Wittgenstein's rejection of the subject, or the "soul", is only announced in the *Tractatus*. There Wittgenstein argues that the use of such forms of expression as "A believes that *p* is the case", and so on suggests a correlation of a fact and a subject; in reality, however, the form is "'*p*' says *p*", which is a correlation of facts by means of a correlation of their objects and shows that there is no soul as conceived in "superficial psychology of the present day" or "modern theory of knowledge".[9] On the other hand, when Wittgenstein gave a sketch of the form of belief-ascriptions in the *Notes on Logic*, the emphasis was on the *bipolarity* of propositions—a feature that Russell had failed to appreciate, according to Wittgenstein—and not on the elimination of the subject. Rather than eliminating the subject, Wittgenstein in fact presupposes it, as when he argues that in order correctly to analyze "A believes *p*", we have to make A stand in a relation to the two poles, *a* and *b*, of the proposition *a-p-b* (Wittgenstein 1913: 97).[10]

Second, when we look at what Russell actually says in *PLA*, we see him continuing to advocate his *mrtj*; or at least we see him continuing to accept its basic underlying idea. This is one of the two key points on judgment that he wants to get through to his audience (*PLA*: 199):

> There really are *two main things that one wants to notice* in this matter that I am treating of just now. The first is the *impossibility of treating the proposition believed as an independent entity*, entering as a unit into the occurrence of a belief... (Italics added.)

Third, Russell not only acknowledged Wittgenstein's influence on him on the topic of judgment but was quite emphatic on the point. But what he says indicates that what he thought is called for is not a renunciation of his *mrtj* but a revision. The previous quotation continues (*PLA*: 199):

> [A]nd the other is the *impossibility of putting the subordinate verb on a level with its terms as an object term in the belief.* That is a point in which I think that the theory of judgment which I set forth once in print some years ago was a little unduly simple, because I did then treat the object verb as if one could put it as just an object like the terms, as if one could put "loves" on a level with Desdemona and Cassio as a term for the relation "believe". (Italics added.)

Russell makes the same point in another context. In "Knowledge by Acquaintance and Knowledge by Description", a paper originally published in 1911, he had explained that if I judge, for example, that A loves B, the judgment is an event consisting of a specific four-term-relation of judging between me and A and love and B, with judging occurring as a relating relation. This is the familiar *mrtj*. When the 1911 essay was reprinted in 1917 in *Mysticism and Logic*, Russell appended the following footnote to the explanation: "I have been persuaded by Mr. Wittgenstein that this theory is somewhat unduly simple [...]" (*KAKD*: 154). Putting these two passages together, we see that Russell's second main thing—that the subordinate verb cannot be put as an object term in a belief complex—came from Wittgenstein; but we also see that the modification has to do with a point on which Russell says his old theory had been somewhat too simple. Admittedly, one may regard this as an "understatement" on Russell's part; but the point remains that what is needed *is* presented as a modification of *mrtj*.[11]

The fourth observation relates to what Russell wrote in *My Philosophical Development* about why he did eventually reject his *mrtj* (*MPD*: 182):

> In my belief [that Socrates loves Plato], the unity of the complex depends upon the relating relation *believing*, where *love* does not enter as a relating relation, but as one of the terms between which the relation of believing holds. [...] I abandoned this theory, both because I ceased to believe in the 'subject', and because I no longer thought that a relation can occur significantly as a term, except when a paraphrase is possible in which it does not so occur.

If we follow up this comment, we should conclude that one of the two reasons for renouncing his *mrtj* (the "Wittgenstein reason", having to do with how a relation occurs in a fact) was in place in the lectures, while the other reason, having to do with discarding the "pin-point Subject", came slightly later: the second reason is asserted in "On Propositions" but is absent from the lectures. If we take the latter reason to have been decisive—as is suggested, I think, by the comment in "On Propositions," that was quoted earlier—then there remains the possibility that Russell's comments in his logical atomism lectures do leave room for a version of his *mrtj*, albeit one that differs from the earlier one on a crucial point; how crucial the point is, is something that I shall consider below.

The fifth and final observation concerns a further point about the passage from *On Propositions* with which we began. Consider carefully how the passage describes his *mrtj*: The multiple relation theory of judgment is the theory that belief consists "in a multiple relation of the subject to the objects constituting the 'objective', i.e. the fact that makes the belief true or false". Now, this is clearly different from his *mrtj* as found in *Philosophical Essays*, the Introduction to *Principia*, *The Problems of Philosophy* or *Theory of Knowledge*. On each of these earlier versions of the theory, a judgment is said to be true when there is a complex corresponding to the judgment and false when there is no such corresponding complex: truth is aligned with the existence and falsehood with non-existence of "complexes" or "facts". For example, Othello's belief that Desdemona loves Cassio is false because there is no complex unity appropriately compounded of Desdemona, loving and Cassio and corresponding to Othello's belief (*PoP*: 75). In *PLA*, on the other hand, Russell embraced positive and negative facts as truth- and false-makers for positive and negative atomic propositions. On this view, an atomic proposition "*d* has the relation *L* to *c*", if false, is false because it is made false by the negative fact that *d* does not have the relation *L* to *c*. This is one well-known and much discussed doctrine in the lectures. But the passage from "On Propositions" suggests that at one time Russell entertained a plan of combining a new version of his *mrtj* with the distinction between positive and negative facts. This is the idea that I shall explore in this chapter.

## 3 Russell's puzzle about the nature of belief

In the logical atomism lectures, the discussion of judgment centers on the "puzzle about the nature of belief" that Russell explains in §3 of Lecture IV. The puzzle is inextricably intertwined with the discovery, which Russell attributes to Wittgenstein, that we cannot make "a map-in-space of a belief" (*PLA*: 198–9) and that, therefore, belief-facts introduce a completely new species into the inventory of logical forms of facts. Here is the key passage (PLA: 198):

> The point is in connection with there being two verbs in the judgment and with the fact that both verbs have got to occur as verbs, because if a thing is a verb it cannot occur otherwise than as a verb. Suppose I take "*A* believes that *B* loves *C*." "Othello believes that Desdemona loves Cassio." There you have a false belief. You have this odd state of affairs that *the verb "loves" occurs in that proposition* and *seems to occur as relating Desdemona to Cassio* whereas *in fact it does not do so*, but yet it does occur as a verb, it does occur in the sort of way that a verb should do. I mean that when *A* believes that *B* loves *C*, you have to have a verb in the place where "loves" occurs. You cannot put a substantive in its place. Therefore it is clear that the subordinate verb (i.e. the verb other than believing) is functioning as a verb, and seems to be relating two terms, but as a matter of fact does not when the judgment happens to be false. That is what constitutes the puzzle about the nature of belief. (Italics added.)

Russell's description of the puzzle is not very helpful, even granting that what he is trying to convey to his audience is his own perplexity. Describing the "odd state of affairs", he speaks of a verb occurring "in that proposition". But which proposition is that? He has just argued in the previous section that the correct analysis of belief does not mention propositions in his old sense. Nor is he talking about linguistic propositions here. But then again we cannot simply substitute "judgment" for "proposition", thinking that Russell is being careless here. For it is far from clear that the verb—relation, that is—"loves" occurs in Othello's judgment that Desdemona loves Cassio, or that it even seems to occur in that judgment as relating her to him.

Of course, we do know *why* Russell was perplexed about belief in his logical atomism lectures. We know this because we know the context in which the perplexity arose. It is this context that we need to consider to see not only what the puzzle was but also whether Russell had something constructive to say about the topic of judgment at the time when he gave the lectures. What is clear enough is that the puzzle about the nature of belief and Wittgenstein's logical discovery about the logical form of belief-facts relate to Russell's old theory of judgment, the one that he now perceived as "a little unduly simple". The lectures do contain further material that is relevant here. It is found in the discussion of *understanding* at the beginning of Lecture III (*PLA*: 181–3); there, as in the discussion of the puzzle, Wittgenstein is mentioned explicitly as the source of important new ideas. Beyond the lectures, further elucidation may be obtained from considering where Russell stood vis-à-vis Wittgenstein and the topic of judgment in 1914, immediately after the event of Wittgenstein's criticism of his *mrtj*, which allegedly destroyed that theory.

## 4 Russell's Struggles with Judgment in 1913

What underlies the puzzle about the nature of belief is Wittgenstein's notorious *nonsense objection* to his *mrtj*.[12] Wittgenstein had argued in *Notes on Logic* that the propositional character of judging, that is, the feature that judging is necessarily judging that so-and-so, is inevitably lost once judging is construed as an act which puts a multiplicity of entities in front of the judging mind. This is, in effect, to treat the entities with which the judgment is concerned as "substantives", and then there is nothing to stop one from putting a substantive for a verb, and the result will be a nonsensical judgment. We find (Wittgenstein 1913: 96):

> When we say A judges that, etc. then we have to mention a whole proposition which *A* judges. It will not do either to mention only its constituents, or its constituents and form but not in the proper order. This shows that a proposition itself must occur in the statement to the effect that it is judged. [...] In "*A* judges (that) *p*", *p* cannot be replaced by a proper name. This is apparent if we substitute "*A* judges that *p* is true and not-p is false". [...] Every right theory of judgment must make it impossible for me to judge that "this table penholders the book" (Russell's theory does not satisfy this requirement). [...] The structure of the proposition must be recognized and then the rest is easy.

Russell was not blind to the fact that judgment is propositional or that with judgment, "the structure of the proposition must be recognized". By the time he went to Harvard in the spring of 1914, however, he had lost his faith in the ability of his *mrtj* to cope with the problem. Before discussing the conclusions that he had drawn by that time, it is useful to have before our minds a rough chronology of Russell's attempts to "recognize the structure of the proposition" (or "propositional structure", as I shall call it) within the framework of his *mrtj*:

1. Propositional structure derives from, or is grounded in, the *sense* or *direction* of the judging relation (before *Theory of Knowledge*).
2. Propositional structure derives from, or is grounded in, *logical form* (in *Theory of Knowledge*).
3. Propositional structure derives from, or is grounded in, *neutral fact* (in Russell's working notes that probably date from late May of 1913).
4. Propositional structure derives from, or is grounded in, the two verbs of a judgment fact; this is the *two verbs solution*, the negative part of which is formulated by the spring of 1914, and of the positive part of which there are indications in *The Philosophy of Logical Atomism* lectures.

I shall not discuss the first two items on the list. It is sufficient to be reminded that the reason why Russell introduced logical forms in *Theory of Knowledge* was precisely that he needed a way of recognizing propositional structure without invoking propositions. This is shown by the following passage which is about understanding rather than judging, but it applies to the latter notion as well (*TK*: 116):

> [W]hen we are concerned with a proposition which may be false, and where, therefore, the actual complex is not given, we have only, as it were, the "idea" or "suggestion" of the terms being united in such a complex; and this, evidently, requires that the general form of the merely supposed complex should be given. More simply, *to understand "A and B are similar", we must know what is supposed to be done with A and B and similarity*, i.e. what it is for two terms to have a relation; that is, we must understand the form of the complex which must exist if the proposition is true. (Italics added)

Apparently, Russell lost faith in logical forms in this specific sense soon after their introduction. This development takes us to stage (3) in our chronology, which consists of a series of working notes which Russell may have composed in late May 1913. Russell wrote (*TK*: 195–9):

> Three objects *x*, *R*, *y* form one or other of two complexes *xRy* or ~*xRy*. The *proposition xRy* points to either indifferently: both contain nothing but *x* and *R* and *y*.
>
> When we *understand* the propositions, what is happening points equally to either of these two complexes—at least it points to whichever there is of the two. […]
>
> It *looks* as if there actually were always a relation of *x* and *R* and *y* whenever they form *either* of the two complexes, and as if this were perceived in understanding. If there is such neutral fact, it ought to be a *constituent* of the positive or negative fact. It will provide a meaning for *possibility*. […]
>
> 1. Call the positive fact +(*xRy*), and the negative fact −(*xRy*).
> 2. Call the neutral *fact* ±(*xRy*), and the *proposition xRy*.
> 3. Call the judgment *J*{+(*xRy*)} or *J*{−(*xRy*)} or *J*{±(*xRy*)}

No, this won't do; it must be + *J*(*xRy*), − *J*(*xRy*), ± *J*(*xRy*). Otherwise we should have to know before judging.

> Judgment involves the *neutral* fact, not the positive or negative fact.
>
> The neutral fact has a relation to a positive fact, or to a negative fact.
>
> Judgment asserts *one* of these. It will still be a multiple relation, but its terms
>
> will not be the same as in my old theory. The neutral fact replaces the *form*.
>
> Call neutral fact "positively directed" when it corresponds to a positive fact, "negatively directed" when it corresponds to a negative fact. […]
>
> There will only be a neutral fact when the objects are of the right types. This introduces great difficulties.

Several important points emerge from a reflection on these notes. First, they introduce the notion of "neutral fact". Talk of the form as being replaced by the neutral fact indicates that Russell conceived the idea *after* rejecting logical forms; or at least that he conceived it as a response to whatever criticisms Wittgenstein directed against Russell's theory of judgment at the time.[13] They suggest, furthermore, that "neutral fact" was introduced for the same purpose as logical form, that is, to capture the propositionality of judgment.

Second, the notes also introduce the distinction between *positive and negative facts* (±-facts). These two innovations, neutral facts and ±-facts, are clearly meant as complementary notions. The idea here, I venture to guess, is to use "neutral fact" to make sense of the propositionality of judgment ("Judgment involves neutral fact"), and to use "±-facts" to find worldly correlates for judgments ("Judgment asserts one of these"). Being *neutral*, neutral facts fall short of being positive or negative, and hence fall short of being *fully actual*, as it were; hence they may occur as constituents in judgments as well in ±-facts. Being *facts*, neutral facts are presumably *unities*. One thing where this might help is in securing the propositionality of judgment and in circumventing Wittgenstein's nonsense objection. But Russell senses difficulties here, witness the final note in the quotation: for there to be a neutral fact, its constituents must be "of the right types", and this "introduces great difficulties".

Third, Russell continues to assert that judgment is a multiple relation. Unfortunately, he does not tell what its terms are supposed to be, and it is easy to see that the working notes are quite unclear on this point. On the one hand, Russell's symbolism, as exemplified by "$+J(xRy)$", suggests that judgment is in fact a binary rather than a multiple relation, as it seems to indicate that judgments operate on something like propositional contents or maybe states of affairs after the manner of the *Tractatus*.[14] Such entities, however, would be excluded by general Russellian principles. On the other hand, that symbolism is inconsistent with Russell's explanation of judgment as involving a neutral fact standing in a relation to a positive or negative fact. And there is the further point that "subject" is nowhere mentioned in the working notes (Carey 2003a: 38). For these reasons, it is difficult, if not impossible, to make anything definitive of these notes on this point.

The notion of a neutral fact, as sketched in Russell's notes, may be a non-starter.[15] What does emerge from the notes, however, is the important new idea that judgment as a multiple relation might be combined with the theoretical innovation of ±-facts.

## 5 Russell on Judgment in 1914

The next step in our chronology takes us to the early 1914 and Harvard, where Russell lectured on logic (with Harry T. Costello as a teaching assistant) and theory of knowledge. Lecture notes by T. S. Eliot and Victor F. Lenzen suggest that, by that time, Russell had come to accept Wittgenstein's criticism of his *mrtj* as definitive. T. S. Eliot's notes on Russell's course on Advanced Logic (edited by Bernard Linsky) from Lecture 26 on April 9, 1914 are illuminating[16]:

> You never can tell whether a proposition is true or false by examining it; you can only find out by examining the fact. True and false propositions are in some sense incomplete symbols; but in a very different sense from descriptions.
>
> My old theory of judgment
>
> 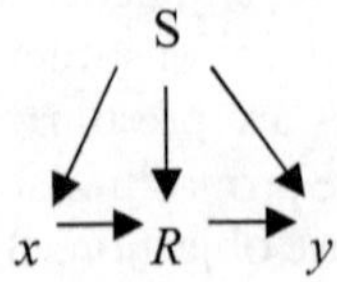
> 
>
> *S* judges that *x* has *R* to *y*
> if *R* was a thing, you could substitute another thing (*z*) for it, and if you do, the judgment is meaningless.
>
> Russell accepts the infinite judgment.

Here we see Russell's formulation of the nonsense objection: if you treat the subordinate relation as a thing, you can substitute another thing for it, "and if you do, the judgment is meaningless". The phrase "My old theory of judgment" here clearly refers to published versions of his *mrtj*, primarily to the explanation in Chap. 12 of *The Problems of Philosophy*. Judgment is there explained to be a relation which has a "sense" or "direction". When a subject (*S*) judges, it puts the objects (*x*, *R*, and *y*) in a certain order; here *R*, too, is "a brick in the structure, not the cement" (1912: 74), and hence it is treated as a thing or a substantive, a feature that gives rise to Wittgenstein's objection.

Arguably and plausibly, Russell spend at least a part of that day's lecture explaining views from Wittgenstein's *Notes on Logic*. As Bernard Linsky's contribution the present volume shows,[17] Russell's next lecture on Advanced Logic, on April 11, was clearly concerned with points from the *Notes*; so Russell may well have explained the nonsense objection against his *mrtj* in the course of discussing Wittgenstein's views.[18] It seems, moreover, that the nonsense or meaninglessness objection was not just a report on Wittgenstein's theory. The quotation from Eliot's note includes the claim that propositions "are in some sense incomplete symbols", and this does not fit Wittgenstein's views but, rather, fits the multiple relation theory. In the *Notes on Logic*, Wittgenstein argues that neither the "sense" nor the "meaning" of a proposition is a thing; these words are incomplete symbols (*TK*: 94). On the other hand, Wittgenstein clearly thinks that a proposition is a *complete* symbol, as when he argues that "a proposition itself must occur in the statement to the effect that it is judged", as in the quotation in §4. So I take it that Russell was explaining what he himself had come think about his "old theory of judgment".

This is in line with what is found in another document relating to Russell's teaching at Harvard 1914: Victor F. Lenzen's notes on Russell's lectures on *Theory of Knowledge*. They are also a little more informative of where Russell stood vis-à-vis the topic of judgment at that time. Sheet [58] of the notes that Victor F. Lenzen took of Russell's seminar on *Theory of Knowledge* at Harvard 1914 (edited by Bernard Linsky) offers the following[19]:

> Take I believe Jones hates Smith. Two verbs-both must occur as verbs. hate not substantive.
>
> Suppose [I] say-Weather is wet today-nice yesterday. Two facts.
>
> I believe Jones hates Smith-single fact-contains two verbs. Constitutes oddity of propositional thought.
>
> Verb occurring in a fact. Jones hates Smith. "hates" points to a different sort of thing than Smith or Jones-Unites them. Gives unity to the fact. Relation between different things different from things related.
>
> In facts believing, disbelieving-two of these verbs coming in-logical form of fact that you believe peculiar.
>
> Remains Judgment-possibility error-not dual relation-Don't simply have-I believe a certain proposition.

Here we see Russell making points that are familiar from the logical atomism lecture. First, in a belief-fact with two verbs, both verbs must occur as verbs; *pace* Russell's old theory of judgment, the subordinate verb really is a verb and not a substantive. Second, a belief-fact is nevertheless a single fact, and not two facts. Third, there is Russell's standard argument that the possibility of error shows that judgment cannot be a dual relation; we do not simply say, "I believe a proposition". Together these points constitute an "oddity of propositional thought": the logical form of a belief-fact is "peculiar", because two verbs come in.

We see, then, that by the spring of 1914, Russell has drawn the conclusions that many interpreters see as an impasse: he has derived from Wittgenstein the lesson about verbs occurring as verbs; nevertheless, he continues to affirm the basic idea of his *mrtj*.

## 6 Russell on Judgment in 1918: A Constructive Suggestion

The chronology given previously mentions the "two verbs solution" to the problem of recognizing propositional structure. But so far we have only seen the negative side of the matter, the "puzzle about the nature of belief", or "oddity of propositional thought", which was in place already in 1914. To make a case for the constructive side, we must enter the realm of speculation. This speculation, though, is not without textual evidence. Some of the speculation to follow makes use of MacBride (2013), an important recent essay on the Russell-Wittgenstein schism. I then add a further speculative layer to MacBride's reading of Russell's 1918 view of judgment.

MacBride is more optimistic than most commentators about the prospects of finding in the lectures on logical atomism a constructive notion of judgment under the general heading of "multiple relation theory". As MacBride himself plausibly puts it, it "seems an unlikely hypothesis that Russell's continued championing of the multiple relation theory during this period was merely a consequence of intellectual inertia" (MacBride 2013: 232).

Prior to 1914/1918, Russell had held, as Landini (2007: 57) puts it, that "universals have both a predicable nature and an individual nature". Accepting a lesson from Wittgenstein, however, Russell came to reject this doctrine. He now argued that predicates and relations can never occur except as predicates or relations, never as subjects (*PLA*: 182).[20] Thus, on the face of it, Russell had come around to a view similar to Frege's distinction between objects and concepts, of which the latter are unsaturated or essentially predicative. In fact, however, Russell's position is rather more involved. For Russell must now recognize *three* kinds of occurrences of constituents in facts:

1. Occurrences as subjects
2. Occurrences as relating relations (or as predicating predicates)
3. Purely predicative occurrences (occurrences "as verbs")

MacBride (2013: 234) argues that recognition of (3) is necessitated not only by the new version of Russell's *mrtj* but also by such innovations as *negative facts*. Our chronology, given earlier, suggests, though, that in Russell's thought there was an intrinsic link between the two. We saw how Russell, in 1913, had the idea that neutral facts are involved in both judgments and ±-facts (although, to be sure, he failed to make full sense of this). In his logical atomism lectures, this dual role is taken over by purely predicative occurrences. Russell continues to accept negative facts, a view which he had defended already at Harvard.[21] He says nothing directly about their structure, but commitment to purely predicative occurrences is nevertheless plausibly read into *PLA*, too. Firstly, there is the general observation that MacBride, too, makes: the fact that *a* and *b* are *not* related by *R*, if it is a fact, can hardly be compounded of *a* and *b*, *actually* related by *R*. Secondly and more importantly, the 1919 essay "On Propositions" argues that the distinction between positive and negative facts is a primitive distinction between two opposing qualities, positive and negative (*OP*: 279–80). Supposing that something like "$[aRb]_{\pm}$" is a reasonable schematic representation of the structure of Russellian atomic facts, this clearly presupposes that universals have purely predicative occurrences. And this view is plausibly read into the logical atomism lectures, too, as it is clearly Russell's version of Wittgenstein's idea of bipolarity.[22] Thirdly, the only way to circumvent the general observation and avoid purely predicative occurrences is in fact to incorporate negativity into the universal itself. Russell, however, is explicit in rejecting negative universals in the logical atomism lectures.[23]

What then of Russell's *mrtj* and purely predicative occurrences? MacBride makes two points about the matter. First, he spells out in a little more detail the conception of judgment toward which Russell was drawn in *PLA*. Predicative occurrences, MacBride (2013: 235) observes, were exactly what Russell needed (1) to cope with the "puzzle of how to deal with error" (*PLA*: 198), and (2) to respond to the nonsense objection. For if an occurrence of *loves* in "Othello believes that Desdemona loves Cassio" is indeed purely predicative, it can occur where it does occur without creating an actual unity; for the same reason, the other objects of the subordinate complex cannot occupy the purely predicative position. The direction of Russell's thought is thus clear: he now conceives of judgment as a family of multiple relations, each possessing its own internal structure, with higher-order argument position or positions reserved for the subordinate verb and lower-order argument positions reserved for the objects of which the relation is predicated.

So far, this adds little by way of a positive construction to the "puzzle about the nature of belief". It is therefore useful that MacBride (2013: 235) adds a further "speculative flourish". He now refers to Arthur Prior's discussion of the conception of judgment that Ramsey sketched in the 1920s, and suggests that "what Russell may have been tentatively edging towards" in logical atomism lectures is an appreciation of a far more radical version of his *mrtj* than hitherto countenanced".

MacBride notes how Prior distances himself from his *mrtj* (Prior 1967: 229). Elaborated, the criticism may be formulated as follows, using Prior's example of Othello's belief that Desdemona is unfaithful.[24] The old version of his *mrtj* is a theory which dispenses with objective falsehoods ("Desdemona's infidelity") and with propositions ("that Desdemona is unfaithful"). It does this by rephrasing "Othello believes that Desdemona is unfaithful" as "Othello ascribes unfaithfulness to Desdemona"; the belief-fact is thus not about any complex entity at all but is about two real entities, Desdemona and infidelity, to which Othello stands "in the complex relation of ascribing the latter to the former" (Prior 1971: 8). Prior objects to this, pointing out that *mrtj*, although rightly dispensing with "Desdemona's unfaithfulness" (and also with "that Desdemona is unfaithful"), still asks us to believe that there is such an entity as *infidelity* (universal) and such an entity as *her fidelity* (a fact), which makes Othello's belief false—Prior's example involves a negative universal, which Russell would not have approved of, but the complication is immaterial here. But propositions and facts, Prior argues, are not the only logical constructions or incomplete symbols that must be paraphrased away (Prior 1971: 9). Consider the identity statement:

> Othello ascribes infidelity to Desdemona = Othello believes that Desdemona is unfaithful.

Prior argues that we must regard the right-hand side of the identity as explanatorily more basic because what needs to be explained (away) is precisely the apparent reference to an abstract object, a universal. Hence Russell's *mrtj* stands condemned, according to Prior.[25]

But if propositions are "logical constructions", as Prior thinks they are, then the right-hand side, too, will have to be detailed accordingly (cf. Prior 1971: Chap. 2.2). To this end, he eliminates the apparent name "that Desdemona is unfaithful" by parsing "Othello believes that Desdemona is unfaithful" as "Othello believes that/Desdemona is unfaithful". Understood in this way, the sentence is not even apparently about a proposition but is about Othello and Desdemona. And to appreciate the way the statement is about them, we note that although "_ thinks that _" is not itself a predicate, it does occur as a part of the predicate "_ thinks that _ is unfaithful", which expresses a compound relation that might hold between Othello and Desdemona and is said by our sentence to hold between them.

As MacBride (2013: 235, fn. 1) notes, it is not really much of a stretch to hold that Russell's new direction of thought in the logical atomism lectures in fact sails very close to Prior's own constructive suggestion regarding the semantics of belief attributions; seeing judgment as involving a compound multiple relation helps us to make sense of how "relating relation" and "purely predicative occurrences" could be reconciled with each other by incorporating the latter into the former.

This is in fact the conclusion that Ramsey drew in 1927, although he did not claim to find it in Russell but presented it as an elaboration of what Russell had at one time held about judgment. Ramsey adds the comment that "it is desirable that we should try to find out more about" judgment construed as a compound multiple relation, and judgment "varies when the form of the proposition believed is varied" (Ramsey 1927: 157).

## 7 Further constructive speculation

In the previous section, I argued that in Russell's thought, there was an intrinsic connection between the new *mrtj* and the distinction between positive and negative facts. In this section, I give a brief elaboration of what was involved in that connection.

At the beginning of Lecture III of *PLA*, Russell explains how understanding and acquaintance for names and particulars differ from understanding and acquaintance for predicates and universals. Here is the relevant passage (*PLA*: 182):

> To understand a name you must be acquainted with the particular of which it is a name, and you must know that it is the name of that particular. *You do not, that is to say, have any suggestion of the form of a proposition, whereas in understanding a predicate you do.* To understand "red", for instance, is to understand what is meant by saying that a thing is red. You have to bring in the form of a proposition. You do not have to know, concerning any particular "this", that "This is red" but you have to know what is the meaning of saying that anything is red. You have to understand what one would call "being red". The importance of that is in connection with the theory of

> types [...]. It is in the fact that a predicate can never occur except as a predicate. When it seems to occur as a subject, the phrase wants amplifying and explaining, unless, of course, you are talking about the word itself. [...] When you understand "red" it means that you understand propositions of the form "*x* is red". So that the understanding of a predicate is something a little more complicated than the understanding of a name, just because of that. Exactly the same applies to relations. (Italics added.)

Understanding predicates and relations, Russell now argues, differs in an important respect from understanding names: hence, acquaintance with a universal must also differ from acquaintance with a particular. And the difference is that understanding a predicate involves ***understanding the form of a position***, while understanding a name does not. Russell notes, moreover, that this is an important point because it connects with the issue of different kinds of occurrences.

This account contrasts with what Russell held in *Theory of Knowledge* regarding logical forms. There he explained—as was noted in Sect. 4—that to understand "*A* and *B* are similar", we must understand "what is to be done with *A* and *B* and similarity" and that this introduces the form of dual complexes (expressed by "something stands in some relation to something"). From the standpoint of *PLA*, this is wrong, because it fails to register the peculiar logical form of predicates. Suppose that *S* understands "A is red". This is now not to be understood as

$$\mathrm{U}\left(S, a, \text{redness}, \alpha\left(\mathbf{x}\right)\right),$$

where $\boldsymbol{\alpha}(\mathbf{x})$ is the logical form of all subject-predicate complexes (*TK*: 113). Russell now holds that understanding must take the following form:

$$\mathrm{U}\left(S, a, \textbf{is red}\right).$$

This is somewhat misleading of what Russell intends, because it might be taken to suggest that understanding involves a relationship between a subject and an entity. What the new construal intends, however, is the following. First, there is no such thing as *redness*. Second, the predicate "red" is not understood through acquaintance with an abstract universal but by understanding what is said when a thing is said to be red; that is, we understand "red" when we understand how things are said to be when a thing is said to be red.[26]

Understanding a predicate thus presupposes understanding a propositional form, according to Russell. But now the question arises: What *is* a propositional form? After all, Russell continues to be committed to his *mrtj*, and hence there aren't any propositions. What there are, or what there could be, though, are *forms of facts*. But what are forms of facts? How can there be a form of a fact if there are no facts of that form?

Russell's conception of fact is one on which this is a legitimate and pressing question: a fact is an actuality, and there are no merely possible facts.[27] The following response is available to Russell, however. Whether a proposition (an atomic proposition) is true or false, there is the positive or negative fact which makes the proposition true or makes it false. And these facts have or can be said to have forms.[28] So I suggest that Russell's new direction of thought—the speculative version of his *mrtj* that has been outlined here—in fact requires there to be positive and negative facts. Thus, Russell was committed to the following line of thought:

*First step:* Understanding words for universals presupposes acquaintance with *propositional forms*; better, involves understanding *propositional forms.*

*Second step:* But there aren't any propositional forms, because there aren't any propositions.

*Third step:* But there are *facts.*

*Fourth step:* So, there could be *forms of facts*, rather than forms of propositions.

*Fifth step:* But what is a form of a fact, if there aren't any facts of that form?

*Sixth step:* This question is answered by introducing *positive and negative facts* corresponding to atomic propositions; there is a fact corresponding to an atomic proposition whether that proposition is true or false.

*Seventh step:* So, the assumption of positive and negative facts guarantees that there will be "sufficiently many" forms of facts.

*Eight Step: mrtj*, as conceived in Russell's lectures on logical atomism, requires there to be positive and negative facts.

If this is along the right lines, it is inaccurate to say, as was said earlier, that when one understands a predicate, one understands how things are said to be using that predicate. Rather, what the new version of *mrtj* really involves is Russell's elaboration of the understanding-bipolarity connection that

Wittgenstein had affirmed in 1913: to understand a predicate is to understand a sentence featuring that predicate, and this in turns means: to know what is the case if the sentence is true *and* to know what is the case if the sentence is false (cf. Wittgenstein 1913: 93–4).

We may conclude, further, that Russell was indeed describing a fresh version of *mrtj*, when he wrote, in *On Propositions*, about the theory of belief which he formerly advocated and according to which a belief consists "in a multiple relation of the subject to the objects constituting the 'objective', i.e. the fact that makes the belief true or false". That version, so I have argued in this chapter, remains largely implicit in Russell's lectures, but it is nevertheless there.[29]

## Notes

1. Russell says they might also be called "attitudes", but he explains that he does not like the term, because there is at least the possibility that not all such verbs are really psychological (*PLA*: 199).
2. Russell uses "verb" ambiguously to mean either a linguistic expression, a member in a word class, or (more often) what is expressed by a word in this class; cf. the discussion of atomic facts and particulars in (*PLA*: 177–8).
3. Russell (*MsN*: 265, 268).
4. Neutral monism—mostly that of William James—is discussed at length in the *Theory of Knowledge* manuscript; see Russell (*TK*: Part I, Chaps. II and III), which were separately published in Russell *OKEW* (1914).
5. One of his prison-letters to his brother Frank shows Russell in the middle of this change: "there seems to me a lot of interesting work to be done on Facts, Judgments, and propositions. I had given up Logic years ago in despair of finding out anything more about it but now begin to see new hope. Approaching the old questions from a radically new point of view, as I have been doing lately, makes new ideas possible" (letter to Frank Russell, dated July 1, 1918; quoted in *CPBR* 8: 248–9).
6. Russell makes the same point in Lecture IV in his brief criticism of the neutral monist theory of belief, which dispenses with belief as an isolated phenomenon and substitutes for it a behavioristic account of propositional attitudes on which the "logical essence" of an attitudinal fact does not involve propositional reference but a causal chain linking a bodily behavior to a suitable external object (*PLA*: 195–6). To resolve the dispute, Russell argues, one would have to plunge deep into psychology, which he is not willing to do.
7. A very clear formulation of the standard story is found, for example, in Candlish (2007: Chap. 3).

8. MacBride (2013) has recently advanced a similar view; I will consider some elements of MacBride's interpretation in Sect. 6.
9. See Wittgenstein (1922: 5.541–5.5422).
10. The manuscript notes that Russell composed in 1918 after the logical atomism lectures strongly suggest that the imperative of discarding the "pin-point subject" came upon him through a continued reflection on neutral monism, as in the following passage: "It must not be *assumed* that believing, wishing, etc. are irreducible phenomena. If this is assumed, it is very hard to avoid the pin-point Subject, which ought to be avoided if possible. This problem, of getting rid of the pin-point Subject, is a vital one in this topic" (*MsN*: 268). The topic is "Propositions".
11. "Understatement" is Landini's (2011: 270) characterization of Russell's comment on his *mrtj* in the logical atomism lectures.
12. This label and my description of Wittgenstein's objection show that I take for granted a simple and straightforward reading of Wittgenstein's criticism of Russell's theory of judgment. Wittgenstein's criticism is formulated in his characteristically brusque and sweeping manner, which gives the conclusion but omits the underlying reasoning (Wrinch 1919: 324–5 gives an equally uncompromising reply to the Wittgenstein-type objection to *mrtj*, chalking it up to a "lingering belief in the unity of a proposition"). What matters here, however, are not the details of Wittgenstein's objection but *Russell's* use of that objection in the spring of 1914 and thereafter and here the straightforward formulation is the gist of the matter; see Sect. 5.
13. Recall that in late May 1913, Russell and Wittgenstein were engaged in a rather heated exchange over the theory of judgment. In a well-known letter to Lady Ottoline, written on May 27, 1913, Russell told about a meeting with Wittgenstein on the previous day: "Wittgenstein came to see me—we were both cross from the heat—I showed him a crucial part of what I have been writing. He said it was all wrong, not realizing the difficulties, that he had tried my view and knew it wouldn't work" (Griffin 1992: 459). There are reasons to think that the "crucial parts" contained Russell's discussion of the notion of logical form, since Russell had composed that part of the manuscript immediately before their meeting (for a discussion, see Connelly 2014). It is not implausible, then, that Russell's working notes were a reaction to this incident. On the other hand, they contain ideas that are clearly reminiscent of Wittgenstein's *Notes on Logic*. (See Carey 2003b.) This observation in itself fixes nothing, but it might be taken to suggest that Russell's notes were composed at some later date. The matter deserves further investigation.
14. The latter suggestion is made by Bonino (2008: 88).
15. Some of the ideas that Russell sketches in the notes are significantly similar to the psychological theory of judgment that he worked out after *PLA*.

With that theory, neutral facts make a kind of return. In the psychological theory, propositions in the fundamental sense are *image propositions* standing in the relation of objective reference to ±-facts, which are truth- and falsehood-makers; see Russell (*OP*: Sects. 3 and 4) and Russell (*AMi*: Lecture 13). Russell (*AMi*: 272) attributes this basic idea to Wittgenstein, and in working it out, he makes further use of Wittgenstein's ideas, including the twin-notions of propositions as facts and as pictures. It is clear, however, that the picture theory was tailor-made to suit Wittgenstein's notion of an atomic proposition, which is quite different from Russell's. Hence, in fact, Russell had difficulties weaving the different ideas together into a coherent whole. Arguably, it is here that "neutral fact" finds room as the neutral element shared by a proposition and a ±-fact. This development, though, lies outside the scope of the present chapter.

16. Philosophy 21: Advanced Logic, Harvard University, 1914; Notes by T. S. Eliot. The material is kept in Harvard University Library. I am grateful to Professor Bernard Linsky for access to transcribed material.
17. See Linsky's chapter in this volume.
18. In the version of *Notes on Logic* that Russell took to Harvard, there is the following remark: "A proper theory of judgment must make it impossible to judge nonsense" (Wittgenstein 1913: 97). This claim is immediately preceded by an explanation that on Wittgenstein's theory of the proposition, "*p* has the same meaning as not-*p* but opposite sense". This latter point is found in Eliot's notes on April 11, although there Russell uses a slightly different terminology, "denotation" instead of "meaning".
19. *Theory of Knowledge*: Philosophy 9c, Harvard 1914; Notes by Victor F. Lenzen, edited by Bernard Linsky. Lenzen's notes are kept in the Bertrand Russell Archives, and can be accessed through Digital Archive at McMaster University Library. Thanks to the Bertrand Russell Archives in the William Ready Division of Research Collections, McMaster University Library, for permission to use unpublished materials.
20. One particularly appealing feature of Russell's new doctrine of universals, also noted by MacBride (2013: 233), was that it promised a way out of the puzzle that F. H. Bradley had formulated for relational thought. Bradley's case against the metaphysical validity of relational thought depends on what looks like an annoyingly simple dilemma: either a relation is something to its terms or else it is nothing to its terms; if it is nothing to its terms, then the terms are not related; but if it is something to them, then that requires a new connecting relation, and we have made no progress in explaining the fact of relatedness (Bradley 1893: 21). Thanks to his "Wittgensteinian turn", Russell is now in a position to argue that Bradley was misled by grammar: "Bradley conceives a relation as something as substantial as its terms, and not radically different in kind", and in so doing he

has been misled by "the fact that the *word* for a relation is as substantial as the *words* for its terms" (*OoP*: 252); see also Russell (*LA*: 332–8).

21. See Russell (*PLA*: 187–90).
22. Wittgenstein (1913: 93–4); cf. Russell (*AMi*: 271–3).
23. See Russell's discussion of Demos and negative propositions: Russell (*PLA*: 187–8).
24. For the criticism, see Prior (1971: Chap. 1.4).
25. Prior (1971: 9) notes Russell's change of mind in the logical atomism lectures from *mrtj* to the two verbs view, but doesn't elaborate on its implications for Russell.
26. Since predicates live in sentences, the view that understanding a name is fundamentally different from understanding a predicate is just one facet of the deep contrast between names and sentences, a doctrine that Russell got from Wittgenstein; see (*PLA*: 167–8).
27. Russell (*CPBR 9*: 8) explains that we cannot say that a false proposition "means the fact which would make it true if it were true, since there is no such fact".
28. See (*OP*: 279–80).
29. I am grateful to participants in the Centenary Celebration of Russell's Lectures for comments and lively discussions. Research for this chapter was supported by a grant from the Alfred Kordelin Foundation.

## References

### Works by Other Authors

Bonino, Guido (2008). *The Arrow and the Point: Russell and Wittgenstein's Tractatus*. Frankfurt: Ontos Verlag.

Bradley, Francis Herbert (1893). *Appearance and Reality: A Metaphysical Essay*. Second edition (revised), with an Appendix. First Published 1897. London: Swan Sonneschein & Co. Ltd. 1908.

Candlish, Stewart (2007). *The Russell/Bradley Dispute and its Significance for Twentieth-Century Philosophy*. Basingstoke: Palgrave Macmillan.

Carey, Rosalind (2003a). "The Development of Russell's Diagrams for Judgment", *Russell: the Journal of Bertrand Russell Studies*, n. s. 23: 27–41.

Carey, Rosalind (2003b). "Wittgenstein on Believing that p." In *Wissen und Glauben*, eds. Winfried Löffler and Paul Weingartner. Beiträge des 26. internationalen Wittgenstein Symposiums, Kirchberg am Wechsel, Österreichische Ludwig Wittgenstein Gesellschaft: 81–83.

Connelly, James (2014). "Russell and Wittgenstein on Logical Form and Judgement: What did Wittgenstein Try that Wouldn't Work?" *Theoria*, Vol. 80: 232–254.

Griffin, Nicholas (1992). *The Selected Letters of Bertrand Russell, Volume 1: The Private Years (1884–1914)*. London: Allen Lane, The Penguin Press.

Landini, Gregory (2007). *Wittgenstein's Apprenticeship with Russell*. Cambridge: Cambridge University Press.

Landini, Gregory (2011). *Russell*. London and New York: Routledge.

MacBride, Fraser (2013). "The Russell-Wittgenstein Dispute: A New Perspective." In *Judgment and Truth in Early Analytic Philosophy and Phenomenology*, ed. Mark Textor. History of Analytic Philosophy. Basingstoke: Palgrave Macmillan. 206–241.

Prior, Arthur (1967). "Correspondence Theory of Truth", in *The Encyclopedia of Philosophy*. Paul Edwards (Editor in Chief), Vol. 2. New York: The Macmillan Company & The Free Press; London: Collier–Macmillan Limited: 223–232.

Prior, Arthur (1971). *Objects of Thought*. eds. P. T. Geach and A. J. P. Kenny. Oxford: Clarendon Press.

Ramsey, Frank P (1927). "Facts and Propositions." *Proceedings of the Aristotelian Society*, Vol. 7: 153–170.

Wittgenstein, Ludwig (1913). "Notes on Logic." In L. Wittgenstein, *Notebooks 1914–16*. Ed. G. H. von Wright and G. E. M. Anscombe. Oxford: Basil Blackwell: 93–106.

Wittgenstein, Ludwig *TLP₁* (1922). *Tractatus Logico-Philosophicus*. Translated by C. K. Ogden and F. P. Ramsey. London: Routledge and Kegan Paul.

Wrinch, Dorothy (1919). "On the Nature of Judgment", *Mind*, Vol. 28 (111): 319–329.

CHAPTER 11

# Russell's Descriptivism About Proper Names and Indexicals: Reconstruction and Defense

*Francesco Orilia*

## 1 Introduction

Since at least the time of "On Denoting," and then throughout his whole philosophical career, Russell endorsed a form of *descriptivism* about *ordinary* proper names and indexicals, that is, the proper names and indexicals that we normally use to communicate in natural language, by means of which we typically refer to ordinary objects.[1] That is, roughly speaking, he took the meanings of such singular terms to be expressible by definite descriptions, and thus characterizable, we may say, as *descriptive contents*. In particular, Russell upholds these ideas in his 1918 lectures on the philosophy of logical atomism (*PLA*). This is interesting, as it shows that this semantic aspect of Russell's thought remains constant in spite of the significant changes in ontological perspective that we find in the logical atomism lectures. Russell now thinks that ordinary objects are reconstructable in terms of classes of sense data and ends up endorsing the neutral

F. Orilia (✉)
Dipartimento di studi Umanistici, Sezione di Filosofia e Scienze Umane,
University of Macerata, Macerata, Italy
e-mail: orilia@unimc.it



L. D. C. Elkind, G. Landini (eds.), *The Philosophy of Logical Atomism*, History of Analytic Philosophy,
https://doi.org/10.1007/978-3-319-94364-0_11

monism that he had seriously considered, but ultimately rejected, in his 1913 *Monist* paper "On the Nature Acquaintance."

In contrast with Russell's loyalty to descriptivism, the current scene in the philosophy of language is hostile to it. Since at least the seventies of the last century, in the light of important works by Donnellan, Kripke, and Kaplan, *Referentialism* about proper names and indexicals has become dominant. According to this view, these singular terms are *directly referential*, that is, their meanings are not descriptive contents, but simply their referents. Despite the triumph of referentialism, in my book, *Singular Reference: A Descriptivist Perspective*, I argued that Russell was right in endorsing descriptivism, as the arguments in favor of referentialism can be rebutted and those in favor of descriptivism are not easily accommodated by referentialism (Orilia 2010). Russell's form of descriptivism is however *subjectivist* in a sense that I shall clarify below, and when descriptivism is combined with this subjectivism there are indeed problems and some of the anti-descriptivist pro-referentialist arguments are successful. What we need to do then is to amend Russell's descriptivism from subjectivism. In this chapter I wish to show how this can be done.

I shall proceed as follows. In Sect. 2, I shall review how Russell presents and motivates descriptivism in the logical atomism lectures, and clarify why his version of this doctrine is subjectivist. In Sect. 3, I shall consider the referentialist arguments that are effective against descriptivism cum subjectivism. Finally, in Sect. 4, I shall first consider the way in which in my book *Singular Reference*, I proposed to free descriptivism from subjectivism, and then illustrated an alternative way, which I think improves on my earlier proposal.

## 2 Russell's Descriptivism and Its Motivations

Russell puts forward descriptivism about ordinary proper names by telling us that any such name is an "abbreviation" of a definite description, say "the *F*," wherein "*F*" is a predicate expressing a contingent property, which in typical cases is commonly attributed to the bearer of the name. We may put this as follows: the meaning of the name, in *one way* of interpreting this name (since a name can be used for different individuals, as "John Smith" paradigmatically witnesses) is provided by a definite description of that sort, which can thus be considered as synonymous with the name. At the same time, Russell admits that there are, as he characteristically says, *logically* proper names, whose meanings are particulars with which the speaker is acquainted. These names, as we may say in current terminology, directly refer to their referents (and thus Russell is, as regards them, a referentialist). In the logical atomism lectures Russell presents these ideas as follows (*PLA*: 178):

> The names that we commonly use, like "Socrates", are really abbreviations for descriptions; not only that, but what they describe are not particulars but complicated systems of classes or series. A name, in the narrow logical sense of a word whose meaning is a particular, can only be applied to a particular with which the speaker is acquainted. You remember, when Adam named the beasts, they came before him one by one, and he became acquainted with them and named them. We are not acquainted with Socrates, and therefore cannot name him. When we use the word "Socrates", we are really using a description. Our thought may be rendered by some such phrase as, "The Master of Plato", or "The philosopher who drank the hemlock", or "The person whom logicians assert to be mortal", but we certainly do not use the name as a name in the proper sense of the word.

The beginning of this passage hints at Russell's view of ordinary objects as "systems of classes or series," which we may set aside for present purposes. Moreover, the part about Adam (as well as other passages at *PLA*: 178 *ff*) seems to tell us that we can be acquainted with ordinary objects and tag them with logically proper names; but this is a misleading suggestion that we should also set aside, since Russell really thinks that the only particulars with which we can be acquainted are private mental occurrences, sense data, or the like (cf. Pears 1985).

I'd rather like to focus on another aspect of Russell's descriptivism about proper names, which the above passage well illustrates: for a typical proper name, there are several very different descriptions with equal claims to provide the meaning of the name in question; for this meaning is captured, depending on what a given speaker has in mind, by an idiosyncratic recourse to a *hook* property, as we may call it: a property that (if all goes well) contingently identifies the alleged referent, but has typically nothing to do with bearing the name in question: being the master of Plato (where of course "Plato" should in turn be understood in terms of some definite description), being a philosopher who drank the hemlock, and the like.[2] This identification occurs, if it does, because it so happens that the referent is unique in having the hook property in question. Thus, the meaning of a proper name is not objectively, or intersubjectively, fixed by the *Langue* of a certain linguistic community; it rather varies from speaker to speaker, or perhaps even for the same speaker at different times. We may thus say that Russell's descriptivism is *subjectivist*.[3] Let us grant, although Russell would not put things in this way, that the meaning of a definite description, "the *F*," is a certain *descriptive content*, [the *F*], within which we can distinguish a *determiner* component, corresponding to the definite article, and a *property* component, that is, the property, *F*, expressed by the predicate "*F*."[4] We can then say that Russell (i) is descriptivist about proper names in that, for him, the meaning of a proper name is a descriptive content and (ii) endorses a subjectivist brand of descriptivism, since the property component of the descriptive content is a hook property that varies from speaker to speaker.

In talking about meaning, however, we need be a bit more precise. As I propose in my book *Singular Reference*, it is in fact appropriate to distinguish between (i) a *semantic* meaning of a certain expression type, (ii) the *contextualized* (semantic) meaning of a certain token of that type, uttered in a certain specific context, and (iii) the *pragmatic* meaning of the token in question. Various semantic meanings are attributable a priori, so to speak, to the expression type on the basis of its being part of the lexicon of a given language. One of them is attributable a posteriori, so to speak, qua contextualized meaning, to the expression token, given the context in which it is used. Moreover, thanks to the information provided by the context, this contextualized meaning is typically "enriched" and thus turned into a pragmatic meaning (it is not ruled out here that no enrichment is needed, so that contextualized and pragmatic meanings coincide). These distinctions are easily illustrated with so-called incomplete descriptions. Consider "the bank." From the point of view of semantic meaning, "bank" could mean bank qua river bank, say $bank_1$, or bank qua building hosting a financial institution, say $bank_2$. Thus "the bank" has a least two semantic meanings, expressible by "the $bank_1$" and "the $bank_2$," respectively. In a context in which it is clear that the latter is to be privileged, a certain token t of "the bank" takes [the $bank_2$] as its contextualized meaning. This meaning is a descriptive content that fails to pick up one specific object, for there are many objects with the property $bank_2$. But in the context in question it may well be clear that a more specific property, which actually identifies a certain object, could and should be invoked in order to identify one object; for example, let us suppose, $bank_2$ and located on Elm Street, in Austin, Texas; or $bank_2$ and larger than any other $bank_2$ in Austin, Texas. We may call the second conjunct in each such conjunctive property, a *t-relevant enrichment* of $bank_2$, where *t*, let us recall, is the token of "the bank" in question. One option is to say that, for any *P* that counts as a *t*-relevant enrichment of $bank_2$, [the ($bank_2$ and P)] is *a* pragmatic meaning of *t*. Another option is to say that there is *one* pragmatic meaning of *t*, call it *[the* $bank_2$**]*. This can be assumed, if we allow for the idea that, given the appropriate context, *one* property, which we have called $bank_2$***, is somehow created, or evoked, as a proxy for all the properties of the kind ($bank_2$ and *P*), where *P* is a *t*-relevant property. We can characterize $bank_2$* as a property that is materially equivalent to any such property of the kind ($bank_2$ and *P*), but not identical to any of them (it is important to note here that all the *t*-relevant properties are taken to be materially equivalent).[5]

In Russell's account, there is no explicit distinction between these three levels of meaning, let alone between the semantic meaning of a proper name type and the contextualized and pragmatic meanings of a proper name token. But the subjectivism that Russell proposes is most obviously interpretable as a *semantic* subjectivism, which regards the level of contextualized semantic meaning. To begin with, we could attribute to Russell the idea that the idiolects of different speakers of a language such as English, or even of the same speaker at different times, differ idiosyncratically as to the meaning of proper name types in such a way that there is no truth of the matter as to which idiolect is right in this respect. In what we attribute to Russell we should make room of course for the idea that a proper name is ambiguous for a certain speaker in the sense that we can illustrate with this example: "Aristotle," qua type, is ambiguously associated, for a certain speaker, to the descriptive content [the husband of Jacklyn], and also to the descriptive content [the Greek philosopher born in Stagira]. However, in a given context, for example, one in which we are discussing philosophy, for this speaker the contextualized meaning of a token of "Aristotle" would be the latter content, rather than the former. Yet, for a different speaker in the same context the very same token could be associated, say, to the descriptive content [the teacher of Alexander the Great]. In this way of seeing the matter, the contextualized meaning typically coincides with the pragmatic meaning[6]: we can hardly differentiate between them, since the definite descriptions that in Russell's view provide the meanings of proper names are not incomplete in the sense in which "the bank" is, and, as we saw above in relation to definite descriptions, the need to differentiate the pragmatic from the contextualized meaning precisely arises with the incomplete descriptions (and more generally with incomplete determiner phrases) (Orilia 2010: Section 5.3).

Could we take Russell's subjectivism to be a *pragmatic* subjectivism, which regards only the level of pragmatic meaning? Perhaps we can do it, by stretching a bit what we actually find in PLA and other Russell's texts. How would it work? Roughly, along these lines. We view a proper name "*N*" as an incomplete definite description, whose semantic meaning is a descriptive content, [the *N**], whose property component, *N**, is a *nominal* property, which is exemplified by an individual by virtue of this: it has been named "*N*" in a baptism and possibly called with that name on other occasions after the baptism.[7] This is of course a very generic property that many individuals may share and thus N* does not succeed in identifying a certain individual. However, in a specific context, depending on the speaker, a token of "*N*" has as pragmatic meaning another descriptive content, [the (*N** and *F*)], whose property component succeeds in identifying a certain individual (at least in typical cases); for example, for a certain speaker, *F* might be a property such as being a philosopher who drank the hemlock, but for another it might the property of being the master of Plato.[8]

There are two main reasons that Russell has in mind in favor of the view that a proper name takes as meaning a descriptive content, rather than its referent: the *co-reference* and the *no-reference problems*.[9]

We can illustrate the former problem with this example. If the meanings of a token of "George Eliot" and a token of "Mary Anne Evans" were simply their referents, a token of "George Eliot is a poet" and a token of "Mary Anne Evans is a poet" should express the same proposition; but this runs contrary to the fact that Tom, who does not know that George Eliot is Mary Anne Evans, may assent to the former, but not to the latter. We need two propositions, one believed and the other disbelieved by Tom; and we get them if the two proper name tokens have different descriptive contents. Following Russell's subjectivism, they could be, say, [the author of *Middlemarch*] and [the woman who was vice-director of the *Westminster Review* in 1951] (perhaps Tom ran into an old 1951 issue of this journal and found there the name "Mary Anne Evans").

As for the no-reference problem, consider this. If, as many experts believe, it was somehow invented that a single poet called "Homer" wrote both the *Iliad* and the *Odyssey*, a token of "Homer did not exist," should express a true proposition. Yet, if the meaning of a proper name token would simply be its referent, it would not be clear which proposition, if any, is expressed by the sentence token in question, unless we assume, *à la* Meinong, that the referent is a non-existent object. In contrast, if the meaning of the token is a descriptive content, we get the proposition we need. Following Russell's subjectivism, the descriptive content could be, for example, the author of the *Iliad* and the *Odyssey*.

As regards indexicals, in the logical atomism lectures we find Russell's view in the following passage (*PLA*: 179):

> One can use "this" as a name to stand for a particular with which one is acquainted at the moment. We say "This is white". If you agree that "This is white", meaning the "this" that you see, you are using "this" as a proper name. But if you try to apprehend the proposition that I am expressing when I say "This is white", you cannot do it. If you mean this piece of chalk as a physical object, then you are not using a proper name. It is only when you use "this" quite strictly, to stand for an actual object of sense, that it is really a proper name.

Apart from asserting that indexicals can be used as logically proper names that directly refer to sense data or the like, this passage suggests that "this," used as an ordinary indexical in order to refer to an ordinary object, since not "a proper name," is a definite description. We may thus take Russell as telling us that a token of an ordinary indexical such as "this" has a descriptive content as pragmatic meaning. But which one? Looking at what Russell has written on these topics before and after his logical atomism lectures we can get a unique perspective. (See Farrell Smith 1989.) We can attribute to him, even in his logical atomism lectures, a descriptivism about indexicals that is even more subjectivist than his descriptivism about proper names: the descriptive contents working as pragmatic meanings of indexical tokens subjectively and idiosyncratically depend on private mental occurrences. For example, suppose you and I look at a statue and I tell you: "I like this." For me the pragmatic meaning of the token of "this" that I have used is something like [the object causing $\text{this}_1$], where $\text{this}_1$ is a certain sense datum, which happens to be in my visual field as a result of the fact that I am looking at the table.[10] But for you it is something like [the object causing $\text{this}_2$], where $\text{this}_2$ is a sense datum in your visual field; it is a different sense datum, though caused by the same object. Moreover, the token of "I" that I have uttered has for me a pragmatic meaning along these lines: [the individual attending to $\text{this}_1$], whereas for you it has a quite different pragmatic meaning, something like this: [the individual causing $\text{this}_3$], where $\text{this}_3$ is a sense datum in your mind caused by my presence in front of you. (See, e.g., the third part of Russell's paper "On the Nature of Acquaintance.")

The idea that indexicals have descriptive contents as meanings can also be motivated by the co-reference and no-reference problems[11] and presumably Russell had these motivations in mind, although I cannot find clear statements of this.

## 3 Criticisms of Descriptivism and the Current Scene

In the second half of the last century various philosophers of language, Donnellan, Kripke and Kaplan in particular, mounted an attack against descriptivism about proper names and indexicals. It is typically believed that this led to a demise of descriptivism and to the affirmation of referentialism. Soames (2010) is paradigmatic in this respect, since it is meant to be a textbook that presents the state of the art in the philosophy of language and takes descriptivism to be unredeemably superseded by referentialism.[12] There are indeed important arguments that have been provided against descriptivism and it is typically taken for granted that they cannot be answered and that accordingly a referentialist turn is necessary. However, once we accept referentialism, to deal with the co-reference and no-reference problems becomes very difficult and these two problems are then either ignored or tackled with various epicycles that make referentialism much less palatable that it might seem at first sight (*Singular Reference*, Section 8.13). In particular, the no-reference problem, as noted above, may take us back to Meinongianism, if we endorse referentialiasm. It is thus not surprising that Soames (2010: 128) acknowledges non-existent objects, even though he does not cite Meinong.

Fortunately, the arguments against descriptivism can be tamed (Orilia 2010: Chap. 8). This is so, however, to the extent that the subjectivism that is part and parcel of Russell's descriptivism is somehow eliminated. Some of these criticisms in fact precisely undermine this aspect of Russell's descriptivism. Let us focus on them. I have in mind in particular the semantic argument and the problems of choice, ignorance and error. (Orilia: Sections 4.11 and 4.4, and references therein.)

Suppose we incorporate semantic subjectivism into descriptivism. Then, if Tom uses a token *g* of "Gödel" in uttering

(1) Gödel is Austrian,

both the contextualized meaning and the pragmatic meaning of *g* would be, for instance, something like [the logician who discovered the incompleteness theorem], and the proposition expressed would accordingly be captured by this sentence:

(1a) The logician who discovered the first incompleteness theorem is Austrian.

Here is the semantic argument. Imagine that, unbeknownst to us all, Gödel stole the proof of the theorem to an unknown German logician, Schmidt. In this case, we should admit, the token g still refers to Gödel, not to Schmidt, and (1a) is false, although (1) remains true. Yet, our theory wrongly predicts that g refers to Schmidt rather than to Gödel, since (i) *g* has as meaning the descriptive content [the logician who discovered the incompleteness theorem], and (ii) Schmidt is the unique individual that exemplifies the property component of this content. Moreover, as a consequence of this, the theory also wrongly predicts that both (1) and (1a) are false, since the individual in question is German, rather than Austrian. We avoid the result that *g* refers to Schmidt, if we assume the pragmatic subjectivism outlined above, and take the pragmatic meaning of *g* to be [the individual who is a Gödel and who discovered the first incompleteness theorem]. We avoid the result, because Schmidt (let us assume) is not called "Gödel" and thus is not a Gödel. But we would still not succeed in securing that *g* refers to Gödel, since, by hypothesis, he did not discover the first incompleteness theorem.

The problem of choice goes as follows. We might imagine that Tom has at his disposal several descriptive contents that he may associate to this token g of "Gödel." For example, in addition to the one we have already considered, there could also be this: [the logician who discovered the second incompleteness theorem]. If so, there is no principled reason to take one rather than the other as contextualized meaning, and as pragmatic meaning, of *g*. Yet, descriptivism cum semantic subjectivism predicts that one specific descriptive content should fulfill these roles. The picture does not change substantially, if we combine descriptivism with pragmatic subjectivism.

Move now to the problem of ignorance. Tom could be so ignorant about Gödel that there is at his disposal no appropriate descriptive content of the sort envisaged by descriptivism cum semantic subjectivism. Hence, he can attach to g no contextualized and no pragmatic meaning. Yet, the theory predicts that one descriptive content of this sort should fulfill these roles. Again, the picture does not change substantially, if we associate descriptivism to pragmatic subjectivism.

Let us finally turn to the problem of error. Assume descriptivism cum semantic subjectivism and suppose now that Tom's ignorance leads him to take, as contextualized and pragmatic meaning of *g*, the following descriptive content: the logician who discovered the undecidability of quantification theory. The theory predicts that *g* refers to Church (since he is the logician who discovered the undecidability of quantification theory) and that (1) is false, but in fact *g* refers to Gödel and (1) is true. As for the semantic problem, the situation slightly improves if we rather assume descriptivism cum pragmatic subjectivism, but not in a decisive manner.

As we shall see in the next section, in order to answer these objections we need a form of descriptivism that avoids subjectivism. It is not obvious that these criticisms are effective against the Russellian account of indexicals. Be this as it may, we shall see that, once we avoid subjectivism about proper names, we shall have the elements for a more satisfactory but still descriptivist account of indexicals.

## 4 Descriptivism Without Subjectivism

In my book *Singular Reference* I have proposed a form of descriptivism that avoids the subjectivism of Russell's approach by putting together causal descriptivism and Reichenbach's token-reflexivity.[13] It works as follows (setting aside details that need not detain us here). It is assumed that, by being baptized with the proper name "*N*," either ostensively or via a description, a given individual acquires the property of being an *N*; and it is further assumed that this very generic property is the property component of the descriptive content [the *N*]. In turn, this content is the semantic meaning of "*N*," as well as the contextualized meaning of any token of "*N*."[14] The causal theory of reference made popular by Kripke takes for granted that there is a causal chain that links a baptism based on a certain name to subsequent uses of the name, a *nominal-causal* chain, as we may call it. Granting this, we can say that, for any token *n* of "*N*" that is used, there is a property, being a source of a nominal-causal chain leading to *n*, that is exemplified by a certain individual *i*, just in case *i* was baptized with "*N*" in a baptism linked by a nominal-causal chain to the circumstance in which *n* is used. This property, we may also grant, is a species of the more general property of being an *N*; that is, whatever has the former property by necessity has also the latter property. With all this is in place, it is then also proposed that the pragmatic meaning of a given token n of "*N*" is simply the token-reflexive descriptive content [the source of the nominal-causal chain leading to *n*].[15] By virtue of what has just been explained, it should be clear that a descriptive content of this sort has a property component that in a typical case identifies one individual and that appropriately enriches the more generic property component of the contextualized meaning of "*N*." Clearly, this approach eschews the semantic argument and the problems of choice, ignorance and error. Let us see how.

Tom, in using a certain token g of "Gödel," may well have in mind a descriptive content such as [the logician who discovered the first incompleteness theorem]. But this does not make it the pragmatic meaning of *g*. The pragmatic meaning is: [the source of the nominal-causal chain leading to *g*]; this is a descriptive content whose property component is exemplified just by Gödel. Thus, even if it turns out that it was Schmidt who actually discovered the first incompleteness theorem, the token g still refers to Gödel. Moreover, it does not matter that in using *g* Tom has in mind several descriptive contents, such as the logician who discovered the first incompleteness theorem or the logician who discovered the second incompleteness theorem. Or that Tom has in mind a descriptive content, such as [the logician who discovered the undecidability of quantification theory], whose property component happens not to be true of Gödel. All these descriptive contents are not meanings of *g*, for the only thing that counts as meaning of *g* (at the pragmatic level) is: [the source of the nominal-causal chain leading to *g*]. And this is true, even if Tom has not this descriptive content explicitly in mind, and indeed if no descriptive content is explicitly in his mind.

This approach has however some shortcomings. A token-reflexive meaning is ephemeral, for its existence is as short-lived as the very token that happens to be one of its constituents. This may be unappealing for those of us who think of propositions as both meanings of sentence tokens and eternal truth-bearers that are true or false independently of the sentence tokens that express them. Moreover, and more importantly, this approach introduces a new form of subjectivism, for it takes any two tokens of the same name to have two distinct pragmatic meanings. For example, suppose that Tom uses a token *g* of "Gödel" in saying

(2) Gödel discovered the first incompleteness theorem,

and Mary uses another token *g*′ in saying

(3) Gödel discovered a theorem.

According to this token-reflexive approach, we are forced to say that *g* and *g*′ have two distinct pragmatic meanings, namely [the source of the nominal-causal chain leading to *g*] and [the source of the nominal-causal chain leading to *g*′]. And yet we may well have the intuition that they have precisely the same meaning. Moreover, we may think that the proposition expressed by Tom logically implies the proposition expressed by Mary. But this can't be true if *g* and *g*′ stand for the two distinct descriptive contents in question, since from a purely logical point of view these two descriptive contents are not equivalent. (Similarly, that the husband of Michelle is a former US president does not logically imply that the 2009 Nobel Peace Prize winner is a former president).

But we can avoid these problems by eliminating token-reflexivity as follows. We assume now that at a baptism performed in a certain specific place *p* at a specific time *t* with the name "*N*," the baptizer *ipso facto* creates (or simply evokes) a property that one has by virtue of being baptized with "*N*" in *p* at *t*. Call a property of this sort a *specific nominal* property. This property is a species of, and thus entails, the more generic nominal property of being an *N*, which one has simply by virtue of being baptized with the name "*N*" (a property shared by all individuals baptized with this name). Thus, "*N*," to the extent that it was ever used in a baptism (thereby counting as a proper name) has as semantic meaning the very generic property of being an *N*, but, ambiguously, it also has as semantic meaning, for any baptism celebrated at a certain time and place, a corresponding specific nominal property. Thus, given that several baptisms with the name "*N*" have taken place, there are corresponding properties $N_1$, $N_2$, $N_3$, and so on. Depending on which baptism is the source of the nominal-causal chain that leads to a given use of a token *n* of "*N*," one of these specific nominal properties, call it $N^*$, happens to be privileged in that [the $N^*$] is the contextualized linguistic meaning of the token *n*, and consequently also the pragmatic meaning of *n*.

It should be clear that this approach steers clear of the semantic objection and the problems of choice, ignorance and error, just like the previous token-reflexive approach. However, contrary to the former, it completely eliminates subjectivism from descriptivism about proper names. To see this, consider again our Gödelian example in which Tom uses a token *g* of "Gödel" in uttering (2) and Mary uses another token *g*′ of "Gödel" in uttering (3). We can of course assume that both *g* and *g*′ can be traced back to the same baptism and thus both have the same contextualized and pragmatic meaning, say [the $G^*$]. Hence, we capture the intuition that the proposition expressed by Tom's sentence token logically implies the proposition expressed by Mary's sentence token. Moreover, for those who care, we make room for the idea that these propositions are not ephemeral like the tokens used by Tom and Mary, so that the former proposition keeps implying the latter independently of the tokens that express them.

In *Singular Reference*, I have also treated indexicals *à la* Reichenbach in a token-reflexive way, but we can provide for them an approach similar to the one that I have just urged here for proper names: an indexical word, qua type, has a very generic property as semantic meaning, but, whenever a corresponding token *t* is used, we have, as it were, a new baptism, which generates (or evokes) a very specific property *P*, which typically identifies a certain individual, and is such that [the *P*] is the contextualized and pragmatic meaning of *t*. If we pick the first-person pronoun "I" and the demonstrative "this" as examples, the corresponding semantic properties can be roughly characterized as follows.[16] For "I," the semantic meaning is the very generic property of being an I, a property which one acquires by virtue of simply uttering something in a given language (whether out loud or in inner speech); for "this," the very generic property of being a this, a property that something acquires simply by virtue of being an item in an area in which utterances occur, which I call an *interdoxastic domain*.[17] Each such property *P* is a genus with respect to a very specific property, *P**, which is generated, or evoked, whenever a token *x* of the indexical in question is used at a specific time and place, which we may call, respectively, the *contextual* time and place of *x*.[18] *P** typically identifies a specific individual and the descriptive content [the *P**] is the pragmatic meaning of *x*. If *t* and *p* are, respectively, the contextual time and place of the relevant token, then (roughly): if *P* is the property of being an I, *P** is that property that one acquires simply by virtue of uttering something in *p* at *t*; if *P* is the property of being a this, *P** is that property that something acquires simply by virtue of being an item that at time *t* is most salient among the items in the proximity of *p*.

Suppose for example that at pastry shop, while pointing at a piece of cake, Tom uses a token *i* of "I," and a token *t* of "this" in saying

(4) I want this.

Then, there are (i) a property *I** that Tom has by virtue of having spoken at that particular time and place and (ii) a property *T** that the piece of cake has by virtue of the fact that Tom's pointing gesture makes it (let us suppose) more salient than anything else at that particular time and place. The descriptive content [the *I**] is the pragmatic meaning of *i*, which, as expected, refers to Tom, since he is the only individual with the property *I**. Similarly, the descriptive content [the *T**] is the pragmatic meaning of *t*, which, as expected, refers to the piece of cake, since it is the only object with the property *T**.

## 5 Conclusion

In his logical atomism lectures Russell propounds a descriptivist account of proper names and indexicals. The standard view in current philosophy of language is that Russell is deeply mistaken about this. Nevertheless, the co-reference and no-reference problems still suggest to us that Russell was right in adhering to descriptivism. His form of descriptivism, however, is subjectivist and this subjectivism exposes it to criticisms such as the semantic argument and the problems of choice, ignorance and error. But this subjectivism can be eliminated from descriptivism and thus these criticisms can be dodged. There are of course other well-known problems that allegedly afflict descriptivism, such as Kripke's modal and epistemic arguments. However, as I argued in SR, they can all be countered. In sum, Russell was basically on the right track in his account of singular reference.

## Notes

1. As we shall see, Russell also acknowledges logically proper names and indexical terms that work as logically proper names, understood as directly referential expressions. In the following, by "proper name" and "indexical" I shall mean ordinary proper names and indexicals, unless otherwise indicated. In the category of indexicals I include both deictic terms such as "I" and "here" and demonstratives such as "this" or "that."
2. I wrote that the hook property has *typically* nothing to do with bearing the name in question, because Russell occasionally considers properties such as being called *N*, where *N* is a certain proper name.
3. As is well-known, Frege holds a similarly subjectivist form of descriptivism.
4. Descriptive contents can be viewed as certain kinds of denoting concepts, which Russell accepted in *Principles of Mathematics*, but rejected in "On Denoting." (See Cocchiarella 1982.) As explained in my book *Singular Reference*, I take descriptive contents to be denoting concepts understood pretty much in Cocchiarella's sense, i.e., as properties of properties.
5. The first of these two options is essentially the one proposed in Bach (1994), which I have followed in Orilia (2000, 2003) (although only in the latter work I explicitly relied on Bach's notion of conversational impliciture). This line is however subject to a *problem of choice* analogous to the one that will be discussed below in relation to proper names. Accordingly, in SR I favored the second option, and then proceeded to characterize in token-reflexive terms the pragmatic meaning of incomplete definite descriptions and more generally of incomplete determiner phrases. (See Orilia 1910: Section 5.3.) I still think that the second option is better, but I would now avoid the recourse to token-reflexivity, as we shall see in more detail in discussing proper names and indexicals.

6. This is the view that in Orilia (2010: Section 3.7), I attribute to Russell.
7. I am assuming that a proper name "*N*" is a phonological word in the sense of Lyons (1968: 69). If we understand "word" in this way, a token of "Aristotle" that is used to refer to the philosopher, and another token that is used to refer to the husband of Jacklyn Kennedy, are tokens of the same word. Similarly, a token of "take" used as a noun and a token of "take" used as a verb are tokens of the same word.
8. This is the view that I proposed in Orilia (2000, 2003).
9. We may add of course other reasons, e.g., anti-essentialist motivations. (See Landini 2011: 211 *ff.*) But we may set them aside for present purposes.
10. The sense datum can be taken to be directly referred to by the token in question.
11. See Orilia (2010: Section 3.3).
12. Even though Soames corrects in his own way the referentialist paradigm and recognizes an element of truth in descriptivism. (See Soames 2010: 171.)
13. Causal descriptivism has been proposed by many authors. See (Orilia 2010: 155, no. 12 for references).
14. This is so to the extent that "*N*" is viewed as a proper name. The theory also acknowledges that a proper name can function as a general term.
15. Frigerio (2017) objects to my descriptivist approach as follows. He notes that a sentence of the form "*N* might not be called *N*," where "*N*" is a proper name, does not have a contradictory *de dicto* reading, and charges that my approach wrongly predicts that it does. It seems to me, however, that it cannot be taken for granted that there is no such *de dicto* reading, unless we beg the question against the descriptivist by ruling out that "*N*" is a definite description. Of course, the *de dicto* reading is most unnatural, but this can be explained by appealing precisely to its contradictory nature.
16. Here I rely on proposals made in Orilia (2010, Chap. 6).
17. See Orilia (2010: 180), for a more precise characterization of an interdoxastic domain, which explains why this term is adopted.
18. See Orilia (2010: Section 2.10), for a more precise characterization of contextual times and places of tokens.

## References

### Works by Other Authors

Cocchiarella, Nino B. (1982). "Meinong Reconstructed versus Early Russell Reconstructed." *Journal of Philosophical Logic*, Vol. 11: 183–214.

Frigerio, Aldo (2017). "Francesco Orilia: Singular Reference: A Descriptivist Approach." *Axiomathes*, 27: 731–733.

Landini, Gregory (2011). *Russell.* London and New York: Routledge.
Lyons, John (1968). *Introduction to Theoretical Linguistics.* Cambridge: Cambridge University Press.
Orilia, Francesco (2000). "The Property Theoretical Performative-Nominalistic Theory of Proper Names." *Dialectica*, 54: 155–176.
Orilia, Francesco (2003). "A Descriptive Theory of Singular Reference." *Dialectica*, 57: 7–40.
Orilia, Francesco (2010). *Singular Reference: A Descriptivist Perspective.* Dordrecht: Springer.
Pears, D. F. (1985). "Introduction." In *The Philosophy of Logical Atomism*, Bertrand Russell, ed. D. F. Pears. Open court: La Salle, Ill. 1–34.
Smith, Janet Farrell (1989). "Russell on Indexicals and Scientific Knowledge," in C. W. Savage and C. A. Anderson, eds., *Rereading Russell: Essays in Bertrand Russell's Metaphysics and Epistemology*, *Minnesota Studies in the Philosophy of Science*, vol. 12, Minneapolis: University of Minnesota Press, 119–137.
Soames, Scott (2010). *Philosophy of Language.* Princeton: Princeton University Press.

PART V

# Epistemology: Acquaintance and Analysis

CHAPTER 12

# The Possibility of Analysis: Convergence and Proofs of Convergence

*David Fisher and David Charles McCarty*

## 1 Ideas of Analysis

Together, Russell and Wittgenstein offered at least three markedly distinct ideas of logico-philosophical analyses of ordinary language sentences. The most notable, and best understood, is that grafted onto Russell's famous treatment of such definite descriptions as 'the present king of France.' According to Russell's 'theory' of definite descriptions, the sentence

The present king of France is bald.

is to be replaced by something akin to

There is one and only one man who now reigns in France and he is bald.

Thus is the phrase 'the present king of France' removed and the sentence reconfigured with the aid of quantifier expressions. The Earl Russell also meant this replacement business to be extended to so-called descriptive

D. Fisher (✉) • D. C. McCarty
Indiana University Bloomington, Bloomington, IN, USA
e-mail: dffisher@indiana.edu



L. D. C. Elkind, G. Landini (eds.), *The Philosophy of Logical Atomism*, History of Analytic Philosophy,
https://doi.org/10.1007/978-3-319-94364-0_12

names, a class of linguistic items thought to include the vast majority of ordinary proper names, so that the sentence

> Moses was a prophet.

gets analyzed, as a first step, into such a sentence as (dodging some obvious questions of tense)

> There is one and only one man who led the Israelites out of Egypt and he was a prophet.

For this, one first applies a replacement rule according to which the presumptively descriptive name 'Moses' is substituted away in favor of the definite description

> the man who led the Israelites out of Egypt

then, there is the further step of definite descriptions analysis carried out so that the last sentence becomes

> There is one and only one man who led the Israelites out of Egypt and he was a prophet.

The second idea of analysis features prominently in Russell's essay 'The Relation of Sense-Data to Physics.' It traces a vaguely specified route to a phenomenalistic-plus-class-theoretic analysis of such utterances as

> My desk is brown.

When the second idea gets applied here, the sentence is replaced, it seems, by the assertion

> There is one and only one equivalence class of sensibilia related to each other (in some desired fashion or other) and, moreover, there are suitable sensibilia in that class that are brown.

A determined Russellian analyst could yet press on, wielding the no-class interpretation from *Principia Mathematica* ($PM_2$: 187–190) and, from this last sentence, produce a sentence with quantifiers ranging exclusively over propositional functions, each of a definite type and order in *Principia*'s ramified hierarchy. Presumably, the final result would resemble this sentence:

> There is a unique propositional function $\phi$ (appropriately typed and ordered) on sensibilia, there are suitable brown sensibilia $b$ that $\phi$ maps to the truth-value **True**, and moreover, $\phi$ maps any sensible to **True** if and only it is related (again, in the desired fashion) to those brown $b$.

Third, there is an—again described only roughly—idea for analysis to be discovered in Wittgenstein's *Tractatus* (*TLP*$_1$: 2.0201, 3.24) and revisited *en passant* in his *Philosophical Investigations.* Here, a sentence about a complex is rendered as a sentence representing pertinent relations in which the components of the complex stand to each other, making a statement about their relata. Wittgenstein's own example (Wittgenstein 1953: §60) leads one to think that this sentence, about a complex 'the broom':

> The broom is in the corner.

is to be analyzed into

> The brush is in the corner, the stick is in the corner, and the stick is stuck in the brush.

This time, the pertinent relation is $\lambda x \lambda y.\ x$ *is stuck in* $y$, its immediate relata are the brush and the stick, and they are, as stated, both to stand in the corner.

## 2 Schemes of Analysis

Each of these informal ideas of analysis is adequately captured as a *scheme* for analysis, conceived as first cousin to formal systems and Turing machines, and operating either on natural language sentences or on formulae from an artificial language. A scheme consists of a decidable collection of symbolic replacement rules, which may be nondeterministic, together with a decidable collection of starting sentences. The idea is that, once a rule is handed a starting or input sentence, it and other rules in the scheme operate recursively and generate, through a series of steps, a putative analysis of the starting sentence. For example, a miniature or toy scheme illustrating the first informal idea, that of Russell's definite descriptions analysis for descriptive names, is given by these two replacement rules:

> Moses ⇒ the man who led the Israelites out of Egypt

and

> The man who led the Israelites out of Egypt is a prophet ⇒ There is one and only one man who led the Israelites out of Egypt and he is a prophet.

A relevant starting sentence would be

> Moses was a prophet.

Following the given rules, we first replace the name 'Moses' by the description 'the man who led the Israelites out of Egypt,' so obtaining

> The man who led the Israelites out of Egypt was a prophet.

Then, proceeding stepwise, the second rule, when applied to the latter sentence, yields

> There is one and only one man who led the Israelites out of Egypt and he was a prophet.

With a small dose of formalism, a nontrivial and general scheme covering a wider but similar territory can be defined. Let

$$n \mapsto P_n$$

be a computable function on linguistic items mapping descriptive names $n$ into predicates $P_n$ such that the latter picks out the former uniquely. Under this rubric, $P_{\text{Moses}}$ might be 'is a man who led the Israelites out of Egypt.' Then, according to the scheme, sentences of the form $Q(n)$ get replaced first using rules that substitute explicit definite descriptions, so that we obtain

$$Q(\iota x.\ P_n x),$$

for example, 'The man who led the Israelites out of Egypt was a prophet,' when *Q* in the generic scheme is instanced by 'was a prophet.' Such scheme output sentences are then fed into rules thereby replacing them with sentences of the form

$$\exists x\left[\forall y\left(P_n y \leftrightarrow y = x\right) \wedge Qx\right],$$

for example,

$$\exists x\left[\forall y\left(y \text{ is a man who led the Israelites out of Egypt} \leftrightarrow y = x\right) \wedge x \text{ was a prophet}\right].$$

A scheme with different rules and different starting sentences suffices to capture Russell's analysis of physical objects, and Wittgenstein's analysis of complexes, as well as any combination of the three informal analytical ideas. All three are symbolic processes that are mechanizable, that is, properly conceived as recursive methods for the replacement of symbols by symbols, and that is certainly the way their progenitors thought of them. In his *Philosophical Grammar*, Wittgenstein wrote:

> Formerly, I myself spoke of a 'complete analysis', and I used to believe that philosophy had to give a definitive dissection of propositions so as to set out clearly all their connections and remove all possibilities of misunderstanding. I spoke as if there was a calculus in which such a dissection would be possible. I vaguely had in mind something like the definition that Russell had given for the definite article. [...]
>
> There could perhaps be a calculus for dissecting propositions; it isn't hard to imagine one. Then it becomes a problem of calculation to discover whether a proposition is or is not an elementary proposition.
>
> The question whether, for example, a logical product is hidden in a sentence becomes a mathematical one. (Wittgenstein, 1974b: 211–212)

Once set into action, a scheme's replacement rules generate a stepwise reduction process on any starting sentence. By means of those processes, sentences get operated on by one or more replacement rules and starting sentences are reduced recursively, applying the necessary rule or rules over and over again to any intermediate outputs of earlier application steps. If such process of recursive replacement ends, a final or *normal form* is produced. This normal form or fully reduced sentence is then the completely analyzed sentence, the intended real or true logical form of the original, starting sentence, as imagined by Russell or Wittgenstein. When that happens on a particular starting sentence, we say that the analytical process *converges* on the starting sentence. Should the analytical process fail to converge, the starting sentence would presumably lack a real logical form, a fully analyzed sentence to which the real logical properties of the starting sentence are supposed to be anchored.

## 3 Analyses as Computable Processes

First, in much the same spirit as the familiar treatment of derivation rules defined over formal languages, schemes for analysis and the replacement rules in them are computable in the sense of (Turing 1937). Even the crudest minds have to be able to recognize them fully and to carry them out, following the replacement rules on the starting sentences, at least in principle. (That is not to say that the crudest minds would be able to *discover* them. More on this point will be discussed later.) Both Russell and the Tractarian Wittgenstein imagined their ideas of analysis capable of being implemented in such fashion, at least in principle, in accord with computable recipes. So, in each scheme, the set of starting sentences ought to be decidable, and the set of replacement rules as well. This does not mean that the replacements cannot be conditional or 'if … then …' in form. After all, some starting sentences will exhibit predictable ambiguities, which the rules will need to reflect.

If Russell's and Wittgenstein's ideas of analysis are not properly interpretable as specifying purely symbolic, strictly computable processes, capable in principle of being digitized and carried out by a Turing (or equivalent) machine unaided, then certainly they could be digitized and carried out by a Turing machine with the aid of an extra collection *D* of data. For example, when analyzing everyday talk of physical objects, a relevant set *D* would contain information on the layout and appearances of tables in the kitchen. Perhaps, for the definite descriptions analysis, *D* would list facts about the current denotations of indexicals in our sentences. The relevant computation would proceed by consulting the data in *D* as a so-called *oracle*, a concept defined by Turing (1939). When *D* serves as an oracle to a computing process, then, during each step of the process, the abstract computing machine is allowed to query the information in *D* at most a finite number of times, even if *D* itself is infinite. In this way, a Turing machine with oracle gets to take cognizance of relevant background knowledge. In the present writing, we proceed as if all the analytical processes governed by schemes are strictly computable, without oracles, but note that there is a fully demonstrative version of what we assert here relative to any data set *D*.

Second, unlike ordinary formal methods familiar from freshman logic, such as that for calculating the dual of a propositional formula or that for substituting a new term for a variable in a quantified formula, the original definite description procedure devised by Russell was not recursively defined on sentences and their components. Ordinarily, one does not start explaining how to carry it out on atomic or simple sentences and then continue to say how the procedure would affect more complex sentences featuring connectives, quantifiers, and other sentential combining expressions, following step-by-step the way in which sentences are constructed 'from below,' up their parsing trees. This is one aspect of what Russell meant to flag by (mis)labeling definite descriptions 'incomplete symbols' (*PLA*: 211). This historical fact is no barrier whatsoever to the treatment of definite descriptions analysis in terms of schemes of analytical replacements.

Third, in alliance with this feature, a scheme of Russellian analysis for definite descriptions can fail to be deterministic. The intermediate outputs obtained at any stage in the reductive process can be produced only via the results of one or more choices of replacement rules along the way. What we have in mind here is exemplified by interactions of the description analysis with the logical operators. For instance, the analysis of 'The present king of France is not bald,' exhibits a scopal ambiguity associated with its negative particle. Famously, an analyst can, at whim, choose the descriptor to be of narrow scope relative to the negation, that is, so that the analysis yields

> It is not the case that there is one and only one man who now reigns in France and he is bald.

or she can opt that it be of wide scope relatively, so obtaining,

> There is one and only one man who rules in France and he is not bald.

We do not mean to suggest that these are the sole possible options for dealing with the 'not' in such sentences. In complicated cases, there seem to be scopes that are neither wide nor narrow, but intermediate (Kripke 1977). Any appearance of a quantifier, connective, modal operator, or other scope-bearing linguistic device in either a starting sentence or a sentence intermediate to the analytical process would in general cause the number of choices to multiply by at least two—one for 'wide scope' and one for 'narrow scope'—for each such appearance. In a full analytical treatment of some starting sentence or sentences, the gradual cumulation of choices—with later choices piled upon earlier ones—can give rise to such panoply of alternative results that computer scientists would call their growing complexity 'exponential in input length.' We warn the reader that, since these choices, even if free or nondeterministic, are at most finite in number, the resulting nondeterministic process is still Turing computable, if the relevant scheme is computable to start with (Lewis and Papadimitriou 1981).

Consequently, logicians prefer to display the complete record of a procedure generated by a scheme on a starting sentence as a tree or root-system of intermediate and normal form outputs that is downward growing with all possible reduction paths running through it. This diagrammatic record is the *tree* of the analysis of a starting sentence. The forks in the tree, if needed, that make for the tree's branching represent possible choices in applying replacement rules, choices resulting in intermediate or normal outputs that are, sometimes, markedly different from one another. Moreover, one may need to apply more than one of the given replacement rules to any sentence at any stage of analysis; after all, a sentence may feature, in addition to definite descriptions, names for physical objects as well as names for complexes. Consequent uncertainty may generate yet further branching. For example, the starting sentence could be

The present king of France sits on a chair.

A rule for physical object analysis could be applied to the term 'chair,' and one for definite descriptions analysis to 'the present King of France.' We emphasize that this further nondeterminism does not lead us outside the realm of the strictly computable. Recall that the procedure for generating a formal proof in a standard Hilbert-style derivational system may be highly nondeterministic, and yet remain computable throughout: it can be encoded and, so encoded, carried out by an unaided Turing machine.

Incidentally, the present authors have no firm idea of how to analyze away, in either Russellian or Wittgensteinian manner, names for such institutional or geographical entities as the nation of France. Is 'France' a descriptive name? Does the word denote a Wittgensteinian complex? Or is it a Russellian physical object—perhaps a landmass conveniently surveyed in its geographic entirety only from outer space—and if so, what sort of sensibilia comprise the equivalence class that is its analysans? When the word is treated as a descriptive name, the relevant replacement rule for 'France' could well create a further range of nondeterministic choices, for example, 'France is the country in which I met my beloved Patricia' or 'France was the home of Jean-Paul Sartre,' so introducing further names and prompting yet more questions, more ambiguities, more choices. In *The Philosophy of Logical Atomism*, Russell seemed to offer his readers little help either in discovering or, once discovered, in choosing among any such prima facie reasonable alternatives. But again, once these replacements are specified, the whole analytical process in which they feature ought to remain mechanical, finitary, and thoroughly computable.

## 4 Further Features of Analyses: Types, Orders, and Sentence Lengths

Russell and Whitehead are justly celebrated for their vision of a two-dimensional, predicative type hierarchy, stratified into types and, within types, ramified into orders in accord with the Vicious-Circle Principle (*PM*$_2$: 37–38). The type theory describing that hierarchy served, after 1910, for the official logical background to Russell's foundational work. Mathematical scaffolding for the hierarchy includes a vertical axis for types—indexed by ever-increasing natural numbers that record the logical heights of the entities over which typed quantifiers range—plus a horizontal axis for numbered orders, marking quantifier interlacing in formulae of the same type. Consequently, all variables or designators appearing in formulae of the theory are required to bear natural number indices recording their types and orders. Neither Russell nor Wittgenstein gave any indication in their analytical works of how to introduce indices for types into analyses of natural language sentences, ones without explicit type indices on their terms or quantifiers to begin with. (Is an expression describing France of higher type, or of the same logical type, as those describing its inhabitants?) Importantly, the replacement rules in the schemes one would naturally use to codify the vaguely described ideas of Russellian or Wittgensteinian analyses do not appear, in any obvious fashion, to decrease type indices associated with the entities under discussion. In the case of definite descriptions for (possible) persons such as the present king of France, replacement rules in the appropriate scheme trade apparent nominal reference to persons for quantifier expressions containing variables ranging over persons. So, there would here be no apparent decrement in type index. In general, there would seem to be no reduction in type (or order) at any stage of an analytical process governed strictly by definite descriptions analysis.

When it comes to physical objects and their Russellian analysis, names and designators for physical objects get replaced by names and designators for classes of sensibilia. If sensibilia are treated as primitive individuals of Type 0, then classes of them are of Type 1 (but there is no order indicated, unless one has implicit recourse to the Axiom of Reducibility (*PM*$_2$: $*12$)). Were Russell's no-classes idea (*PM*$_2$: $*20$) applied to these classes, the outcome of the analysis would reference propositional functions of Type 1. In the case of the Wittgensteinian analysis of designators for complexes, there is a proposed replacement of brooms by brushes and sticks. Hence, so far as one can see, there is no apparent reduction in the logical types on variables or quantifiers here: terms for physical objects are replaced by yet more terms for other physical objects, parts of those mentioned in the analysandum, all presumably of the same type. Officially, Wittgenstein rejected the type theory of *Principia*; hence, this may not present any pressing problem for his analyses *in se*, yet it may make even more pressing the problem of determining if and proving when Wittgenstein's analyses converge (*TLP*$_1$: 3.331–3.333). *Vide infra.*

As mentioned, fully legal expressions of the predicate type theory will also feature indices for orders, but neither Russell's nor Wittgenstein's informal ideas of analysis for physical objects or complexes tell us how to introduce or handle such indices.

These observations stand markedly at variance with Russell's own assertion about the relative type differences between complexes and simples that are discovered in his analyses. In *Logical Atomism*, he wrote,

> To speak loosely, I regard simples and complexes as always of different types. (*LA*: 173)

Perhaps Russell meant his savvy readers to put some special linguistic spin onto the phrase 'to speak loosely.' The sample analyses he offers us in *Logical Atomism* do not seem intuitively to reflect any manner of reduction in type. Nor do we receive counsel on how such reduction might be justified. Why would a complex whole consisting of physical objects, for example, a chess set, be of strictly higher logical type than the individual physical objects—the chessmen—of which the set is composed?

One should also note that none of the intermediate outputs at any stage of any analytical process needs to be, as a sentence, shorter or even of the same length—crudely counting the individual words in it—as the input sentence.

> The present king of France is bald.

turns into

> There is one and only one man who now reigns in France and he is bald.

This is a transformation of a sentence containing seven words into one containing 16. (We count 22 symbols in the output when explicit formal quantifiers are introduced.) For Wittgenstein,

> The broom is in the corner. (six words)

gets replaced by

> The brush is in the corner, the stick is in the corner, and the stick is stuck in the brush. (20 words in all)

These features matter in signal fashion to the issues, introduced in the section to follow, of the *convergence* of schemes of analysis on their starting sentences and of *proofs of convergence.*

## 5 Questions of Convergence

Do the analytical procedures above specified, based on the appropriate schemes or others of that ilk, always converge, always come to an end, always produce a sensible output, anything rightly deemed the normal form or complete logical form of the starting sentences? If so, can one prove in general that they do converge? When processes of analysis for sentences of a class ***S*** do not converge, the logical forms, atomic facts, and logical objects that Russell and Wittgenstein imagined to undergird the logic of ***S*** may not be known to exist. Are atomic facts the sentence-sized chunks of reality that are meant to accord with the ultimate sentential components of the logical forms of ***S*** sentences? When logical forms fail to exist, what ken do we then have of atomic facts? If the processes producing complete analyses do not converge, the entire doctrine of logical atomism may be nothing but vague fantasy. Commentators have already raised these kinds of convergence questions. Here is David Pears on the issue:

> An atom is something indivisible or not further analyzable. A logical atomist, therefore, needs to show not only that the divisions traceable in logic correspond to real divisions in the nature of things, but also that the two corresponding processes of analysis do not continue indefinitely. If Russell is right, there must be a point at which words and things will be found to be not further analyzable. But why should we believe that? (Pears 1985: 2)

Can one prove that all individual paths through every process tree, individual attempts at full analysis, fail to loop, that is, fold back upon themselves and start repeating intermediate outputs? Can one prove that they fail to produce ever longer and more involved intermediate sentences without end? If, as we have seen, the outputs of certain schemes of analysis grow ever longer with respect to their inputs, can they not keep going to infinity, producing ever more definite descriptions or names for complexes or physical objects that must be eliminated, in their turns, in favor of yet further descriptions, or class or relation terms? There is no mathematical proof—or even very good reason to believe—that different ideas of analysis never, when applied as schemes to the same starting sentence, come into conflict, perhaps via a complicated series of interactions. Russell certainly had it in mind that analyses for definite descriptions must be used in tandem with ideas of analysis applicable to physical objects and complexes. He asserted,

> The names that we commonly use, like 'Socrates,' are really abbreviations for descriptions; not only that, but what they describe are not particulars, but complicated systems of classes or series. (*PLA*: 178)

There is no proof that schemes for analysis cannot in practice clash on a single starting sentence, making the reductive process grind to a halt without yielding any reasonable output.

It is nowise anachronistic to raise questions like these. Russell faced them himself, if not in the very terms employed in this essay. At the close of Lecture II in *The Philosophy of Logical Atomism*, Russell responded to an objector who feared that stepwise processes of analysis could continue forever without converging:

> I think that it is perfectly possible to suppose that complex things are capable of analysis *ad infinitum*, and that you never reach the simple. I do not think that it is true, but it is a thing that one might argue, certainly. (*PLA*: 180)

In the present writing, this is precisely what we shall argue. In fact, this is what we shall demonstrate.

Following Kripke (2005), Grabmayer et al. (2011) offer an example of a process of definite descriptions analysis that does not converge on examples formulated in a language of first-order predicate logic. It is child's play to come up with perfectly clear examples, drawn from natural language, that lead to endless analytical processes. Here is a simple scheme plus starting sentence obviously exhibiting nonconvergence. Let the starting sentence be

> Moses wore a beard.

Let the relevant replacement rules of the scheme be

> Moses ⇒ the younger brother of Aaron

and

> Aaron ⇒ the older brother of Moses.

Plainly, under this scheme, the analytical process is unambiguous—the process tree never branches—but it continues endlessly, from the starting sentence, yielding, under stepwise analysis, first

> The younger brother of Aaron wore a beard.

then

> The younger brother of the older brother of Moses wore a beard.

and then

> The younger brother of the older brother of the younger brother of Aaron wore a beard.

and so on forever. This (single) branch in the process will never end.

An obvious objection to this toy nonconverging scheme proves ineffective: 'One can easily tell that this scheme generates an infinite branch. In other cases and in general, all you have to do is to check for such plainly circular schemes and eliminate them.' Using techniques to be introduced, it becomes straightforward to devise schemes of analysis for which infinite, nonconverging branches cannot be detected by any computable method. In addition, if, as Russell suggested in *PLA*, the replacement rules for a scheme of analysis reflect the route by which the meanings of such words as 'Moses' were, as a matter of fact, learned initially by the analyzer, it would be impossible to rule out, either a priori or a posteriori, such nonconverging schemes as the foregoing. Simple counting arguments suffice to show that the routes by which children in fact learn the meanings of names must be awash with circularities like that in the Moses-Aaron scheme, if perhaps not so readily discerned.

More sophisticated and telling examples of nonconvergent schemes can be constructed. Logicians will remember that for $m$ the number of a Turing machine and number $n$ an input to it,

$$I(m,n,p)$$

is a code representing the ***instantaneous state description*** or ISP of machine $m$ on input $n$ at precisely number $p$ steps in the computation of $m$ on $n$. In effect, $I(m, n, p)$ is a numerical 'aerial photograph' of what machine $m$ is doing at step $p$ in its computation on input $n$. In any of those state descriptions, either $m$ is converging with an expected output at step $p$ or it is not: that is a computably decidable matter. The question of convergence at $p$ (but not necessarily convergence in general, i.e., convergence for some $p$ or other) is answered easily via consultation of the relevant ISP $I(m, n, p)$.

Here is a scheme, based on Russell's idea for eliminating definite descriptions. For each triple of natural numbers $\langle m, n, p\rangle$, let

$$N_{m,n,p}$$

be a designator. The replacement rules are, for each triple,

$$N_{m,n,p} \Rightarrow \text{the predecessor of } N_{m,n,p+1},$$

for $p \geq 1$, provided that state description $I(m, n, p)$ does not show convergence, together with rules for the standard, narrow scope Russellian analysis of

$x$ precedes the predecessor of $N$.

The latter rules are, for any variable or designator $x$ and designator $N$,

$$x \text{ precedes the predecessor of } N \Rightarrow \exists y\left[\forall z\left(z \text{ precedes } N \leftrightarrow z = y\right) \wedge x \text{ precedes } y\right].$$

The starting sentences are $N_{m,n,0}$ precedes $N_{m,n,1}$. We assume here that every item that has a predecessor has only one.

It is easy to see that, for any $m$ and $n$, the analytical process generated by the scheme is perfectly unambiguous—there will be no branching at all—and that there are never any plain failures of the 'Moses and Aaron' type. The set of starting sentences is clearly computable, as are the scheme's rules for replacement. Also, the collection of logically proper names of the scheme is computable as well: one could write a simple program, using instantaneous state descriptions, to determine which of the names $N_{m,n,p}$ do not admit of further analysis: these are the 'logically proper' designators. However, it is not true that every process of analysis generated by the scheme converges: just start the process on a designator $N_{m,0,0}$ such that machine $m$ never gives an output on input 0. Such a designator is given by a Turing machine implementation of a computation rule like this:

Step 1: Set the current number to 2.
Step 2: Add one to the current number.
Step 3: If the current number is 1 or more, go to Step 2.

A crucial feature of the above scheme is that it is impossible, by any computable procedure or program, to select out and eliminate (or even to list computably) those designators $N_{m,n,0}$ on which, under this scheme, the process of analysis does not converge. This elimination would require a solution to the famous Halting Problem (Davis 1958: 70), known to be computably unsolvable.

## 6 Proofs of Convergence

What serious reasons can latter-day analytical philosophers provide for thinking that Russell's and Wittgenstein's ideas of analysis, once bugs and unclarities get cleared away and they are presented as schemes, actually work, actually converge on a wide range of ordinary, scientific, and mathematical sentences to produce complete logical forms? For the Wittgenstein of *Tractatus*, the sole conception of a logical atom seems to be that of the referent of a logically proper name. And the sole conception of that in turn seems to be as linguistic element in a completely analyzed sentence, a normal form, the final result of a convergent analytical process. Therefore, to show that logical atoms exist, one must, to use the words of Wittgenstein above cited, solve the 'mathematical' problem. In Russell, there was what seemed to be a conception of logical atom independent of logical form, that is, as item of acquaintance. If there is such an independent conception, a committed Russellian analyst will have to show that the required full analyses converge on sentences all the names in which refer to items of direct acquaintance only. Hence, there is here a higher standard to meet: to prove that the analyses of English sentences—scientific, ordinary, and mathematical—all not only converge but also do so in a normal form of the right sort. If the analyst has a conception of logical atoms independent of the details of convergent analyses and fully analyzed sentences, perhaps derived from the old ideas of sense data or of phenomenal primitives, for example, 'Pain here now,' it remains to prove that carefully formulated processes of analysis accept as inputs a wide range of sentences and succeed in reducing them all to sentences whose denoting terms stand for sense data or phenomenal primitives exclusively.

Some commentators maintain that Wittgenstein offered an argument a priori for the convergence of the analyses of complexes presupposed in his ($TLP_1$). In our terms, that argument has, as premises, the claims

1. the meanings of the starting sentences in any scheme of analysis are determinate, and
2. the sole way in which those meanings can be determined is via the convergence of the relevant Wittgensteinian analyses. *Vide* (Pears 1985).

Under that idea of analysis, as we have seen, the reduction of such a sentence as 'the broom is in the corner' requires the sentence, exposed at step one of the replacement process, 'the stick is stuck in the brush' to be true. Wittgenstein staunchly maintained that this could not, in general, be the case with every sentence arising in any procedure of analysis, at least if the starting sentence has a determinate meaning. In *Tractatus*, he wrote

> If the world had no substance, then whether a proposition had sense would depend on whether another proposition was true. It would then be impossible to form a picture of the world (true or false). (*TLP*$_I$: 2.0211, 2.0212)

That is, Wittgenstein held that no analytical process in accord with his ideas for reducing complexes to their components, once started on a meaningful sentence, could fail to converge, since this would require that the meaning of every sentence produced in the course of analysis would depend upon the truth of yet another, different sentence spelling out some relation crucial to the constituents mentioned in the analyzed sentence.

Certainly, this line of argument, particularly its main premise, is hard to square with simple facts of ordinary language. That the sentence

> My dog has fleas.

is meaningful does seem to depend upon the truth of several other sentences, among them,

> 'Fleas' is a word of English.

Often, proving convergence is, even in cases of schemes that appear trivial on the face of it, a tall order. Convergence proofs for programs that look relatively clear and simple can be extremely difficult. For example, metamathematical proofs of the consistency of formal systems for real analysis are proofs of convergence for algorithms of notable conceptual simplicity. Yet, they can call for infinitary ordinal numerations that are often mind-boggling in complexity (Schütte 1977; Buchholz et al. 1981). Surely, many readers already know that, in both practical and theoretical computing, a proof of termination or convergence of a process, computable or otherwise, is serious business. Most such proofs succeed, if they succeed at all, against a background that features at least one well-ordering, most often the well-ordering that is the standard 'less than' order on the natural numbers. In numerous cases, the proofs work by noting that each pass through the program (in the present essay, each application of a replacement rule) reduces in value some index or pointer to elements of the well-ordering. That accords with what 'well-ordering' means: a well-ordering can contain no infinitely order-descending path. It is important to emphasize that there appears to be no such order in play in the philosophical analyses now under examination. As we have just seen, sentence length will not do, since it is, in general, increasing as replacements are made. One prima facie obvious choice of well-ordering would be the joint lexicographic order on pairs of type and order indices for variables and terms in formulae from the ramified language of *Principia Mathematica*. If each application of the rules of an analytical scheme always reduced indices under the joint order, one could prove at once that the processes generated from that scheme converge in a finite number of steps. Unfortunately for that suggestion, the ideas of Russell and Wittgenstein seem neither to introduce nor, once introduced, to reduce *Principia* indices.

High among our aims is to demonstrate categorically that there can be no general proof of convergence, even for Russell/Wittgenstein analyses whose trees show no branching and whose single processing path converges. We offer our argument, first, in strictly mathematical terms and then explain how these examples can be sashayed into plainly natural language. Parenthetically, we must remark that mathematical language—in the present case, mathematics expressed in English—is an example of a perfectly natural language.

## 7 Schemes That Converge, But Not Provably

The set of formal proofs or derivations in standard set theory—each one finite with numbered lines—is enumerable computably: there is a program such that, for each $n$, it prints out a formal proof $P_n$ and, given $m$, follows that proof out to its step numbered $m$. Since a formal proof is finite in length, there is always a step numbered $m$, for some $m$, at which it ends. A Hilbert-style, linear, formal proof from the axioms of set theory, printed out by a computer, is also a complex à la Wittgenstein. It is a relational conglomerate of individual formulae arrayed on numbered lines, the relevant relations being spatial, as governed by the logical rule of *detachment*. One can therefore imagine an analysis of complexes that takes, as inputs, formalized metamathematical descriptions of the individual formal proofs. It then works algorithmically, line-by-line, through the descriptions of the proof, separating out component lines and pointing out their relations to previous lines. If a line contains a formal contradiction, we can imagine that the process is determined to loop without end, thereby failing to converge. If not, the process produces, computably from the input, a full analysis of the proof into its individual lines and formulae on those lines. This is certainly an analysis of obvious complexes into their significant parts and their interrelations. Furthermore, if set theory is consistent, the process will converge on every input: it never finds a contradiction on a line. However, set theory itself is quite incapable of proving that this process converges, and produces a normal form or full analysis on every starting complex, for example, formal proof.

Such a procedure is equally easy to capture using descriptions analysis. For each pair $\langle m, n \rangle$ of natural numbers, let $D_{m,n}$ be a descriptive name. Let the replacement rules for descriptions be, for each $m$ and $n$,

$$D_{m,n} \Rightarrow \text{the father of } D_{m,n+1}$$

provided that $P_n$ has at least $m$ steps or $P_n$ has a formal contradiction on a line appearing prior to step $m$. The starting sentences are

$$D_{m,0} \text{ is the father of } D_{m,1},$$

for each $m$. This set of schemes is clearly computable. There is no ambiguity, no branching in the processes. The collection of normal forms and that of the logically proper or unanalyzable names are computable as well. Since, as far as one can tell, set theory is consistent, no formal contradiction will ever be encountered in scanning its formal proofs. Therefore, all analytical processes on this scheme converge and normal forms are, in each case, eventually produced. However, there is no proof from the laws and deducible truths of set theory that the processes on the given sentences all converge. This arises from the application, to set theory, of Gödel's Second Incompleteness Theorem (Gödel 1931); set theory cannot prove formally the standard statement of its own consistency, that is, the statement that none of its formal proofs ends in a contradiction.

There is no obstacle to recasting these schemes in wholly natural language terms. The Tweedledees are a royal family, with each Tweedledee a king or a prince. Assume that every king Tweedledee has an eldest son, as does every male descendent, a prince, of every king Tweedledee. Here is the scheme. You see easily that it is equivalent to the above.

$$\text{Tweedledee } \alpha, \beta \text{ is the father of Tweedledee } \alpha, \gamma$$

provided that formal proof $\alpha$ in set theory has a least $\gamma$ steps or contains a formal contradiction before step $\gamma$. Here, all schematic variables $\alpha$, $\beta$, and

$\gamma$ are to be replaced by Roman numerals such that $\gamma$ always gets the Roman numeral for the successor of the number denoted by the Roman numeral inserted for $\beta$. So, one of the rules tells us, for example, that

$$\text{Tweedledee I, CI} \Rightarrow \text{the Father of Tweedledee I, CII.}$$

The starting sentences are all sentences of the following form, wherein $\alpha$ gets replaced by suitable Roman numerals.

$$\text{King Tweedledee } \alpha\text{, I is the father of Tweedledee } \alpha\text{, II.}$$

By the way, one could well imagine a determined (and set-theoretically obsessed) subject in the kingdom of Tweedledee learning the names, past and present, of the members of his ruling royal family by working precisely the replacement rules that the foregoing scheme would have him do. He would be able, in truth and in principle, to grasp in that way the meaning of each and every such royal name but would, from his set-theoretic mathematics, be unable to prove that fact.

Finally, we note that a mild variation on the above scheme would yield a new scheme in which every attempted process of analysis from it converges, but the analyzer is wholly incapable, in general, of picking out those names that are not further analyzable, that is, the names that appear in a normal form and are proper logically.

## 8 General Results on the Convergence of Schemes

*Definition* (*rewriting system*) Given a finite alphabet, a rewriting system is any decidable collection of consistent rewriting rules of the form

$$X\alpha Y \Rightarrow X\beta Y,$$

where $X$ and $\Upsilon$ are schematic variables for any strings from the alphabet, including the null string, while $\alpha$ and $\beta$ any fixed strings from it.

*Example* Over the alphabet of binary digits 0 and 1,

$$X11Y \Rightarrow X00Y$$

is such a rewriting rule, licensing the replacement of the substring 11 by 00 in any word.

Naturally, when set to work on any starting string of the background alphabet, a rewriting system will generate an obvious computable process, moving step-by-replacement-step.

*Theorem 1* Every (Turing) computable procedure is faithfully mimicked stepwise by some rewriting system, where the rewriting system can be discovered, from the original procedure, via a provably computable and correct transformation.

*Proof 1* The proof is standard in the literature, making use of the fact that every Turing machine can be reconceived as a replacement system that operates on the alphabet in which the instantaneous state descriptions of the original Turing machine can be written (see, e.g., Davis 1958: 94). ■

*Theorem 2* Every rewriting system can be mimicked in its stepwise processing by some scheme of analysis, where the scheme can be discovered, from the original rewriting system, via a transformation that is provably computable and correct.

*Proof 2* Given a rewriting system, redescribe the abstract strings over its alphabet using sentences of standard English describing the original strings. For instance, the displayed string

$$ABC$$

is rendered as the English sentence

$$A \text{ precedes } B \text{ precedes } C.$$

Then, think of the rewriting rules in terms of correlative replacements, such as

$$A \text{ precedes } B \Rightarrow C \text{ precedes } D.$$

In this example, we think of the expression

$$A \text{ precedes } B$$

as describing a complex that is to be analyzed into the complex described as

$$C \text{ precedes } D$$

with components $C$ and $D$ standing in the

$$\lambda x \lambda y. x \text{ precedes } y$$

relation. ■

*Theorem 3* (Gödel) If set theory is consistent, there are (Turing) computational procedures that converge or halt on all relevant inputs, but cannot be proven, even in set theory, to converge on all relevant inputs.

*Proof* There is a computable procedure that checks formal proofs from the axioms of Zermelo–Fraenkel set theory to see if they are proofs of a formal contradiction, and converges on an input proof whenever it fails to lead to a contradiction. By Gödel's Second Incompleteness Theorem, if set theory is consistent, then this procedure converges on all relevant inputs but cannot, in set theory, be proved thus to converge. ■

*Corollary* There are schemes of analysis, applying to recognizably English sentences, that converge (i.e., produce a complete logical form) on all relevant inputs, but cannot be proven, even in set theory, to converge, that is, to produce complete logical forms for all the allowable starting sentences. ■

These example schemes cannot be barred from consideration on the grounds that their replacement rules are imaginary or contrived rather than naturally occurring. Since analysis was intended to be a logical process, it must apply to bizarre starting sentences and replacement rules as well as more familiar ones. It must apply to every meaningful sentence of a language, and yield a logical form—whether that sentence reports on the real or on the unreal, for example, 'The present King of France is bald.' After all, the traditional ideas of analytical philosophers were meant to expose the true logical forms undergirding *all thought*.

## 9 Conservativeness: A Note

So far, we have conceived inputs and outputs to ideas and schemes of traditional analysis as strictly and individually syntactical: items for mindless programs to manipulate. We have not thought of them either as bearers of truth or falsity, or even as representing assumptions in derivations from a formal or informal theory. The progenitors of analytical philosophy, however, conceived them as both: as yielding true principles auxiliary to theories ruled by sound deductions. *Vide* (*$PM_2$*: 173ff). Official expressions of ideas for analysis were conceived to take on the status of definitions. As such, they need to satisfy constraints ordinarily imposed upon definitions. At a minimum, they must, over the background theory in the original language to which they are appended, be provably *conservative*: under deduction, they must yield no theorem in the original language that the background theory could not have proven on its own (Enderton 1972: 154–163). Therefore, even if one could prove the convergence of analytical schemes, one has also to prove that the stated results of the replacement processes are conservative over the original background theory. Here, we can only assert that, in general, this will not be possible. It is easy to construct example schemes that are demonstrably nonconservative over reasonable and well-known background theories. Alas, we must postpone these demonstrations for another time.

## 10 Objections and Replies

### *10.1 Objection 1*

*It is unfairly anachronistic to endeavor to refute a philosophical view by employing mathematics or science unknown at the time at which the view was originally promulgated.*

*Reply 1* First, this manner of objection is absurd, since it suggests that we cannot refute, and soundly, Hegel's opinion that there are no more than five planets by referring to the basic astronomy school children have learned since Hegel's day. We need not restrict ourselves to the moral thinking of Aristotle's time to show that his views on human slavery are utterly mistaken.

Second, appeals to natural number functions that are computable, to those that are noncomputable, and to their respective mathematical properties are not, in the cases of Russell and Wittgenstein between 1900 and

1922, anachronistic. Noted German mathematician and philosopher Paul du Bois-Reymond (1882) distinguished plainly between natural number sequences that can be computed and those that, on pain of violating laws of physics, cannot. Du Bois-Reymond (1875) gave a proof that employs, at its heart, an argument that is plainly a diagonalization, that is, plainly of the same sort as those underwriting the famous incompleteness and uncomputability results of Gödel and Turing.

### *10.2 Objection 2*

*In order to be fair to Russell, you must submit his views to intellectually generous interpretation. This you have not done.*

*Reply 2* There is nothing at all ungenerous about our work in this article. On the contrary, we have performed a definite service. Such a philosopher as Russell would have known, and known well, how great a service it is to be refuted capably. We leave it to the objector to show how a business of reinterpretation, without adding further and significant mathematical constraints to the ideas of analysis here described, would reveal that our arguments are not probative or do not properly apply. If we have committed a definite error in our statements about Russell's many views, and that error affects the present mathematical reasoning, it is up to the champions of Russellian analyses to tell us, and plainly, what that error is.

### *10.3 Objection 3*

*The analysis of a sentence is a process that requires hard philosophical work, insight, effort, and sometimes luck. It will not, therefore, submit to mechanization.*

*Reply 3* The treatment we employed above, the representation of vague ideas of analysis by mathematically delineated schemes of analysis, no more applies to the discovery of the analysis of a sentence or range of sentences than does the completed formalization of the proof of a mathematical theorem represent the process by which the proof was originally discovered. The first discovery of a proof or an algorithm that solves a mathematical problem may indeed call for hard work, insight, effort, and luck, but its later, completed formalization—or the implementation of that formalization—will most likely not. Formalizations and their implementations treat the proofs or algorithms as already discovered. More to the point, the analyses of Russell and Wittgenstein were meant, once discovered, to be applicable to the same range of sentences over and over again in fixed, established patterns, as with the definite descriptions analysis. It is this unchanging pattern that our schemes are intended to capture.

Moreover, as the world has already seen in the cases of automated theorem provers and robotic chess masters, an undertaking that can seem the very epitome of human intellection, effort, and insight can be formalized within a high-level program language and implemented.

Third, Russell clearly intended his analytical ideas, as applied to scientific and ordinary sentences of English, to lead to computable processes (or processes computable relative to a delimitable set of data). To admit that and then question the suggestion that these same symbolic processes cannot be carried out on abstract computing devices such as Turing machines (or Turing machines with oracles) would be to call into question, at least in this case, the Church–Turing Thesis: that every humanly computable symbolic procedure can be represented on a Turing machine tape and carried out, on that tape, by a suitable Turing machine.

## *10.4 Objection 4*

One can argue for the convergence of methods of analysis as follows:

1. It is indisputable that thought occurs.
2. All thought is either direct thought or indirect thought.
3. Indirect thought is only possible given that direct thought is possible. To deny this leads immediately to an infinite regress.
4. Finally, how direct thoughts relate to indirect thoughts so that the former can make possible the latter is via analysis. Any indirect thought must somehow be analyzable into direct thoughts.

Therefore, analyses *must* converge or else thought itself would not be possible.

*Reply 4* Experience with the literature leads one to suspect that the notion of direct thought here intended could be explicated only with the help of such concepts as *logically proper name* or *epistemically proper name*. How, then, does one show that these latter concepts are at all coherent and really apply to something without providing full, convergent analyses for a large range of sentences? Even if that hurdle is overcome, it has yet to be shown that it is analyses of the sorts that Russell and Wittgenstein once had in mind, rather than some other, yet to be discovered kind of analysis, that throws the required bridge across the chasm between indirect and direct thought.

**Acknowledgments** An early version of this article was presented at the Obermann Center conference on Logical Atomism in Iowa City, Iowa, 12–16 June 2017. The authors wish to thank the conference patrons, organizers, and especially those participants who so graciously, and often forcefully, offered us comments. We also wish to thank Wade Munroe and Andrew Smith who also endured an earlier version.

> *Yet it would perhaps be thought to be better, indeed to be our duty, for the sake of maintaining the truth even to destroy what touches us closely, especially as we are philosophers or lovers of wisdom; for, while both are dear, piety requires us to honor truth above our friends.* (Aristotle 2009: I.6)

## References

### Works by Other Authors

Aristotle, tr. W. D. Ross (2009). *The Nichomachean Ethics.* Oxford: Oxford University Press.

Buchholz, Wilfried, Solomon Feferman, Wolfram Pohlers, and Wilfried Sieg (1981). *Iterated Inductive Definitions and Subsystems of Analysis: Recent Proof-Theoretical Studies.* Berlin: Springer-Verlag.

Davis, Martin (1958). *Computability and Unsolvability.* New York: McGraw-Hill Book Company.

du Bois-Reymond, Paul (1875). Ueber asymptotische Werthe, infinitäre Approximationen und infinitäre Auflosung von Gleichungen. *Mathematische Annalen*, Vol. 8: 363–414.

du Bois-Reymond, Paul (1882). *Die allgemeine Functionentheorie.* Tübingen: Verlag der H. Laupp'schen Buchhandlung.

Enderton, Herbert B. (1972). *A Mathematical Introduction to Logic.* New York: Academic Press.

Gödel, Kurt (1931). *On Formally Undecidable Propositions of* Principia Mathematica *and Related Systems I.* In *Collected Works: Volume I, 1929–1936*, eds. Solomon

Feferman, John W. Dawson, Jr., Stephen C. Kleene, Gregory H. Moore, Robert M. Solovay, and Jean van Heijenhort. Oxford: Oxford University Press: 1986. 144–195.
Grabmayer, Clemens, Joop Leo, Vincent Van Oostrom, and Albert Visser (2011). "On the Termination of Russell's Description Elimination Algorithm." *The Review of Symbolic Logic*, Volume 4, Number 3: 367–393.
Kripke, Saul (1977). "Speaker Reference and Semantic Reference." In *Contemporary Perspectives in the Philosophy of Language*, eds. French, Peter A. and Theodore E. Uehling Jr., and Howard K. Wettstein. Minneapolis, Minnesota: University of Minnesota Press. 1983: 6–27.
Kripke, Saul (2005). "Russell's Notion of Scope." *Mind*, Vol. 114: 1005–1037.
Lewis, Harry R. and Christos H. Papadimitriou (1981). *Elements of the Theory of Computation.* First edition. Englewood Cliffs, NJ: Prentice-Hall. "Nondeterministic Turing Machines": 204–211.
Pears, D. F. (1985). "Introduction." In *The Philosophy of Logical Atomism*, Bertrand Russell, ed. D. F. Pears. Open court: La Salle, Ill. 1–34.
Schütte, Kurt (1977). *Proof Theory.* Grundlehren der mathematischen Wissenschaften. Band 225. Berlin: Springer-Verlag.
Turing, A. M. (1937). "On Computable Numbers." *Proceedings of the London Mathematical Society* (Series 2), Vol. 42: 320–265.
Turing, A. M. (1939). "Systems of Logic Based on Ordinals." *Proceedings of the London Mathematical Society* (Series 2), Vol. 45: 161–228.
Wittgenstein, Ludwig (1953). *Philosophical Investigations*, Tr. G. E. M. Anscombe. New York: The Macmillan Company.
Wittgenstein, Ludwig *TLP*$_2$ (1974a). *Tractatus Logico-Philosophicus*, Second edition. Tr. D. F. Pears and B. F. McGuinness. London: Routledge & Kegan Paul Ltd.
Wittgenstein, Ludwig (1974b). *Philosophical Grammar.* ed. R. Rhees, tr. A. Kenny. Berkeley, California: University of California Press.

CHAPTER 13

# The Underlying Presuppositions of Logical Atomism

*Richard Fumerton*

## 1 Introduction

There are all kinds of interesting questions concerning what ultimately motivates logical atomists to conclude that it is both important and possible to analyze all meaningful statements into an ideal language whose key terms refer to logical atoms. It is tempting to think that Russell's interest in logical atomism is tied closely to his conviction that direct acquaintance is not only the key to understanding direct knowledge but also the key to understanding what I call direct thought. I think a version of that view is correct, but that one should keep distinct the question of how to understand the distinction between direct and indirect thought from more specific ideas about how to locate the objects of direct thought. I'll try to suggest that there are good reasons for divorcing the idea of direct thought from what I take to be the view that Russell never gave up, the view that the analysis of direct thought involves the idea of direct acquaintance.

R. Fumerton (✉)
University of Iowa, Iowa City, IA, USA
e-mail: richard-fumerton@uiowa.edu



L. D. C. Elkind, G. Landini (eds.), *The Philosophy of Logical Atomism*, History of Analytic Philosophy,
https://doi.org/10.1007/978-3-319-94364-0_13

## 2 Logical Atomism

Before discussing what is and isn't central to the doctrine of logical atomism, one might suppose that we should get clear about what precisely the tenets of logical atomism are, but that is not an easy task. Indeed, it might be futile to suppose that the question has a definitive answer. Like most "isms," logical atomism might be best construed as involving a number of theses, no one of which is critical to embracing a version of the view. Consider, for example, another famous "ism," empiricism: Which doctrines are critical to embrace if one is to fall into the class of empiricists? Certainly, some paradigmatic empiricists are committed to the Humean view that all simple ideas are "copies" of prior impressions. But, even Hume allowed that one might be able to form the idea of a "missing" shade of blue. His thought was that one might have experienced a smooth continuum of shades of blue, notice an odd "gap," and fill in the gap with an idea—still a simple idea—that would smooth out the continuum. Furthermore, while Hume embraced some version of the view that simple ideas are copies of prior impressions, he was unequivocally committed to the conclusion that if true, this "cornerstone" of his philosophy was only contingently true. And the only grounds he could consistently recognize for believing it at all are inductive. I presume he surveyed his own simple ideas and seemed to remember some prior impression to which each of them corresponded. The induction in question is particularly problematic in that it would seem to involve hasty generalization. After all, just because he could not find any ideas of his that were simple but not copies of prior impressions, why would he think it reasonable to conclude that no one else had such ideas? What's even worse, I take Hume's skepticism with respect to induction to be genuine. If so, the cornerstone of his empiricism is a doctrine that he himself should believe that he has no reason to accept.

Are empiricists committed to the view that all necessary truths are analytic? It's hard to say so, of course, because the most well-known empiricists didn't employ that terminology. To take Hume as an example, he did seem to embrace the conclusion that all necessary truths have as their truthmakers relations between ideas, but some of those necessary truths don't naturally fall into our contemporary category of analytic truths. One of those relations between ideas that can ground a necessary truth for Hume is the relation of difference. The difference between the idea of being blue and being red would ground the necessary truth that being blue is different from being red. But if an analytic truth is one the statement of which can be reduced to a tautology through substitution of synonymous expressions, this isn't a very plausible example of an analytic truth.

Rather than trying to give necessary and sufficient conditions for being a logical atomist, perhaps it would be much more profitable to spend our time and energy trying to figure out what specific views were endorsed by this or that historically influential philosopher, and, more importantly trying to figure out what motivated that philosopher to embrace those views. Russell was one of the most famous philosophers who endorsed some version of atomism; we might look at his commitments as a way of identifying at least the sort of commitment that would be usefully characterized as a commitment to logical atomism. There is, however, significant disagreement as to just what Russell's view was and whether he remained resolute over time in his defense of it. When trying to understand a philosopher's position over time, however, it has always seemed to me rather natural to pay attention to that philosopher's own reflections on his or her past views. In *My Philosophical Development* (1959), Russell says the following:

> I have maintained a principle, which still seems to me completely valid, to the effect that, if we can understand what a sentence means, it must be composed entirely of words denoting things with which we are acquainted or definable in terms of such words. It is perhaps necessary to place some limitation upon this principle as regards logical words—e.g. *or, not some, all.* We can eliminate the need of this limitation by confining our principle to sentences containing no variables and containing no parts that are sentences. In that case, we may say that, if our sentence attributes a predicate to a subject or asserts a relation between two or more terms, the words for the subject or for the terms of the relation must be proper names in the narrowest sense. (*MPD*: 125–26)

It is clear that the proper names in question are what Russell sometimes called *logically* proper names—names whose meaning for a person S is something with which S is directly acquainted. The view is explicitly stated in terms of conditions for understanding language, but it seems obvious to me that for Russell, it is *thought* that breathes life into language. And one could just as easily express the heart of the view without appealing to features of our understanding of language at all, appealing instead only to the conditions required for the possibility of thought. The idea is that all complex thought is composed of simpler thought that succeeds in being about some object only in virtue of the object being an object of direct acquaintance. Put more straightforwardly still, thought of X, at least what I call *direct* thought of X, just is our standing in a relation of direct acquaintance to X. There is certainly a sense in which I can think about your favorite color and form the belief that it is not pink even when I don't know what your favorite color is. I can think of that color indirectly as the color (whatever it is) that you like more than any other. I can think about Jack the Ripper and form the thought that he was truly depraved, even though there is also a *sense* in which I don't know who Jack the Ripper was. I can think of that person (whoever he was) as the person who committed those atrocities in London near the end of the nineteenth century.

The commitment stated above captures, it seems to me, at least one form of logical atomism. And it also explains why the atomist is so concerned with employing the method of analysis as a tool for improving one's understanding of a given assertion. The understanding one gets from reducing an ordinary sentence to a sentence in an ideal language will often enable one to solve philosophical puzzles and avoid philosophical error. As Landon D. C. Elkind plausibly suggests in his thesis, *The Search for Logical Forms: In Defense of Logical Atomism*, Russell's theory of definite descriptions and its employment in understanding ordinary sentences as a good illustration of how analysis can dissolve philosophical puzzles. It is surely *tempting* to adopt the so-called Millian view of ordinary proper names[1] according to which the meaning of a name is its referent. But, on reflection the view has problematic implications. The sentence "Zeus does not exist" is obviously meaningful, but how could it be if to be meaningful "Zeus" must have a referent? "Mark Twain was Samuel Clemens" is surely meaningful, true, and *informative*, but how could it be informative if understanding the sentence involves grasping the meaning of the proper names and the meaning of the proper names just is the one single referent of the two names. I can believe that Mark Twain was an author without believing that Samuel Clemens was an author, and how is that possible if the very same individual is the one object of thought partially constitutive of both belief states? Though hardly universally accepted these days, Russell's suggestion was that we can solve all of these problems by understanding ordinary proper names as disguised definite descriptions, and by in turn offering his famous analysis of the meaning of definite descriptions. If "Zeus" just means something like "The divine being recognized by Greeks as the father of the Olympians" and sentences involving that description assert the existence of one and only one being exemplifying the relevant property, then we can understand the assertion that Zeus doesn't exist as simply the assertion that there isn't a unique thing exemplifying the relevant properties. Problems solved, and solved without a voyage into the exotic land of Meinong where we search for non-existent "objects" to serve as the referent for ordinary names that refer to no existing objects.

The lessons learned from an application of the theory of descriptions to our understanding of ordinary names can also be applied to the use of most indexicals. If we solve the problem of informative identity claims involving ordinary names by construing the names as "disguised" definite descriptions, then we will presumably find attractive the same solution to the problem of informative identity claims involving demonstratives. I see two ends of a tangled piece of rope and assert that this rope is the same rope as that rope. I see a person and a reflection in a mirror and think that he (pointing at the person) is he (pointing at the person in the reflection). Such claims are informative and, in my view, contingent. The best way to explain how this can be so is to treat the terms flanking the identity claim as disguised definite descriptions.

Indeed, if we employ this criterion to decide when an expression is a disguised description, we'll probably be led eventually to the conclusion that subject terms in a natural language are almost always disguised descriptions. On Russell's view, the exception might be subject terms denoting properties. Some predicate expressions and some names for properties (like "being phenomenally black") are, Russell thought, plausibly construed as expressions whose meaning is the property they pick out. But only some are like this. For just as we can pick out things that are not properties using descriptions that denote those entities, so also we can pick out *properties* using descriptions that denote those properties. We can think of a property indirectly as the property, whatever it is, that uniquely exemplifies certain higher-order properties.

It is a bit misleading to offer the theory of definite descriptions as a paradigm of atomistic commitments working at their best. One can obviously accept the theory of definite descriptions and its application to ordinary proper names and most uses of indexicals, and extoll its success at solving philosophical puzzles, without embracing logical atomism. A great many philosophers did precisely that. The theory of descriptions itself doesn't tell you how far along the reduction should continue, after all. Russell's own examples of the theory at work virtually never achieved the *atomist's* goal of reducing the meaningful statements to statements whose terms all referred directly to items with which we are directly acquainted. Indeed, a great many philosophers with widely differing views about meaning all thought that analysis, broadly construed, is a tool for solving philosophical puzzles that arise from the often misleading "surface grammar" of ordinary sentences. Ryle, for example, was no logical atomist. But, he did try to lead you away from philosophical error by "translating" ordinary sentences in ways that allow us to grasp more easily their meaning.

Russell himself (*LA*: 164) advanced a more generic maxim than the one tying direct reference (and, I would argue, direct thought) to acquaintance: "Wherever possible, substitute constructions out of known entities for inferences to unknown entities."[2] The maxim is cryptic enough that it allows for a number of different interpretations of the key terms: "known entities," "unknown entities," and "constructions." But if the known entities he refers to are items directly known, then the maxim could easily lead us right back to direct acquaintance. For Russell, direct knowledge that some X exists requires that we be directly acquainted with X. The constructions in question would be complex propositions built out of simple atoms known through direct acquaintance. Notice, however, that the maxim concerning reduction seems a bit more cautious than the principle I quoted from *My Philosophical Development*. The maxim only enjoins one to carry out the reductions "wherever possible." And that might at least contextually imply that he was acknowledging that it might not always be possible. This raises the obviously critical question concerning criteria for successful reduction. Here Russell scholars might well disagree. I don't have anything particularly original to suggest other than that a reduction will be successful only if the analysans captures the meaning of the analysandum. For Russell's theory of definite descriptions to succeed, for example, his reduction of the sentence containing the definite description to a claim involving quantifiers must succeed in capturing what we were trying to say using the definite description. The reduction is philosophically helpful because we don't always understand what we are saying and the reduction can reveal to us a meaning that was, in some sense, "hidden."[3]

The point I want to stress here is only that one can embrace the value of analysis, and even insist that analysis must terminate without embracing the details of Russell's own view about how to identify the end point of analysis. And even if one is convinced that acquaintance plays a critical role in identifying the point at which ideal analysis will end, there are at least two importantly different ways in which one might conceptualize that role. Let me explain.

## 3 Direct Reference and Direct Thought

The foundationalist in epistemology argues that all knowledge and justification is parasitic upon foundational knowledge and justification. The most famous argument for that view is the regress argument. If the only way to know a given proposition is to infer that proposition from another proposition, problematic regress threatens. One can form a justified belief that P by inferring P from E only if one has justification for believing E. Much more controversially, one might also argue that to be justified in inferring P from E one must also have some sort of epistemic access to a relation of making probable holding between E and P (where entailment can be viewed as the upper limit of making probable). But if the only way to justify a belief is to infer what is believed from some other different proposition, then one could be justified in believing E only if one in turn justifiably inferred E from some other proposition F, and, so on, ad infinitum. If we also need justification for believing that the relevant evidential connections hold, the regresses proliferate indefinitely. If all justification is inferential, then to be justified in believing anything one would need to complete not just one, but an infinite number of infinitely long chains of reasoning. And it doesn't even make sense to talk about *completing* an infinitely long chain of reasoning. Just as God couldn't finish counting the natural numbers, so also even God couldn't *complete* an infinitely long chain of reasoning. So either there are beliefs that are noninferentially justified or the most radical of all skepticisms is true—one has no justification for believing anything at all to any extent whatsoever.

Just as one might use the threat of vicious regress to argue that there must be a foundation for knowledge and justified belief if there is to be any knowledge or justified belief at all, so also one might appeal to a structurally similar threat of regress to argue that reference and thought must have foundations if there is to be any reference and thought at all.

This dramatic claim about a foundation for thought is plausible because of the following prima facie compelling idea. As we saw earlier, much of what we think about we think about only *indirectly*—only through our thought of something else. I can think of your favorite color without knowing what it is because I can think of you, I can think of being colored, and I can think of preferences. It may be that I can think of you, in turn, only by thinking of the bearer of certain properties. It may be that I can think of some properties only by thinking of those properties as the bearers of still higher-level properties (e.g. the property whose exemplification causes certain things to happen). But, this can't go on forever. There must be some things I can think of directly. There must be thoughts that have the capacity to correspond to some aspect of reality—thoughts that have an intrinsic nature that allow them to correspond to one and only one aspect of reality. My thought of pain, it seems to me, is a good candidate for *foundational* thought. When I think of a certain sort of searing pain, I'm not thinking of the pain only as whatever it is that plays a certain functional role, or whatever it is that causes a certain sort of behavior, or whatever it is that is typically caused by damage to a body. I probably do think that pain has a functional role to play, that pain does cause behavior, and that it is usually caused by damage to bodily tissue, but my access in thought to the pain isn't through any of these other thoughts I have. It is not like my thought of your favorite color when I don't know what that color is.

So, there is a kind of regress argument for foundational thought that resembles in some respects the regress argument for foundational justification. There is such a thing as indirect thought. As I noted earlier, I can think of some things only through thinking of others. I can think of Jack the Ripper and I can form the thought that he was insane. When I think of Jack the Ripper, I am thinking of that person (whoever it was) who committed those atrocities in London in the late nineteenth century. But it can't be the case that all thought is like that. If for all x, thinking of x always involves thinking of something else y, then every thought would be infinitely complex—complex in a way that precludes its very possibility.

Just as the conclusion that there is a foundation for justification and knowledge leaves open the question of how to understand such foundations and leaves open the question of which beliefs are foundational justified/known, so also the conclusion that there is foundation for thought leaves open the question of how to understand foundational thought, and leaves open the question of which thoughts are foundational. And as we noted above, it is here that Russell and others think that the relation of acquaintance plays a fundamental philosophical role. It seems plausible to some of us to suppose that the extension of foundational thought is restricted to items with which we are, or have been, directly acquainted. So to return to the example of pain, I am able to think of pain directly, but only because I have been directly acquainted with pain. I am able to think of phenomenal redness, but only because I have been directly acquainted with phenomenal redness. An extreme version of empiricism, might restrict the class of entities about which one can think directly to those that have been introspectively accessible. But if, like Russell, one concludes that one can be directly acquainted in thought with such abstract entities as numbers, one might also suppose that one can have direct thought of them.

If one can form a direct thought of some particulars and properties and also form a direct thought of causal connection, then one can "build" indirect thought out of direct thought. I can think not only of pain, but of whatever it is that causes pain. I can think of being appeared to in that phenomenally red way, and also think of whatever it is that plays a certain causal role in producing that phenomenal redness. But, this already oversimplifies the way in which direct thought can be about something. The immediate problem is that there is no one event or state of the world that causes pain or an appearance of phenomenal redness. There are causal chains leading to these effects, and each link in the causal chain has as good a claim to being a cause of the relevant effect as do the others. So, if we hope to pick out surface properties of an object when we think about the cause of a given visual experience, we will need to "triangulate" on the relevant property.[4] We'll need to think of the common cause of both this and other experiences of red. That common denominator (whatever it is) might be the property we are trying to think about what we think about the redness of a physical object.

I have so far talked about acquaintance with X as a condition for having a direct thought of X. I still haven't said anything by way of characterizing what direct thought is. There are at least two possibilities. One is that acquaintance figures into the very *analysis* of direct thought. As I said, I think that this was always Russell's view.[5] To think of X directly just *is* to stand in a relation of direct acquaintance to X. To be plausible, such a view would probably require, among other things, that there be universals, and that we can be directly acquainted in thought at a given time with such universals even without our being directly acquainted at that time with the exemplification of the universals. So, I can still think directly of the searing pain I experienced five years ago, because I can still hold "directly" before my mind the property of being in such pain. The only alternative for the acquaintance theorist is to suppose that something like that property of being in pain is exemplified in the very imagination of the pain. If Hume were right, for example, and simple ideas are "pale copies" of what they are ideas of, then in forming the idea of pain we might experience something very much like the pain—just less vivid. But phenomenologically that just seems wrong to me. Fortunately for us, we can remember (and thus think about) severe pain without experiencing anything remotely like the severe pain.

If we analyze direct thought of, say, a property, as direct acquaintance with the property, we might still suspect that it would be causally impossible to stand in such relations with the relevant properties without having at least once been directly acquainted with the exemplification of the property.[6] The intuition that Jackson's color-deprived Mary would never have been able to think directly of phenomenal redness is shaped largely by the idea that without having actually had the experience in question Mary would never have been able to think of that property in the *direct* way she could after having had the experience. But, I take it that having had the experience once, Mary would continue to be able to think directly of phenomenal redness as long as she was able to remember what that experience was like. The intuition pushed by Jackson's thought experiment is perfectly compatible with its being the case that all along, even before her color experiences, Mary's was able to think indirectly of phenomenal redness. She was able to think of that experience (whatever it is) that people have introspective access to as they look at apples, as their brain undergoes certain changes, that causes them to say such things as "That looks red," and so on. But, we don't think she could think of phenomenal redness *directly*—think of it in the way that those not color-deprived could.

As David Lewis (2004) pointed out, however, the subjunctive conditional that we think is true doesn't seem to be necessarily true in any strong sense of "necessarily." We can conceive of a world in which color-deprived Mary did form the thought of a phenomenally red experience even without ever having had it. We can conceive of a world in which a pain-deprived Mary was able to think directly of pain even without ever having experienced pain. We—at least ones who reject externalist accounts of thought and its object—just don't think that as a matter of contingent fact Mary would form such thoughts.

So, when we think that there is a connection between having been directly acquainted with something and being able to think of it directly, we might only be thinking that the former is something like a *causally* necessary condition for the latter. Or perhaps, we might be thinking only that the former, given certain conditions concerning how our brains currently work and the current state of neural science, is causally necessary for the latter.[7] If the relation between direct acquaintance and direct thought is contingent in this way, then we will need to search elsewhere for an analysis of what direct thought is. That search might take us to a view similar to one embraced by Gustav Bergmann (1964, 1967) and his student Laird Addis (1989), a view that inspired in some respects my own view (Fumerton 2002) about the nature of thought. On their view to have a thought of X is to exemplify a certain property, where the exemplification of that property has the capacity to correspond (in the case of true thought) to some aspect of the world. There are thoughts that don't correspond to anything, but they still have the *capacity* to "reach" out to some feature of the world. The property, exemplification of which just is thinking, is unanalyzable. One can identify properties of the property, but one can't *define* the relevant property. But, we all know what it is by virtue of our acquaintance with our exemplification of the property.

The above view leaves open the conditions that are causally relevant to the formation of such thought. Neither Bergmann nor Addis embrace my conception of acquaintance and thus, trivially, would not share speculations about direct acquaintance being a causally necessary condition for the formation of direct thought. They both treat direct acquaintance as just another species of thought (every thought has both a content and is of a given species for them), where on my interpretation of their views, to have a thought is just to exemplify a non-relational property, albeit one that has the capacity to correspond to some feature of the world.[8]

Which "acquaintance" conception of direct thought is more plausible—one that analyzes direct thought in terms of direct acquaintance, or one that embraces the much more cautious thesis that direct acquaintance might be a contingently necessary condition for direct thought? The candid answer is I've never been sure. It is one of those questions in philosophy the answer to which is closely tied to the answer one should give to a host of other difficult philosophical questions. This much seems clear to me. As I said earlier, one can now think directly of properties the exemplification of which one is not *currently* acquainted. Furthermore, if that is possible it seems equally plausible to suppose that one can think directly of a property that has *never* been exemplified—Hume's missing shade of blue, for example. That, by itself, is no obstacle to the Russellian conception of direct thought provided that one can identify properties with universals, provided that universals can exist even if they have never been instantiated, and provided that one can be directly acquainted with such universals. Although there are serious dialectical pressures to embrace the theory that there are uninstantiated universals, I am still attracted to the view that at least some properties are best treated as tropes, where what makes a class of tropes all red tropes is some sort of resemblance between them. We might need to acknowledge certain *relations* as universals, but the question is complicated. In any event, until I become convinced that there are uninstantiated universals with which one can be acquainted in thought, I'm still inclined to the adverbial theory of thought I described above—a theory that doesn't analyze direct thought in terms of direct acquaintance with its object, but allows that past direct acquaintance might be a causally necessary condition for formation of the thought.

## 4 Externalist Accounts of Thought

Acquaintance theories of direct thought are not the only, nor are they now the most popular, attempts to understand how thought reaches out to the world. These days content externalism is coming close to being the received view. Kripke, Putnam, Dretske, and many others are convinced that to understand how at least some language and thought succeeds in being about some entity X we need to look to the causal history of language use and neural activity. After all these years, it still seems to me that the basic idea behind the causal theory is little more than a sketch. But the sketch is just this. When I use a term like "Josephus" in the sentence "Josephus was an historian" I am referring to that person (whoever it is) that figured in the initial link of a long causal chain culminating in my use of the token "Josephus" in that sentence. When I think of Josephus there is some neural activity that is the ultimate effect of that same sort of causal chain that begins with some tokening of Josephus. Kripke (1980) always denied that he was giving an overall theory of reference, though his explanation of his reticence was always a bit cryptic. One suggestion is that he had no real account of the initial "baptism" that secured the initial reference of a term or a thought token. And here, one might suggest that acquaintance still has a role to play. One could (though one needn't) insist that it is only in virtue of being acquainted with some item that one can one fix a label to it or secure a thought about it; it is only in virtue of acquaintance that one can secure the first critical link in a causal chain that enables people thousands of years later to refer to the same item. But, even if one conceded this much, the resulting theory would be a far cry from the kind of acquaintance theory that Russell advanced. That acquaintance theory requires acquaintance that is cotemporaneous with direct reference and direct thought.

This is not the place to rehearse arguments for and against content externalism. Elsewhere (Fumerton 1989) I have argued that one can "steal" all that is plausible from a causal theory and incorporate into a

Russellian descriptive theory of ordinary proper names. The resulting theory will have the advantages of two philosophical worlds. It will solve the problems of informative identity and referentially opaque sentences while accommodating the externalist's insight that traditional descriptivists don't have the resources to secure for many people plausible definite descriptions that can capture the meaning of the ordinary proper names they use. Furthermore, though the issue is complex, I have argued (2003) that content externalism can't handle the datum that we sometimes have unproblematic access to the content of our thoughts.

## 5 Conclusion

There must be foundations for knowledge and thought if knowledge and thought are possible. This, I believe, is one of the fundamental insights of logical atomism. Furthermore, direct acquaintance enters into the most plausible story of what makes foundational thought possible. But, there are two quite distinct theories that identify quite distinct roles for acquaintance to play. On the Russellian view, direct thought is identified with direct acquaintance with what the thought is about. On the Fregeian view, past direct acquaintance with X is a plausible candidate for a causally necessary condition for direct thought of X.

## Notes

1. It is not at all clear to me that Mill was a Millian. See Fumerton (2009: 185–9).
2. He stated the same principle in a slightly different way in (*RSDP*: 11).
3. Stevenson (1937: 15) offered the following criterion for an analysis being successful: "Those who have understood the definition must be able to say all that they then wanted to say by using the term in the defined way. They must never have occasion to use the term in the old, unclear sense."
4. See Fales (1990: Chap. 12).
5. Again, Russell didn't explicitly state the view in terms of the idea of direct thought. As we noted, he often raised it in the context of his related distinction between genuine names and disguised descriptions. Terms of the conditions required for understanding a sentence, but it is clear from the corpus of his work that he believes that thought doesn't conceptually require language, and that the view would be extended to any sort of complex thought. To be possible that thought would need to be constructed out of simple thoughts each of which relates the thinker to an object with which the thinker is directly acquainted.

6. Or to accommodate something like Hume's idea of the missing shade of blue, something very much like the property in question.
7. Perhaps, we will sometime be able to routinely manipulate the brain so as to produce the direct thought of phenomenal redness even in unsighted people.
8. This is, I believe, the view that Addis settled on. Bergmann always included in his analysis of thought a "meaning relation. "The thought that P means P whether there is a fact that P or not. But I put "meaning relation" in scare quotes precisely because Bergmann denies that the intentional relation of meaning requires any *existing* relata. But he also doesn't seem to want to embrace a Meinongian realm of subsistence, so I'm not sure with what we are left.

## References

### Works by Other Authors

Addis, Laird (1989). *Natural Signs.* Philadelphia: Temple University Press.

Bergmann, Gustav (1964). *Logic and Reality.* Madison: University of Wisconsin Press.

Bergmann, Gustav (1967). *Realism: A Critique of Brentano and Meinong.* Madison: University of Wisconsin Press.

Fales, Evan (1990). *Causation and Universals.* New York: Routledge.

Fumerton, Richard (1989). "Russelling Causal Theories of Reference." In *Rereading Russell: Essays in Bertrand Russell's Metaphysics and Epistemology*, eds. C. Wade Savage and C. Anthony Anderson. Minneapolis: University of Minnesota Press: 108–118.

Fumerton, Richard (2002). *Realism and the Correspondence Theory of Truth.* Boston: Rowman and Littlefield.

Fumerton, Richard (2003). "Introspection and Internalism", *New Essays on Semantic Externalism, and Self-Knowledge,* ed. Susana Nuccetelli. Cambridge, Massachusetts: MIT Press: 257–276.

Fumerton, Richard (2009). *The Philosophy of John Stuart Mill* (with Wendy Donner). Blackwell Publishing.

Kripke, Saul (1980). *Naming and Necessity.* Cambridge, Massachusetts: Harvard University Press.

Lewis, David (2004). "What Experiences Teaches." In *There's Something About Mary: Essays on Phenomenal Consciousness and Frank Jackson's Knowledge Argument*, eds. Peter J. Ludlow, Yujin Nagasawa, and Daniel Stoljar. Cambridge, Massachusetts: MIT Press. 2004: 77–103.

Stevenson, C. L. (1937). "The Emotive Meaning of Ethical Terms." *Mind*, Vol. 46: 14–31.

CHAPTER 14

# Russell and Wittgenstein on Occam's Razor

*James Levine*

## 1 Introduction

Given that Russell prefaces the published version of his lectures titled "The Philosophy of Logical Atomism" by writing that they "are very largely concerned with explaining certain ideas which I learnt from my friend and former pupil Ludwig Wittgenstein" (*PLA*: 160), it is understandable that much of the commentary on *PLA* concerns how it reflects Wittgenstein's influence on Russell. My focus here, however, is on a central topic of PLA—namely, Occam's razor and its role in analysis—concerning which Russell expresses views that pre-date his interaction with Wittgenstein and that are sharply opposed to Wittgenstein's position in the *Tractatus*.

In the final *PLA* lecture, Russell claims that one "purpose which runs through all that I have been saying" throughout the lectures "is the purpose embodied in the maxim called Occam's razor" (*PLA*: 235). By following Occam's precept that "entities are not to be multiplied without necessity", Russell seeks to avoid assuming "metaphysical entities"—"things which are supposed to be part of the ultimate constituents, but not to be the kind of thing that is every empirically given". The way he does so is by "constructing" a "logical fiction" out "of empirically given

J. Levine (✉)
Trinity College, Dublin, Ireland
e-mail: jlevine@tcd.ie



L. D. C. Elkind, G. Landini (eds.), *The Philosophy of Logical Atomism*, History of Analytic Philosophy,
https://doi.org/10.1007/978-3-319-94364-0_14

things" that "can be substituted for your supposed metaphysical entity and will fulfill all the scientific purposes that anybody can desire" (*PLA*: 236). However, Russell emphasizes that in applying Occam's razor, he only refrains from asserting—and thus does not actually *deny*—that there are "metaphysical entities" that he has avoided assuming (*PLA*: 237):

> I want to make clear that I am not *denying* the existence of anything; I am only refusing to affirm it. I refuse to affirm the existence of anything for which there is no evidence, but I equally refuse to deny the existence of anything against which there is no evidence. Therefore I neither affirm nor deny it, but merely say that is not in the realm of the knowable…

For Russell, that is, Occam's razor is a principle designed to "diminish your risk of error" (*PLA*: 243). Thus, for Russell, we diminish such "risk" if, instead of holding that an ordinary physical object, such as a desk, is a "persistent substance underlying its appearances", we "construct" the desk out of its appearances. But for Russell, having constructed the desk out of appearances, we also diminish such risk if, instead of "*denying* the metaphysical desk", we neither affirm nor deny that there is such a metaphysical entity (*PLA*: 243).

Russell similarly applies Occam's razor in 1914 in *Our Knowledge of the External World*, where he calls it the "the maxim which inspires all scientific philosophizing" (*OKEW*: 107). There he contrasts the absolute theory of space and time, which "assumes that, besides the things which are in space and time, there are also entities, called 'points' and 'instants', which are occupied by things" (*OKEW*: 146), with the relative theory, on which there are no such metaphysically primitive "points" and "instants". For Russell, "[t]here is, so far as I can see, no conceivable evidence either for or against" the absolute theory; while it is logically possible and consistent with the facts, so too is the relative theory (ibid.). Given this situation, Russell writes (*OKEW*: 146–7):

> Hence, in accordance with Occam's razor, we shall do well to abstain from either assuming or denying points and instants. This means, so far as practical working out is concerned, that we adopt the relational theory; for in practice the refusal to assume points and instants has the same effect as the denial of them. But in strict theory the two are quite different, since the denial introduces an element of unverifiable dogma which is wholly absent when we merely refrain from the assertion. Thus, although we shall derive points and instants from things, we shall leave the bare possibility open that they may have an independent existence as simple entities.

Thus, for Russell, the proper way to apply Occam's razor is not to hold that since we need not assume there are entities that are points or instants, we should deny that there are such entities; rather, we should abstain from either assuming or denying points and instants and "leave the bare possibility open" that there are such entities.

In his *Notebooks*, in a 1915 entry that was apparently written in response to Russell's comments on Occam's razor in *Our Knowledge of the External World*[1] and that appears almost *verbatim* in the *Tractatus*, Wittgenstein writes (Wittgenstein 1979b: 42):

> "Occam's razor" is, *of course*, not an arbitrary rule nor one justified by its practical success. It says that unnecessary units of a sign-language mean nothing.

(See also Wittgenstein *TLP*$_2$: 5.47321.) From this remark, it would seem that if an apparent predicate (e.g., "is a point" or "is an instant") is, as Russell contends, dispensable, then the proper way to apply Occam's razor is not to hold that we should neither affirm nor deny that there are such entities to which that predicate applies while, as Russell put it, "leav[ing] the bare possibility open", but rather to hold that that apparent predicate is meaningless, so that there is no question at all—even a question that we are in no position to answer—as to whether there are such entities.[2]

My purpose here is to identify views that lead Russell and the early Wittgenstein to their different interpretations of Occam's razor. I argue, first, that Russell's interpretation of Occam's razor, which derives from the philosophy of mathematics he develops in his 1903 book *The Principles of Mathematics*, depends on his accepting not only a view of analysis as making "precise" what was previously "vague" but also a view of quantification, according to which we can quantify over entities with which we are not acquainted; second, that the early Wittgenstein rejects both these views and is thereby committed to rejecting Russell's interpretation of Occam's razor; and, third, that in his later philosophy, Wittgenstein moves in the direction of Russell with regard to the issues involved. In defending this reading of Wittgenstein, I refer, at various points, to writings of Ramsey and suggest in the Appendix that through Ramsey, Russell indirectly influenced Wittgenstein to change his view of analysis.

## 2 Russell on Analysis, Generality, and Occam's Razor

Influenced by Moore, Russell broke with Idealism toward the end of 1898; however, he characterizes his attending the International Congress of Philosophy in Paris in August 1900, at which he encountered Peano and his students, as "the most important event" in "[t]he most important year in my intellectual life" (*MMD*: 12). Accordingly, I distinguish the "Moorean" philosophy—specifically, the Moorean view of analysis—that Russell accepts immediately following his break with Idealism from the "post-Peano" philosophy—specifically, the post-Peano view of analysis—that he develops following the Paris Congress and that surfaces at various points in *The Principles of Mathematics.* I argue that while Occam's razor plays no role in the Moorean conception of analysis, Russell's post-Peano conception of analysis together with his view of generality provides the basis for Russell's mature view of Occam's razor and its role in analysis.

### 2.1 *Russell's Moorean Conception of Analysis*

For the Moorean Russell, each word has a definite meaning; the meaning of a word is an entity, "either simple or complex", that that word stands for; the proposition expressed by a sentence is a complex entity, whose constituents are the meanings of the words in that sentence; and understanding a sentence requires apprehending the proposition it expresses, which in turn requires being acquainted with each constituent of that proposition.[3] This combination of views determines a number of features of the Moorean conception of analysis.

On this view, where a word stands for a complex entity, "analysis" requires identifying its simple constituents; where a word stands for a simple entity, that entity (and derivatively, that word) is "philosophically indefinable". Hence, for Russell, analyzing proposition $P$ expressed by sentence $S_1$ involves identifying the simple constituents of $P$—in particular, requires replacing each word in $S_1$ that stands for a complex entity by words that stand for the simple constituents of that entity. Where analysis is complete, $S_1$ will be transformed into $S_2$, each of whose words stands for a simple (i.e., ultimate) constituent of $P$. Thus, while $S_1$ and $S_2$ both express $P$, $S_2$ does so perspicuously: there will be exactly as many words in $S_2$ as there are ultimate constituents in $P$, so that $S_2$, as opposed to $S_1$, mirrors the ultimate constitution of $P$.

Further, on this view, since understanding a sentence requires being acquainted with each constituent of the proposition it expresses, analysis makes explicit what one was already aware of, prior to analysis, in understanding the sentence initiating the analysis. Thus, where $S_1$ expresses P, understanding $S_1$ requires being acquainted with each constituent of *P*, so that even where analysis involves transforming $S_1$ into a sentence $S_2$ that perspicuously represents *P*, the analysis does no more than make explicit the simple entities with which one acquainted in understanding $S_1$ prior to analysis.

This view of analysis is reflected in the Moorean Russell's defense of "absolute", as opposed to "relative" theories of time, magnitude, and number. In each case, the "absolute" theory admits indefinables of a sort of not countenanced by the corresponding "relative" theory: moments in addition to events; heights in addition to physical quantities; cardinal numbers in addition to classes. More specifically, in each case, a central issue concerns the analysis of propositions expressed by sentences of the form

$$(\mathrm{A}_1): E(\alpha,\beta),$$

where "*E*" is replaced by an expression signifying an equivalence relation. In particular, Russell has to decide whether instances of ($\mathrm{A}_1$) are perspicuous representations of the propositions they express or whether those propositions are perspicuously represented by corresponding instances of.

$$(\mathrm{A}_2): (\exists x)\big(R(\alpha,x)\,\&\,R(\beta,x)\big),$$

where "*R*" is replaced by an expression signifying an appropriate many–one relation. To accept the former view is to accept a "relative" theory of order (time, height, cardinal number), according to which the equivalence relation in question (*simultaneity*, *equality in height*, *equality in number*) is indefinable and can obtain between the relevant entities (events, physical quantities, classes). To accept the latter view is to accept an "absolute" theory of order, according to which to say that that equivalence relation obtains between two entities is, when fully analyzed, really to say that those two entities stand in the same many–one relation (*occurring at*, *having*, *possessing*) to an indefinable third entity (moment, height, cardinal number) of a sort not countenanced on the relative theory.

In his pre-Peano draft of *The Principles of Mathematics*, in discussing the choice between relative and absolute theories of magnitude, Russell writes (*dPoM*, 57):

> The [relative] theory is designed to avoid the inconvenience of holding every magnitude to be simple and indefinable. ... The kernel of the difference between the present [relative] theory and the former [absolute theory] is, that now equality is taken as indefinable, whereas formerly each magnitude was indefinable. ... [T]he present theory is simpler than the former, since it does not require so many indefinables.

However, while Russell regards the relative theory of magnitude as "simpler" and thus less "inconvenien[t]" than the absolute theory, in that "it does not require so many indefinables", he does not regard these considerations as relevant for deciding between the two theories. Thus, he writes that it "would ... be a grave philosophical error" to "regard definition as subject to convenience" and continues (*dPoM*, 57):

> Every concept is necessarily either simple or complex, and it is not in our power to alter its nature in this respect. If it is complex, it should be analyzed and defined; if simple, it should be used in defining other terms, without itself receiving a definition. Thus equality either may be analyzed into sameness of magnitude, or it may not be so analyzed. ... It does not lie with us to choose what terms are to be indefinable; on the contrary, it is the business of philosophy to discover these terms.

For the Moorean Russell, that is, what is at issue here is a question of fundamental metaphysics, in which case, considerations of "convenience" or minimizing ontological commitments are irrelevant. Are there indefinable magnitudes (such as heights) in addition to quantities? If so, then *equality in height* (for example) is to be defined as "sameness" of height. If not, then *equality in height* is indefinable. And it is "the business of philosophy" to "discover" which are the genuine indefinables.

Moreover, for the Moorean Russell, expressions such as "equal in height", "simultaneous", or "equal in number" have definite meanings with which we are acquainted in understanding them. Accordingly, in defending absolute theories of order, he holds that they are "common sense" theories that we can recognize to be true by "mere inspection". (*ONOP*: 233). Thus, he writes (*dPoM*: 58):

> … when we consider what we mean when we say that two quantities are equal, it seems preposterous to maintain that they have no common property not shared by unequal quantities. All relations of inequality are unchanged when quantities respectively equal to the two in question are substituted, and this seems to indicate that inequality is rather a relation between magnitudes than between quantities.

Similarly, in considering absolute position in time, Russell writes (*PAR*: 227–228):

> A direct consideration of the question, again, makes it very difficult to hold that that simultaneous events have absolutely nothing in common beyond the common qualities of all events.… If a common quality were once admitted, this would have to be temporal position, and it would then follow that there is such a thing as a position [that is, the moment], distinct from all the events which have the position, by relation to which the events become simultaneous. Thus the admission of a common quality among simultaneous events seems unavoidable, but is fatal to the relational theory of time.

Again, Russell says that "equality [in number] plainly consists in possession of the same number", so that "the absolute theory [of number] is plainly correct" (*PAR*: 225).

Thus, for the Moorean Russell, Occam's razor plays no role in analysis. The task of analysis is not to find analyses that minimize ontological commitments, but rather to identify the simple entities referred to or quantified over by the words in the sentences in question, as ordinarily used and understood.

Not only does the Moorean Russell indicate generally that considerations of ontological parsimony are irrelevant to his concerns; he also defends absolute theories or order, despite acknowledging that relative theories require fewer indefinables.

### 2.2 *The Post-Peano Russell on Analysis and Occam's Razor*

On the Moorean Russell's "absolute" theory of number, each cardinal number is a simple, or indefinable, entity—an ultimate constituent of the universe—metaphysically distinct from any class that has that number of elements. In the *Principles of Mathematics*, in contrast, Russell defines the cardinal number of a class $\alpha$ as the class of classes "similar to" $\alpha$, where two classes are "similar", or equal in number, if and only if the members of those classes can be put into a one–to–one correspondence with each other. Thus, as opposed to the "absolute" theory of number, numbers are no longer indefinable entities in terms of which the relation of *similarity* (or *equality in number*) between classes is defined; now, as on the "relative" theory he previously took to be obviously incorrect, the relation of *similarity* is used to define number.[4]

In thus defining what he previously took to be indefinable, the post-Peano Russell has not merely rejected his Moorean view of cardinal number; he has, more fundamentally, ceased to apply his Moorean conception of analysis. Russell does not justify his definition of the cardinal number of a class by arguing (implausibly) that while he previously took it to be obvious "by inspection" that the meaning of a numeral is an indefinable, he now takes it to be obvious "by inspection" that it is a class of similar classes. On the contrary, he acknowledges in *Principles* that "[t]o regard a number as a class of classes must appear, at first sight, a wholly indefensible paradox[5]" (*PoM*: 115). Similarly, in *Our Knowledge of the External World*, he grants that his definition "does not *seem* to be what we have hitherto been meaning when we spoke of 2 and 3", before adding that "it would be difficult to say *what* we had been meaning" (*OKEW*: 204).

Accordingly, rather than presenting his definitions of the cardinal numbers as reflecting what our words, as ordinarily used, mean and hence as obvious by "inspection", he characterizes these definitions (along with his definitions of other sorts of numbers) as "giv[ing] precision to … notion[s] which had hitherto been more or less vague" (*PoM*: xix; see also $PM_2$ vol. 1: 12). Likewise, in *Our Knowledge of the External World*, in discussing his definition of the cardinal number of a class, Russell writes (OKEW: 205)

> [T]he real desideratum about such a definition as that of number is … that it should give us objects having the requisite properties. Numbers, in fact, must satisfy the formulae of arithmetic; any indubitable set of objects fulfilling this requirement may be called numbers. So far, the simplest set known to fulfill this requirement is the set introduced by the above definition. In comparison with this merit, the question whether the objects to which the definition applies are like or unlike the vague ideas of numbers entertained by those who cannot give a definition, is one of very little importance.

In the logical atomism lectures, he presents a general account of analysis that is in accord with his post-Peano practice in the philosophy of mathematics (*PLA*: 161–2):

> The process of sound philosophizing, to my mind, consists mainly in passing from those obvious, vague, ambiguous things, that we feel quite sure of, to something precise, clear, definite, which by reflection and analysis we find is involved in the vague thing that we start from.... Everything is vague to a degree you do not realize till you have tried to make it precise, and everything precise is so remote from everything that we normally think, that you cannot for a moment suppose that is what we really mean when we say what we think.

No longer does Russell hold that each word has a definite meaning—the entity, simple or complex, that it stands for; that understanding a sentence containing that word requires being acquainted with the entity that is its meaning; and that the task of analysis is simply to "inspect" what that meaning is. Instead, he holds that analysis begins with sentences that we take to be obviously true but that are "vague" or "ambiguous" and that the task of analysis is to assign precise meanings to those sentences that will render them true. Since those sentences have no definite meaning prior to analysis, there is no claim that the precise meaning that is assigned is what we really meant when we originally uttered it. Nor is there any question of there being a single correct analysis; instead, there are a variety of precise meanings we could assign to render the sentences in question true. Thus, in analyzing arithmetic, the task is to assign precise meanings to numerical expressions that "satisfy the formulae of arithmetic", in which case "any indubitable set of objects fulfilling this requirement may be called numbers".

It is in the context of his post-Peano conception of analysis that Occam's razor comes to play a central role for Russell. In particular, it provides guidance for choosing among different ways of making precise the original vague sentences that initiate analysis (while preserving the truth-values we take those sentences to have): of all the candidate analyses, choose one that involves a commitment only to indubitable entities—that is, to entities with which we are, or at least may be, acquainted; and of candidate analyses that countenance only indubitable entities, choose the simplest. Accordingly, Russell indicates that of all the indubitable sets of objects that satisfy the formulae of arithmetic, the set of objects he has defined the cardinal numbers to be is "[s]o far, the simplest set known to fulfill this requirement" (*OKEW*: 205; see also $PM_2$ Vol. 2: 4). Moreover, to apply Occam's razor in this way is not thereby to deny that that there are entities besides those used in a given analysis that others have taken to be involved in the analysis of the sentences in question. Thus, in *Our Knowledge of the External World*, in discussing how the same "principle of abstraction" that he uses to avoid assuming indefinable cardinal numbers can also be used to avoid assuming, for example, indefinable magnitudes, colors, and instants, thereby "clear[ing] away incredible accumulations of metaphysical lumber" (*OKEW*: 42), Russell writes that in applying this principle, he "do[es] not deny" that there are the indefinables he has avoided assuming "but merely abstain[s] from asserting them" (*OKEW*: 126).

In *Principles*, in justifying his definitions of the cardinal numbers, Russell does not cite Occam's razor; however, he makes the same points he later does in appealing to that maxim. Thus, after commenting that his definition of a cardinal number as a class of similar classes "allows the deduction of all the usual properties of numbers" (*PoM*:116), Russell considers whether there are indefinable "predicates" common to similar classes—that is, whether there are indefinables of the sort he previously took the cardinal numbers to be (*PoM*: 116):

> [I]f we can find, by inspection, that there is a certain class of such common predicates, … then we may, if we see fit, call this particular class of predicates the class of numbers. For my part, I do not know whether there is such a class of predicates, and I do know that, if there be such a class, it is wholly irrelevant to Mathematics. … For the future, therefore, I shall adhere to the above definition, since it is at once precise and adequate to all mathematical uses.

Thus, Russell indicates that his account of the cardinal numbers not only does all that is required of it since it is "precise and adequate to all mathematical uses" but also avoids an unnecessary metaphysical assumption—namely, that in addition to the classes of similar classes that he has defined as the cardinal numbers, there are indefinable predicates common to similar classes. Moreover, in accord with his later applications of Occam's razor, Russell does not deny that there are such indefinables; instead, he claims merely not to know whether there are such entities.[6] By lectures in 1910 as well as in 1911, Russell explicitly characterizes his definitions of the cardinal numbers as an application of Occam's razor.[7]

### 2.3 *Generality and the "Bare Possibility" of "Metaphysical Entities"*

While Russell's (post-Peano) view of analysis as beginning with vague sentences—which thereby admit of different, truth-preserving, precisifications—makes it possible for Occam's razor to play a central role in analysis, his view of generality enables him to hold that in avoiding commitment to unnecessary "metaphysical entities", he leaves open the "bare possibility" that there are such entities.

In the logical atomism lectures Russell characterizes "metaphysical entities" not merely as entities that *are* not empirically given but rather as entities of a kind that *cannot* be empirically given (*PLA*: 236). Accordingly, Russell distinguishes the issue as to whether there are, in addition to the sense-data with which he is acquainted at a given time, sense-data with which others are acquainted as well as "sensibilia[,] … which have the same metaphysical … status as sense-data, without necessarily being data to any mind" (*RSDP*: 7) from the issue as to whether, in addition to "appearances" (i.e., sense-data of all minds along with unsensed sensibilia), there are "metaphysical and constant" physical objects distinct from and underlying their appearances. (See *PLA*, 243.) For Russell, while sense-data of other people along with unsensed sensibilia are entities with which I am not now acquainted, they are entities of the same kind as those with which I am acquainted; in contrast, "metaphysical and constant" physical objects are entities of a kind with which I cannot ever be acquainted. Hence, although Russell holds that admitting any entities beyond those with which he is currently acquainted requires an inference, he is willing to countenance the sense-data of other people and unsensed sensibilia as inferred entities, since they are "similar to [entities] whose existence is given" (*RSDP*: 12), but refrains from either affirming or denying that there are "metaphysical entities" such as "metaphysical and constant" physical objects.

However, what enables Russell to raise the question as to whether there are *any* entities beyond those with which he is currently acquainted—whether those additional entities are sense-data of other people and unsensed sensibilia or whether they are "metaphysical entities"—is a view of quantification according to which the entities over which a given variable in my language ranges are not confined, in principle, to entities with which I currently am acquainted and thus, for Russell, to entities that I am now in a position to name. For Russell, understanding a name "in the narrow logical sense" (*PLA*: 178) requires being acquainted with the particular designated by that name; hence, if I understand a sentence of the form

1. I am not now acquainted with *a*,

where "*a*" is to be replaced by a name "in the narrow logical sense", that sentence will be false. That is, no sentence of form (1) that I can understand—and thus no sentence of that form in my language—will be true. Hence, in order for Russell to hold that it is coherent to suppose that the general sentence

2. $(\exists x)$(I am not now acquainted with $x$)

is true—something he needs to maintain in order not only to affirm that there sensibilia other than those he is currently experiencing but also "to leave open the bare possibility" that there are "metaphysical entities"—he cannot accept a substitutional view of quantification, according to which (2) will be true (in his language) if and only if it has a true "substitution-instance"—a true sentence (in his language) of the form (1).[8] For Russell, that is, if the truth of (2) depended on the truth of a sentence of form (1), then since there can be no true sentence in his language of that form, then neither could (2) be true (in his language).

Accordingly, Russell makes clear that he accepts a view of generality, according to which the truth of (2) does not depend on its having a true "substitution-instance" (in the speaker's language), but rather on the variable "$x$" including in its range an entity with which Russell is not acquainted (and hence, for which he has no name). Thus, in his 1913 manuscript *Theory of Knowledge*, in discussing the issue of solipsism, Russell writes (*TK*: 34):

> I can never give an actual instance of a thing not now within my experience, for everything I can mention otherwise than by a description must lie within my present experience,

Russell indicates in this passage, in effect, that no sentence of the form (1) can be both understood and true. And in order to explain how it is nevertheless coherent for him to suppose that (2) is true, he writes (*TK*: 11–12):

> [T]he logical possibility of the knowledge that there are things which we are not now experiencing ... depends upon the fact that we may know propositions of the form: "There are things having such-and-such a property", even when we do not know of any instance of such things. In the abstract mathematical world, it is very easy to find examples. For instance, we know that there is no greatest prime number. But of all the prime numbers that we shall have ever thought of, there certainly is a greatest. Hence there are prime numbers greater than any that we shall have ever thought of.

Here, Russell presents a case in which he holds that we can know—and not merely coherently suppose—of a sentence of the form "$(\exists x)Fx$" that it is true, even though we can produce no true substitution-instance of that sentence. Thus, for Russell, I can know that "$(\exists x)$($x$ is a prime number greater than any I shall have ever thought of)" is true, even though any of its substitution-instances that I can understand will be false (since understanding it will require thinking of the number that has been claimed to be greater than any I shall have ever thought of). For Russell, cases in which I can know (by description) that there is an entity as having a certain property but can name no entity that has that property are cases in which I can understand a sentence of the form "$(\exists x)Fx$" and know it to be true even though "no *experienced* object"—and thus no object I am in a position to name—has the property in question (*TK*: 34). In such cases, the quantified sentence will be true in virtue of containing a variable that ranges over an object that is "not experienced" and hence that I cannot now name.

Likewise, much later, in 1940, Russell writes (*IMT*: 236–7):

> The question of truth which transcends experience may be put as follows: suppose $a_1$, $a_2$, …, $a_n$ are all the names in my vocabulary, and that I have named everything I can name. Suppose $fa_1$, $fa_2$, …, $fa_n$ are all false, is it nevertheless possible that "there is an $x$ for which $fx$" should be true?

And shortly thereafter he adds (*IMT*: 237–8):

> It is clear that, unless our knowledge is very much more limited than there seems any reason to suppose, there must be basic existence-propositions [propositions of the form "There is an $x$ for which $fx$"], and that, in regard to some of these, every instance "$fa$" that we can give is false. The simplest example is "there are occurrences which I do not perceive". I cannot in language express what makes such statements true, without introducing variables; the "fact" which is the verifier is unmentionable.

Thus, as in 1913, Russell indicates that it is coherent for me to suppose that (2) is true even though it has no true substitution-instance in my language. And for Russell, in regarding it as coherent to suppose that (2) is true, we regard the individual variable in our language as ranging over all individuals absolutely, regardless of whether we have or even can, in principle, name each of them.

The central point here is that the same view of quantification that allows Russell to coherently suppose that solipsism (of the present moment) is false and also to countenance some entities—such as sense-data of other minds and unsensed sensibilia—with which he is not acquainted at a given moment also allows him "to leave the bare possibility open that" there are "metaphysical entities", entities with which he will never, in principle, be acquainted. By rejecting the "substitutional" interpretation of the variable in question, Russell is able to hold that the variable includes in its range not only entities with which I am acquainted (and so can now name), but also entities, if any, with which I could be acquainted but have not yet been (so have not yet named), as well as "metaphysical entities", if any, with which I cannot, in principle, be acquainted (and so cannot name).

## 3 The Early Wittgenstein on Analysis, Generality, and Occam's Razor

Russell's view of Occam's razor, I have argued, depends both on his view of analysis as moving from the "vague" to the "precise" and on his view of generality. I argue now that the early Wittgenstein rejects both these aspects of Russell's position and is thus in no position to accept Russell's interpretation of Occam's razor.

### 3.1 The Early Wittgenstein on Analysis

By 1919, Russell ceases to hold that we are capable of attaining precise language; hence, while he continues to maintain that analysis begins with vague language, he now holds that the most that analysis can achieve is a movement from the more to the less vague.[9] This view is reflected in his 1921 Introduction to the *Tractatus*, where Russell presents Wittgenstein as "concerned with the conditions for *accurate* Symbolism, i.e. for Symbolism in which a sentence 'means' something quite definite" (CP 9, 101), adding that "[i]n practice, language is always more or less vague, so that what we assert is never quite precise" (ibid.) and similarly writes that Wittgenstein is "concerned with the conditions for a logically perfect language—not that any language is logically perfect" (ibid.).

However, as commentators from Ramsey on have recognized, it seems clear that the early Wittgenstein does not regard himself as characterizing language that is more precise than ordinary language but rather as characterizing language generally, including ordinary language. Thus, in his 1923 review of the *Tractatus*, Ramsey writes that, as against Russell's characterization of "the problem with which [Wittgenstein] is concerned",

Wittgenstein "seems to maintain that his doctrines apply to ordinary languages in spite of appearances to the contrary (see especially 4.002ff)" (1923: 270). Similarly, in notes on the *Tractatus* apparently written shortly after his review, Ramsey writes: "Not dealing with a perfect but any language [—] Russell wrong" (Ramsey Papers, Box 2, Folder 29).

More specifically, Wittgenstein indicates that logical notation is not more precise than ordinary language, but rather helps "reveal" (see 4.002) what is expressed by ordinary language. Thus, in a 1922 letter to Ogden, in commenting on his remark in 5.5563, that "all the propositions of our everyday language, just as they stand, are in perfect logical order", Wittgenstein writes (Wittgenstein 1973: 50):

> By this I meant to say that the prop[osition]s of our ordinary language are not in any way logically *less correct* or less exact or *more confused* than prop[osition]s written down, say, in Russell[']s symbolism or any other "*Begriffsschrift*". (Only it is easier for us to gather their logical form when they are expressed in an appropriate symbolism.)

Given these remarks, Wittgenstein's Tractarian view of analysis has more in common with that of the Moorean Russell than the post-Peano Russell. Unlike the post-Peano Russell, Wittgenstein does not regard analysis as a matter of finding more precise interpretations of vague sentences we take to be true; instead, like the Moorean Russell, he regards it a making perspicuous—making "it easier for us to gather"—the "real" forms ($TLP_2$: 4.002, 4.0031, 5.5563) of "the propositions [*Sätze*] of our everyday language just as they stand".[10] Hence, whereas the post-Peano Russell holds that we face a choice among different precisifications of the same vague sentence, the early Wittgenstein holds, with the Moorean Russell, that "[a] proposition [*Satz*] has one and only one complete analysis" ($TLP_2$: 3.25). And hence, whereas Occam's razor plays a role for the post-Peano Russell as a principle for choosing among different ways to make given vague sentences more precise, it can play no such role for the early Wittgenstein, for whom analysis is a matter not of choosing among different alternatives, no one of which is more ultimately "correct" than the others, but rather of revealing "the" (one and only) complete analysis of what is expressed by a given sentence.

### 3.2 *The Early Wittgenstein on Generality and Occam's Razor*

Further, whereas Russell accepts a view of generality that enables him to leave open "the bare possibility" that there are "metaphysical entities"—

entities of a sort that with we are not, and never will be, acquainted—the early Wittgenstein does not. For while Wittgenstein holds, with Russell (and as opposed to Frege), that understanding a sentence containing a name requires being acquainted with (*kennen*) the object designated by that name,[11] he also accepts, as against Russell, a substitutional view of quantification. Thus, he writes (*TLP*$_2$: 3.313–3.314):

> [A]n expression is presented by means of a variable whose values are the propositions [*Sätze*] that contain the expression. ... I call such a variable a 'propositional variable'. ... All variables can be construed as propositional variables. (Even variable names.)

What has been translated here as "proposition" is the German "*Satz*", which for Wittgenstein is an expression or symbol (and hence is not what Russell in *Principles* calls a "proposition"). Thus, for Wittgenstein, given a sentence (proposition) of the form "*Fa*", where "*F*" occupies the place of a predicate and "*a*" the place of a name, replacing "*a*" by a variable yields a "propositional variable" of the form "*Fx*", whose values are the sentences obtained by replacing "*x*" with a name. In contemporary terminology, he is treating the variable "x" substitutionally, as having as its "values" names (in the relevant language). Since the names in the elementary sentences in my language refer to objects with which I am acquainted (or which I "*kenne*"), the general sentences in my language are incapable of reaching beyond the objects I am thus "given".

In accord with that view of variables, Wittgenstein holds that each sentence—including each general sentence—is a truth-function of elementary sentences. In particular, for Wittgenstein, a sentence of the form "$(\exists x)Fx$" (in my language) is a logical sum of the corresponding sentences of the form "*Fx*" (in my language). Given Russell's view of the "variable name", it is coherent for me to suppose that (2) is true, even though no sentence (in my language) of form (1) can be true; in contrast, for Wittgenstein, *because* no sentence of form (1) that I understood can be true, it is not coherent for me to suppose that (2) is true. Not only is Wittgenstein in no position to grant with Russell the bare possibility that there are metaphysical entities with which we cannot, in principle, be acquainted; he is not even in a position to agree with Russell that it is coherent to suppose—let alone reasonable to believe—that there are objects of the same kind with which I am not acquainted but with which I am not now acquainted.

For Wittgenstein in the *Tractatus*, each sentence with sense is such that in understanding it, we do not thereby know its truth-value, and so can coherently conceive it as true and can coherently conceive it as false. Thus, for Wittgenstein, since no instance of (1) nor (2) itself can be coherently conceived as true, no instance of (1) nor (2) itself is a sentence with sense; and, given that he cannot regard them as "senseless" contradictions, he regards them as "nonsensical".[12] Thus, while Russell regards "There are metaphysical entities with which I can never be acquainted" as expressing a bare possibility that he neither affirms nor denies, Wittgenstein regards it as a nonsensical pseudo-sentence. Like Russell, he neither affirms nor denies it, but he does so, not because he holds with Russell that it is a meaningful sentence whose truth-value "is not in the realm of the knowable" (*PLA*: 237), but rather because he does not regard it as a meaningful sentence at all.

Russell's and Wittgenstein's differing views of Occam's razor are closely tied to their differing views of generality. Given his view of generality, Russell holds that where "*F*" is replaced by a predicate (such as "is a persistent substance", "is an indefinable point", "is an indefinable moment") that will be true only of "metaphysical entities", we can coherently suppose that there are *F*s; hence, he faces a choice as to whether such entities should be invoked in the interpretation of relevant discourse (regarding, for example, ordinary physical objects, or spatial or temporal relations between objects), and he regards Occam's razor as recommending that we interpret that discourse in such a way that does not require assuming either that "There are *F*s" is true or that it is false. Since Wittgenstein does not regard it as coherent to suppose that there are "metaphysical entities", he faces no choice as to whether such entities should be invoked in the analysis of meaningful sentences; hence, he regards Occam's razor as concerned not with avoiding a commitment to such entities but rather with avoiding expressions that contribute nothing to the expression of sense. In particular, for Wittgenstein, "the point of Occam's maxim" is that "if a sign is *useless*"—that is, makes no contribution to the expression of sense—"it is meaningless" (*TLP*$_2$: 3.328).[13] Thus, for Wittgenstein, if an apparent predicate that is to be true only of "metaphysical entities", then Occam's razor prescribes that since it does not contribute to any sentence with sense, that apparent predicate is meaningless.

In notes unpublished in his lifetime, Ramsey recognizes, in effect, how Wittgenstein's view of generality bears on Russell's understanding in the logical atomism lectures of Occam's razor. Ramsey begins by writing (Ramsey Papers: Box 4, Folder 21):

> Russell says LA [inserted above line] no reason to believe in metaphysical entities, i.e. ones different in kind from those given in experience; really in type (W's form).
>
> Therefore nonsense to believe in them.

By "LA", Ramsey is apparently referring to passages from Russell's logical atomism lectures I have cited at the outset[14]; and he indicates here that while Russell holds that "we have no reason to believe in metaphysical entities", Wittgenstein holds that because such entities would have to be different "in type", and not merely "different in kind", from "those given in experience", it is "nonsense to believe in them".[15] Shortly thereafter, in the same note, he adds (Ramsey Papers:Box 4, Folder 21):

> W. says nonsense to believe anything not given in experience not merely different in kind. For to be mine, i.e. to be given in experience is [a] formal property…

By writing that Wittgenstein regards it as "nonsense to believe anything not given in experience not merely different in kind", Ramsey is indicating that for Wittgenstein, it is just as nonsensical for Russell to suppose that there are entities of the same "kind" as those are given (to him) in experience but which, in fact, he is not experiencing as it is for him to suppose that there are "metaphysical entities" different "in kind" from those given in experience. Moreover, by writing that "to be given in experience is [a] formal property", Ramsey suggests that Wittgenstein identifies the formal concept "is an object" with "is given in experience", or "is an object of acquaintance".[16] For Wittgenstein, formal concepts are pseudo-concepts that are not properly represented by predicates but rather by variables. Thus—in accordance with interpreting Wittgenstein as accepting a "substitutional interpretation" of a "variable name", according to which its "values" are the names in my language, which (for Wittgenstein) thereby stand for objects with which I am acquainted—Ramsey suggests that, for Wittgenstein, the domain of objects over which a "variable name" ranges in my language consists of the objects with which I am acquainted. And, as I have argued, accepting that view of a "variable name" requires holding not merely (with Russell) that no instance of (1) can be both understood and true, but also (as against Russell) that (2) cannot be both understood and true, in which case it is incoherent to suppose that there are any objects that I am not now given—be they "metaphysical entities" or entities of the same kind as those which I am now given.

Similarly, in another note, Ramsey connects Wittgenstein's view of generality with his view of solipsism. In particular, he writes (Ramsey Papers Box 5, Folder 16):

> [I]n $(x).\varphi x$, $\varphi a$ is asserted though $a$ is not explicitly mentioned... That I must be acquainted with it in a transcendental sense is W's solipsism.

Thus, for Ramsey, Wittgenstein's solipsism[17] is related to his view that I must be acquainted (at least "in a transcendental sense") with each object that a variable name ranges over. Accepting this view of quantification precludes Wittgenstein from holding that it makes sense to suppose that there are metaphysical entities—entities with which I cannot, in principle, be acquainted—and thereby precludes him from understanding Occam's razor in the same way that Russell does.

## 4 The Later Wittgenstein on Analysis and Generality

While the early Wittgenstein accepts views of analysis and generality that are incompatible with Russell's understanding of Occam's razor, in his later writings he rejects those views in favor of others that are, in some respects, closer to Russell's.

### *4.1 The Later Wittgenstein Against the Moorean View of Analysis*

First, in the *Philosophical Investigations*, Wittgenstein criticizes the Moorean view of analysis, a version of which, I have indicated, he accepted in the *Tractatus*. Thus, for example, in discussing the view that a sentence in which only the parts of a composite objects are named is "an analyzed form" (Wittgenstein 1953: §60) of a sentence in which the composite objects are named, he writes that it "seduces us into thinking that the former is the more fundamental form; that it alone shows what is meant by the other, and so on" (Wittgenstein 1953: §63). And while he writes that the sort of "grammatical" investigation he carries out in PI "may be called an 'analysis' of our forms of expression" (Wittgenstein 1953: §90), he cautions that by so labeling his investigation "may come to look as if there were something like a final analysis of our forms of language, and so a *single* completely resolved from of every expression" (Wittgenstein 1953: §91).

Further, in *Philosophical Investigations*, Wittgenstein indicates that ordinary language is in many respects "inexact", so that the concepts expressed have "blurred" or "vague" boundaries, in which case in giving words "sharp boundaries", we are not making explicit what we had previously meant, but are rather replacing one concept with another that is related, in certain ways, to the original concept. Thus, in discussing the concepts of *number* and of *game*, he writes (Wittgenstein 1953: §68):

> I *can* give the concept 'number' rigid limits…, but I can also use it so that the extension of the concept is *not* closed by a limit. And this is how we do use the word "game". For how is the concept of a game bounded? What still counts as a game and what no longer does? Can you give the boundary? No. You can *draw* one; for none has so far been drawn.

Again, in discussing the concept of *game*, he writes (Wittgenstein 1953: §76):

> If someone were to draw a sharp boundary I could not acknowledge it as the one that I too always wanted to draw, or had drawn in my mind. For I did not want to draw one at all. His concept can then be said to be not the same as mine, but akin to it. The kinship is that of two pictures, one of which consists of color patches with vague contours, and the other of patches similarly shaped and distributed, but with clear contours. The kinship is just as undeniable as the difference.

Moreover, Wittgenstein indicates that part of what he now does in philosophy is to "eliminate misunderstandings by making our expressions more exact" (Wittgenstein 1953: §91).

Thus, like Russell, Wittgenstein now holds that since ordinary language is in many respects inexact and vague, associating those words with concepts that have sharp boundaries does not make explicit what we had "in mind" all along in using those words, but rather introduces new concepts that while similar to, are distinct from, the blurred concepts originally expressed. Moreover, in writing that "several … sharply defined rectangles can be drawn to correspond to the [original] indefinite one" (Wittgenstein 1953: §77), Wittgenstein indicates, like Russell, that there are various different ways to give more exact meaning to a given inexact word. Thus, Wittgenstein now allows for a kind of flexibility in analysis that is not only opposed to his earlier Moorean view of analysis but is also akin to the post-Peano Russell's view.

That is not to say that the later Wittgenstein's view of analysis is no different from Russell's post-Peano view. Thus, immediately after indicating that he eliminates misunderstandings by making our expressions more exact, Wittgenstein warns that to so characterize his work "may [make it] look as if we were moving towards a particular state, a state of complete exactness; and as if this were the real goal of our investigation" (Wittgenstein 1953: §91), thereby rejecting the standard of precision and the view of analysis as a movement from the vague to the precise that Russell presents in his logical atomism lectures. Instead, for Wittgenstein, there is no one standard of precision—"no *single* ideal of exactness" (Wittgenstein 1953: §88)—at which analysis aims, so that drawing "sharp boundaries" that had not previously been drawn is done "for a special purpose" (Wittgenstein 1953: §69) in response to a specific concern, not to meet a single, absolute standard of precision. My only point here is while the early Wittgenstein is committed to rejecting Russell's view of Occam's razor by accepting a broadly "Moorean" conception of analysis that Russell had abandoned years before, the later Wittgenstein also rejects the Moorean conception of analysis in ways that are anticipated, in at least some respects, by the post-Peano Russell. Moreover, in the Appendix, I suggest that even if Russell did not directly influence Wittgenstein's later view of analysis, he indirectly influenced it, by influencing Ramsey, who, in turn, influenced Wittgenstein.

### 4.2 *The Later Wittgenstein on Concepts Whose Extension "Is* Not *Closed by a Limit"*

Finally, by indicating that the concept of *number*, and likewise of *game* (and hence, by implication, of *proposition*), that we typically use is such that its extension "is *not* closed by a limit", the later Wittgenstein allows that we may come to apply these concepts to cases that we are not now in a position to anticipate. Hence, as opposed to his earlier view, and broadly in accordance with Russell's view of generality, it becomes coherent to ask whether there are numbers (or propositions) different in kind from any I can now conceive of.

For the early Wittgenstein, *number*, like *object* and *proposition*, is a formal concept,[18] and as such, it is properly represented by a variable, whose values are the expressions in my language that can be substituted for it. Thus, the extension of a formal concept is determined by the expressions that can replace the variable by which that formal concept is properly represented. And since the expressions that can replace that variable all belong to the language I now speak, there is no possibility of anything belonging to the extension of a formal concept that I do not now recognize as belonging to its extension. Just as the early Wittgenstein's view of a "name variable" as the proper representation of the formal concept *object* precludes my coherently being able to suppose that there objects besides those I have been given, so to his view of *number* and of *proposition* as formal concepts precludes my coherently being able to suppose that there are any numbers or propositions that I am not now in a position to express. On this view, which he continues to maintain in writings in the early 1930s, "[i]t is inconceivable to discover ... a new number" (Wittgenstein 1979a: 216) and likewise "we cannot discover a propositional form" (Wittgenstein 1979a: 217), since we can already foresee the form of any proposition that can be constructed. (See *TLP*$_2$: 4.5, and Wittgenstein 1979a: 217.)

However, by indicating in *Philosophical Investigations* that we may use the word "number", or "game" or "proposition", so that "the extension of the concept is *not* closed by a limit", Wittgenstein allows that we may come to recognize "numbers" that we are not now in a position to recognize. In a late handwritten addition to the *Big Typescript*, Wittgenstein suggests the sort of case where that might occur. For there he asks us to "[t]hink about the relationship of complex numbers to the older concept [or in a variant of this remark, 'the concept'] of numbers" (Wittgenstein 2005: 53). In his *Philosophical Investigations* §67, after introducing phrase "family resemblance" to characterize our use of the term "games", he writes (Wittgenstein 1953: §67):

> And for instance the kinds of number form a family in the same way. Why do we call something a "number"? Well, perhaps because it has a—direct—relationship with several things that have hitherto been called number; and this can be said to give it an indirect relationship to other things we call the same name. And we extend our concept of number as in spinning a thread we twist fibre on fibre. And the strength of the thread does not reside in the fact that some one fibre runs through its whole length, but in the overlapping of many fibres.

And, in the following section, he continues (Wittgenstein 1953: §68):

> I *can* give the concept 'number' rigid limits, that is, use the word "number" for a rigidly limited concept, but I can also use it so that the extension of the concept is *not* closed by a limit. And this is how we do use the word "game".

Thus, insofar as our ordinary use of "number", like our ordinary use of "game", is such that the extension of the associated concept is not closed by a limit, then on our ordinary use of "number", we may come to recognize kinds of numbers that we do not now recognize, just as we have previously come to recognize kinds of numbers—for example, irrational numbers, complex numbers, transfinite numbers—that we did not recognize when the word "number" was originally introduced. On this understanding of number—as opposed to his earlier view, according to which it signifies a formal concept—it is coherent to raise the question as to whether we will come to recognize kinds of numbers that we do not now recognize and that we are not now in a position to formulate.

Hence, like Russell, but as opposed to the early Wittgenstein, the later Wittgenstein makes room for asking a question of the form "Are there any *Fs*?", even in cases in which I cannot formulate in my current language any true sentence of the from "*Fx*" and thereby makes room for the sort of view of generality that Russell needs in order to formulate his agnosticism regarding metaphysical entities that he emphasizes when he applies Occam's razor. Again, this is not to say that the later Wittgenstein would endorse the view of metaphysics that underlies Russell's conception of analysis and hence his understanding of Occam's razor, or even that he would accept Russell's view of generality. Rather, it is to say that just as the early Wittgenstein accepts a Moorean view of analysis that precludes him from accepting Russell's understanding of Occam's razor, but later abandons it in favor of views of analysis that bear some similarities to Russell's, so too the early Wittgenstein accepts a view of generality that precludes him from accepting the agnosticism regarding metaphysical entities that is central to Russell's understanding of Occam's razor, but later abandons that view of generality in favor of views that bear some similarities to Russell's. In that case, identifying views that are central to Russell's understanding of Occam's razor along with the principles that commit the early Wittgenstein to reject that understanding of Occam's razor and considering the fate of those principles in Wittgenstein's later philosophy reveals ways in which Russell's logical atomism lectures incorporate views that, far from merely reflecting the influence of the early Wittgenstein on Russell, are opposed to Wittgenstein's early views and anticipate aspects of Wittgenstein's later views.

## Appendix: Ramsey's (and Russell's) and Wittgenstein's Changing Views of Analysis

In the early 1920s, Ramsey followed the early Wittgenstein in accepting a Moorean conception of analysis (that he criticizes Russell for failing to recognize Wittgenstein accepted); however, by the late 1920s Ramsey accepts a view of analysis very close to (and plausibly influenced by) that of the post-Peano Russell. Moreover, it is plausible that Ramsey not only played a role in influencing Wittgenstein to reject his earlier Moorean view of analysis but also provided the model of analysis against which Wittgenstein reacts when he denies that in analysis we are "moving towards a particular state, a state of complete exactness" (Wittgenstein 1953: §91).

At the outset of Ramsey's essay "Philosophy", written in 1929 (the last full year of his life), he writes (Ramsey 1929: 263–4):

> I do not think it is necessary to say with Moore that the definitions explain what we have hitherto meant by our propositions, but rather that they show how we intend to use them in future. ... [P]hilosophy should clarify and distinguish notions previously vague and confused, and clearly this is meant to fix our future meaning only. ... I used to worry myself about the nature of philosophy through excessive scholasticism. I could not see how we could understand a word and not be able to recognize whether a proposed definition of it was or was not correct. I did not realize the vagueness of the whole idea of understanding...

In the final paragraph, he adds (Ramsey 1929: 269):

> The chief danger to our philosophy, apart from laziness and woolliness, is *scholasticism*, the essence of which is treating what is vague as if it were precise and trying to fit it into an exact logical category. A typical case of scholasticism is Wittgenstein's view that all our everyday propositions are completely in order...

Thus, Ramsey indicates that while he once accepted the Moorean conception of analysis, he now regards it as a form of scholasticism; further, he attributes that conception of analysis to Wittgenstein; and he advocates a view of analysis according to which we replace "notions previously vague and confused" by ones that are exact or precise. That is to say, these comments indicate that Ramsey has changed his view of analysis in the same sort of way that Russell had done almost three decades earlier, when he abandoned the Moorean view of analysis that was central to his rejection of Idealism in favor of the style of analysis he presents in his post-Peano philosophy of mathematics.

Moreover, it is plausible that Ramsey was led to reject the scholastic Moorean conception of analysis at least in part as a result of the influence of Russell. Ramsey ends his 1927 paper "Facts and Propositions" by writing that "[e]verything that I have said [in the paper] is due to [Wittgenstein], except the parts which have a pragmatist tendency", adding that "[m]y pragmatism is derived from Russell" and that "the essence of pragmatism I take to be this, that the meaning of a sentence is to be defined by reference to the actions to which asserting it would lead, or, more vaguely still, to its possible causes and effects" (Ramsey 1927: 170). In that paper, Ramsey cites Russell's 1921 book *The Analysis of Mind*; and there Russell writes that "[t]he relation of a word to its meaning is of the nature of a causal law governing our use of the word and our actions when we hear it used" (*AMi*: 198); that "a person understands a word when (a) suitable circumstances make him use it, (b) the hearing of it causes suitable behavior in him" (*AMi*, 197); and that "the meaning of a word is not absolutely definite: there is always a greater or lesser degree of vagueness" (*AMi*, 197–8).[19] And remarks such as these license the view that analysis in the style of Moore (and the early Wittgenstein) is "scholastic".

Further, it seems clear that Ramsey's critique of "scholasticism" had an impact on Wittgenstein. Thus, for example, in 1941, Wittgenstein writes (quoted in Misak 2016: 247):

> Ramsey was right in saying that in philosophy one should be neither 'woolly' nor scholastic. But yet I don't believe that he has seen how this should be done; for the solution is not: being scientific.

(In this connection see also Wittgenstein 1953: §81). While the later Wittgenstein follows Ramsey (and Russell) in rejecting the "scholastic" Moorean conception of analysis and in accepting a form of analysis that can involve a movement from the "vague" or "inexact" to the more precise, he does not follow Ramsey (or Russell) in accepting a single ideal of precision or a conception of analysis in the service of "scientific philosophizing". That is, while Wittgenstein's view of analysis moves away from the Moorean conception of analysis in a direction similar, in certain respects, to Russell's, he does not accept Russell's (or Ramsey's) conception of how analysis involves a movement from "the vague to the precise"—and hence does not accept the conception of analysis that provides the context for Russell's understanding of Occam's razor.[20]

## Notes

1. In January 1915, Wittgenstein wrote to Keynes: "I'm very interested to hear that Russell has published a book [*Our Knowledge of the External World*] lately. Could you possibly send it to me and let me pay you after the war? I'd so much like to see it" (Wittgenstein 1995: 79). The entry in the *Notebooks* on Occam's razor occurs on a day (April 23, 1915) in which Wittgenstein is commenting on other topics that arise in *Our Knowledge of the External World*, including the "law of conservation", and principles of "sufficient reason" and "of continuity in nature". See (*OKEW*: 105 (on conservation), and 108–9, 147–50 (on continuity and sufficient reason)); in *Our Knowledge of the External World*, Russell mentions Occam's razor twice (*OKEW*: 107, 146).). On May 1, Wittgenstein makes other remarks apparently stimulated by discussions in *Our Knowledge of the External World* of skepticism (compare (*OKEW*: 67)), of the distinction between hard and soft data (compare (*OKEW*: 70ff)), and of the "scientific method in philosophy" (a theme throughout *Our Knowledge of the External World*).
2. See Sect. 2.2 below (including note 13) for some discussion of what Wittgenstein means here by an "unnecessary units of a sign-language".
3. In my (Levine 2016: 168–72), I defend attributing these views to the Moorean Russell.
4. Although on his *Principles* view, as opposed to the relative theory of number he describes in his Moorean period, *similarity* is not indefinable but is rather defined in terms of one–to–one correspondence. However, the central point is that Russell has reversed the order of definitional (and metaphysical) priority that he accepted during his Moorean period.
5. In *Principles*, what is "paradoxical" is counterintuitive; what is now called "Russell's paradox," he there regards as a "contradiction" (see *PoM*: Chap. X).
6. Similarly, in *Principia Mathematica*, in presenting his interpretation of sentences using class symbols, according to which such symbols do not stand for entities that are classes, Russell does not "assert dogmatically that there are no such things as classes" but only takes himself to have shown that his interpretation of such sentences "yield[s] all the proposition for the sake of which classes might be thought essential" ($PM_2$ Vol. 1: 72; see also *IMP*: 184). However, as Richard Fumerton pointed out in discussion at the workshop, Russell is not always so "undogmatic" in rejecting various metaphysical views, as, for example, when he criticizes Meinong for a "failure of [a] feeling for reality" in maintaining that there are "unreal objects" designated by such phrases as "the round square" or "the golden mountain" (see *IMP*: 169–70). While the point merits further discussion, one reason for Russell's view of Meinong's view may be that in "On Denoting", Russell argues that Meinong's view leads to "intolerable" violations of "the law of contradiction" (*OD*: 418).

7. Thus, in 1910, Sheffer records Russell as saying (Sheffer 1910):

   > Perhaps all couples have a predicate, and all trios have another predicate. But: 1) We don't know this, 2) Calling numbers "classes" fulfills the same for properties of cardinal numbers. Therefore, by Occam's razor, disregard the philosophical predicates and call numbers "classes of classes".

   Sheffer's notes are in the Bertrand Russell Archives: I thank the Bertrand Russell Archives in the William Ready Division of Research Collections, McMaster University Library, for permission to use unpublished materials.

   Similarly, in 1911, Moore records Russell as indicating generally that by Occam's razor "entities are not to be multiplied beyond necessity" but also that "[y]ou oughtn't to assert dogmatically that there aren't those entities" and, more specifically, cites Occam's razor in justifying his definitions of the cardinal numbers, which "[g]ive [an] interpretation to symbols, which will be true, and which involves [the] least possible assumption" (Moore 1911–1912). The originals of Moore's notes are in Cambridge University, MS. Add. 8875. I thank the Estate of G. E. Moore for permission to quote from them.
8. In his contribution to this volume, Klement argues that Russell accepts a substitutional theory of quantification, albeit one in which the substitution-instances are sentences in an "ideal language", which contains a name for each simple object, and hence contains names (and sentences) which a given speaker cannot understand (and moreover, if there are "metaphysical" entities with which no human can be acquainted, it will contain names (and sentences) with which no human can understand). I think it implausible that this is Russell's considered view of quantification. Thus, while Klement emphasizes that in the logical atomism lectures Russell writes that "a logically perfect language" will have one and only one a name for "every simple object" (*PLA*: 176), in that same passage, Russell writes that "[a] logically perfect language ... would be very largely private to one speaker", since "all the names that it would use would be private to that speaker and could not enter into the language of another speaker" (*PLA:* 176). This suggests, that for Russell, there are different "logically perfect" languages for different speakers, that a logically perfect language for a given speaker will only contain names with which that speaker is acquainted, in which case (in the absence of any omniscient speakers), there is no one "logical perfect" language containing a name for absolutely each simple object that provides the basis for our understanding of quantification. However, even on the substitutional theory that Klement attributes to Russell, I can quantify over simple entities with which I am not acquainted (given that the "ideal language" contains names for simples with which I am not acquainted); and that is the central point here. For the remainder of the chapter, in denying that Russell accepts a "substitutional"

theory of quantification, I am denying that he accepts the view (which I take to be the standard understanding of a "substitutional" theory of quantification) that an existential quantification will be true in a given language that a speaker understands if it has a true substitution-instance in that language.

9. See Levine (2016: 195–201); see also Levine (2018: 139–141) for discussion of this change in Russell's philosophy.
10. In attributing the Moorean view of analysis to the early Wittgenstein, I am not thereby attributing to him all aspects of Moore's (or Russell's Moorean) view of analysis, such as that every word has as its meaning an entity, simple or complex. Rather, I am only attributing to him the view that analysis makes explicit the meaning that a sentence has prior to analysis, the meaning that we must already, in some sense, be aware of prior to the analysis.
11. See, for example, *TLP*$_2$: 4.243, 6.232, and 6.2322, where Wittgenstein indicates, with Russell but as against Frege, that there can be no informative identity sentence (where two names flank the identity sign), since anyone understanding those names will ***kennen*** their meaning. For a fuller defense of the view that Wittgenstein agrees with Russell that understanding a sentence containing a name requires being acquainted with (***kennen***) the meaning of that name—while also differing significantly with Russell regarding what acquaintance with an object designated by a name requires—see Levine (2013: §3.1).
12. I defend this claim in my (2013: §3.3).
13. In indicating in *TLP*$_2$: 5.47321 that the "the point" of Occam's razor is that "***unnecessary*** units in a sign-language mean nothing", Wittgenstein is indicating that if, in a given sentence, a "unit in a sign-language" contributes nothing to the sense of that sentence (and so is unnecessary for the expression of that sense by that sentence), then that "unit" in the sign-language is meaningless. Thus, even though a logical connective, such as "$\supset$" may be defined in terms of other connectives, and thus may be regarded as dispensable in a logical notation, in given sentences it contributes to the expression of sense by those sentences and so is not to be regarded as meaningless by Occam's razor, as Wittgenstein construes it.
14. Ramsey seems not to be referring here to Russell's 1924 essay "Logical Atomism" in which there is no discussion of metaphysical entities.
15. In particular, he suggests that, if "metaphysical entities" are different "in type" than those given in experience, we would be unable to quantify over them by means of the variables we can use, in which case it would be "nonsense" to suppose that there are such entities.
16. See in this context Wittgenstein (2005: §107), entitled "Color, Experience, etc., as Formal Concepts".

17. In my (2013: §3.4), I discuss Wittgenstein's view of "the truth of solipsism". In particular, I argue that given his view of quantification, Wittgenstein cannot hold that it makes sense to suppose that there are objects with which he is not acquainted; but I argue as well that he does not thereby hold that he is acquainted with all the objects there are. Nor, I argue, does he hold that while he cannot *say* that he is acquainted with all the objects there are, that is nevertheless *shown*. Rather, I argue that, for Wittgenstein, while he is shown the objects he is given, nothing can be said or shown as to whether they are all the objects that there are.
18. For *number* and *object* as formal concepts, see *TLP*$_2$: 4.12721. That Wittgenstein regards *proposition* as a formal concept follows from 4.1271 ("Every variable is a sign for a formal concept") and *TLP*$_2$: 4.53 ("The general propositional form is a variable").
19. Russell's remarks here reflect changes that occur in his views after 1918 (to which I allude in note 9). For some discussion as to how these changes are the product of Russell's application of his post-Peano method of analysis to issues in the philosophy of mind, see my (2018).
20. Thanks to the participants at the Iowa Seminar on *The Philosophy of Logical Atomism* for helpful comments.

## References

### Works by Other Authors

Levine, James (2013). "Logic and Solipsism." In *Wittgenstein's Tractatus: History and Interpretation*, eds. P. Sullivan and M. Potter. Oxford: Oxford University Press. 2013: 170–238.

Levine, James (2016). "The Place of Vagueness in Russell's Philosophical Development." In *Early Analytic Philosophy—New Perspectives on the Tradition*, ed. S. Costreie. Dordrecht: Springer. 2016: 161–212.

Levine, James (2018). "Russell, Pragmatism, and the Priority of Use over Meaning." In *Pragmatism and the European Traditions: Encounters with Analytic Philosophy and Phenomenology Before the Great Divide*, eds. M. Baghramian and S. Marchetti. New York: Routledge: 110–154.

Misak, Cheryl (2016). *Cambridge Pragmatism*. Oxford: Oxford University Press.

Moore, G. E. (1911–1912). *Notes on Russell's Lectures on the Philosophy of Arithmetic*. Cambridge University Library.

Ramsey, Frank P (1927). "Facts and Propositions." *Proceedings of the Aristotelian Society*, Vol. 7: 153–170.

Ramsey, Frank P (1929). "Philosophy." In *The Foundations of Mathematics and other Logical Essays*, ed. R. B. Braithwaite. Paterson, NJ: Littlefield, Adams & Co. 1960: 263–269.

Ramsey, Frank P (1923). "Critical Notice of L. Wittgenstein's 'Tractatus Logico-Philosophicus'." In *The Foundations of Mathematics and other Logical Essays*, ed. R. B. Braithwaite. Paterson, NJ: Littlefield, Adams & Co. 1960: 270–292.

Ramsey, Frank P. (1929–1930). Frank Plumpton Ramsey Papers. ULS Archives and Special Collections, University of Pittsburgh. https://digital.library.pitt.edu/islandora/object/pitt%3AUS-PPiU-asp198301/viewer#ref27.

Sheffer, Henry (1910). "Notes on Russell's Lectures." Bertrand Russell Archives. Hamilton, Ontario: McMaster University.

Wittgenstein, Ludwig (1953). *Philosophical Investigations*, Tr. G. E. M. Anscombe. New York: The Macmillan Company.

Wittgenstein, Ludwig (1973). *Letters to C. K. Ogden*, ed. G. H. von Wright. Oxford: Basil Blackwell.

Wittgenstein, Ludwig (1979a). *Wittgenstein and the Vienna Circle*, conversations recorded by Friedrich Waismann, ed. B. F. McGuinness, tr. J. Shulte and B. F. McGuinness. New York: Rowman & Littlefield Publishers.

Wittgenstein, Ludwig (1979b). *Notebooks 1914–1916*, second edition, G. H. von Wright and G. E. M. Anscombe (eds.), tr. G. E. M. Anscombe. Chicago: The University of Chicago Press.

Wittgenstein, Ludwig (1995). *Wittgenstein in Cambridge: Letters and Documents 1911–1951*, ed. B. McGuinness. Oxford: Blackwell Publishing.

Wittgenstein, Ludwig (2005). *The Big Typescript: TS 213*, eds. C. Grant Luckhardt and Maximilian A. E. Aue. Oxford: Blackwell Publishing.

# References

## Works by Russell


*AIT*. "On the Axioms of the Infinite and of the Transfinite" (1911). In *CPBR* 6: 43–53. 1st published. *Société mathématique de France, Comptes rendus des séances*, No. 2: 22–35.

*AMi*. *The Analysis of Mind* (1921). London: Allen & Unwin.

*AMa*. *The Analysis of Matter* (1927). New York. Harcourt, Brace & Co. Inc.

*AR*. "Analytic Realism" (1992). In *CPBR 6*: 132–146. 1st published. French "Le Réalisme Analytique" (1911). *Bulletin de la société de philosophie*, Vol. 11: 282–291. French original in *CPBR 6* (Appendix V): 409–432.

*Aut*. *The Autobiography of Bertrand Russell* (1951). Boston: Little Brown and Company.

*BoR*. "The Basis of Realism" (1911). In *CPBR 6*: 125–131. 1st published. *The Journal of Philosophy, Psychology, and Scientific Methods*, Vol. 8, No. 6: 158–161.

*CPBR 3*. *The Collected Papers of Bertrand Russell*, vol. 3, *Towards "The Principles of Mathematics, 1900–02*, ed. Gregory Moore. London: Routledge, 1993.

*CPBR 4*. *The Collected Papers of Bertrand Russell*, vol. 4, *Foundations of Logic 1903–05*, ed. Alasdair Urquhart. London: Routledge, 1994.

*CPBR 5*. *The Collected Papers of Bertrand Russell*, vol. 5, *Towards "Principia Mathematica," 1905–08*, ed. Gregory Moore. London: Routledge, 1994.

*CPBR 6*. *The Collected Papers of Bertrand Russell*, vol. 6, *Logical and Philosophical Papers, 1909–13*, eds. J. G. Slater and Bernd Frohmann. London: Routledge, 1992.




L. D. C. Elkind, G. Landini (eds.), *The Philosophy of Logical Atomism*, History of Analytic Philosophy,
https://doi.org/10.1007/978-3-319-94364-0

*CPBR 7*. *The Collected Papers of Bertrand Russell*, vol. 7, *The Theory of Knowledge: the 1913 Manuscript*, eds. E. R. Eames and K. Blackwell. London: Routledge, 1984.
*CPBR 8*. *The Collected Papers of Bertrand Russell*, vol. 8, *The Philosophy of Logical Atomism and Other Essays. 1914–19*, ed. J. G. Slater. London: Routledge, 1986.
*CPBR 9*. *The Collected Papers of Bertrand Russell*, vol. 9, *Essays on Language, Mind and Matter, 1919–26*, ed. J. G. Slater. London: Routledge, 1988.
*CPBR 10*. *The Collected Papers of Bertrand Russell*, vol. 10, *A Fresh Look at Empiricism 1927–42*, ed. John G. Slater with Peter Köllner. London: Routledge, 1996.
*CPBR 11*. *The Collected Papers of Bertrand Russell*, vol. 11, *Last Philosophical Testament 1943–68*, ed. John G. Slater with Peter Köllner. London: Routledge, 1997.
*EIP*. "The Existential Import of Propositions" (1905). In *CPBR 4*: 486–489. 1st published. *Mind*, 14: 398–401.
*HK*. *Human Knowledge: Its Scope and Limits* (1948). New York: Simon and Schuster.
*HWP*. *A History of Western Philosophy* (1946). London: George Allen and Unwin LTD.
*IL*. "Is Mathematics Purely Linguistic?" (1950–1952). In *CPBR 11*: 352–362. 1st published. *Essays in Analysis*, ed. Douglas Lackey. London: George Braziller. 1973: 295–306.
*IMP*. *Introduction to Mathematical Philosophy* (1919). London: George Allen & Unwin.
*IMT*. *An Inquiry into Meaning and Truth* (1940). London: Allen & Unwin.
*KAKD*. "Knowledge by Acquaintance and Knowledge by Description" (1911). In *CPBR 6*: 147–161. 1st published. *Proceedings of the Aristotelian Society*, 11: 108–128.
*LA*. "Logical Atomism" (1924). In *CPBR 9*: 162–179. 1st published. *Contemporary British Philosophy*, ed. J. H. Muirhead. London: Allen & Unwin.
*LO*. "Logic and Ontology" (1957). In *CPBR 11*: 625–629. 1st published. *The Journal of Philosophy*, Vol. 54: 225–230.
*LP2*. "Logical Positivism" (1950). In *CPBR 11*: 155–167. 1st published. *Revue internationale de philosophie*, Vol. 4: 3–19.
*MaL*. *Mysticism and Logic, and Other Essays* (1917). London: Longsman, Green, and Co.
*MG*. "The Meaning of Good" (1904). In *CPBR 4*: 571–575.
*MI*. "Mathematical Infinity" (1958). In *CPBR 11*: 363–364. 1st published. *Mind*, Vol. 67: 385.
*ML*. "Mathematical Logic" (1908). In *CPBR 5*: 585–625. 1st published. *American Journal of Mathematics* Vol. 30: 222–262.
*MMD*. "My Mental Development" (1944). In *CPBR 11*: 5–17. 1st published. *The Philosophy of Bertrand Russell*, ed. P. A. Schilpp: 3–20. LaSalle, Illinois: Open Court.
*MPD*. *My Philosophical Development*. London: George Allen & Unwin, 1959.

*MsN*. "Manuscript Notes [1918]" (1986). In *CPBR 8*: 247–271.

*NTF*. "On the Nature of Truth and Falsehood" (1910). In *CPBR 6*: 115–124. 1st published. *Proceedings of the Aristotelian Society*, n.s., 7: 28–49.

*OKEW*. *Our Knowledge of the External World: As a Field for Scientific Method in Philosophy* (1914). Chicago: Open Court Publishing Company.

*OM*. "On Matter," *CPBR* 6: 77–95. 1912.

*ONA*. "On the Nature of Acquaintance," *CPBR 7*: Chs. 1–3: 5–52. 1st published. *The Monist*, Vol. 24, No. 1: 1–16; No. 2: 161–187; No. 3: 435–453. 1914.

*ONOP*. "On the Notion of Order and Absolute Position in Space and Time [1901]" (1993). In *CPBR* 3: 234–258.

*OoP*. *An Outline of Philosophy* (1927). London: Allen & Unwin.

*OP*. "On Propositions: What They Are and How They Mean" (1919). In *CPBR 8*: 278–306; 1st published. *Proceedings of the Aristotelian Society*, Supplementary Volume 2: 1–2.

*PAR*. "Is Position in Time Absolute or Relative (1900)?" (1993). In *CPBR* 3: 219–233.

*PaM*. "Physics and Metaphysics" (1928). In *CPBR 10*: 271–278. 1st published. *The Saturday Review of Literature*, 4: 910–911.

*PLA*. "The Philosophy of Logical Atomism [1918]" (1918–1919). In *CPBR 8*: 160–244; 1st published. *The Monist*, 28: 495–527; and *The Monist*, 29: 32–63, 190–222, 345–380.

$PM_1$. Whitehead, A. N. and B. A. W. Russell. *Principia Mathematica* (1910, 1912, 1913). Cambridge: Cambridge University Press. 3 volumes.

$PM_2$. Whitehead, A. N. and B. A. W. Russell. *Principia Mathematica*, Second edition (1925, 1927). Cambridge: Cambridge University Press. 3 volumes.

*PoL*. "The Paradoxes of Logic" (1906). In *CPBR 5*: 273–296. 1st published. *Revue de Métaphysique et de morale*, 14: 627–650.

*PoM*. *The Principles of Mathematics*, Second edition (1937). London: Allen and Unwin.

*dPoM*. "*The Principles of Mathematics*, Draft of 1899–1900" (1993). In *CPBR 3*: 9–212.

*PoP*. *The Problems of Philosophy* (1912). London: Home University Library.

*PoU*. "The Problem of Universals." (1946). In *CPBR 11*: 257–273. 1st published. *Polemic*, no. 2: 21–35.

*RC*. "Reply to Criticisms." In *CPBR 11*: 18–63. 1st published. *The Library of Living Philosophers Volume 5: The Philosophy of Bertrand Russell*, ed. Paul Arthur Schilpp. Evanston and Chicago: Northwestern University. 1944: 679–742.

*RIWFM*. "Recent Italian Work on the Foundations of Mathematics" (1993). In *CPBR 3*: 350–362.

*RPsyL*. "The Relevance of Psychology to Logic" (1938). In *CPBR 10:* 360–370.

*RSDP*. "The Relation of Sense-Data to Physics" (1914). In *CPBR 8*: 3–26. 1st published. *Scientia* 16: 1–27 and supp. 3–34.

*RUP*. "On the Relation of Universals and Particulars" (1911). In *CPBR 6*: 167–182. 1st published. *Proceedings of the Aristotelian Society*, 12: 1–24.

*RWPM*. "Recent Work on the Principles of Mathematics" (1901). In *CPBR 3*: 363–379. 1st published. *The International Monthly*, Vol. 4: 83–101. Reprinted as "Mathematics and the Metaphysicians" (1917). In *MaL*: 74–97.

*SHM*. Russell, Bertrand and Woodrow Wyatt. *Bertrand Russell Speaks His Mind* (1960). London: A. Baker. https://archive.org/details/BertrandRussellDiscussesPhilosophy

*SMP*. "On Scientific Method in Philosophy." In *CPBR*.

*STCR*. "On the Substitutional Theory of Classes and Relations." In *CPBR 5*: 243–261. 1st published. *Essays in Analysis*, ed. Douglas Lackey. 1973: 165–189.

*T*. "Types." [undated; circa 1906–1907] In *CPBR 5*: 498–514. 1st published. *CPBR 5*: 2014.

*TK*. *The Theory of Knowledge: the 1913 Manuscript* (1984). In CPBR 7: 1–178. 1st published. CPBR 8.

*TNI*. "The Nature of Inference" (1920). In *CPBR 9*: 82–86. 1st published. *The Athenaeum*, No. 4, 694: 514–515.

*ToI*. "The Theory of Implication" (1906). In *CPBR 5: 14–61*. 1st published. *American Journal of Mathematics*, 28: 159–202.

*V*. "Vagueness" (1923). In *CPBR 9*: 145–149. 1st published. *The Australasian Journal of Psychology and Philosophy*, Vol. 1: 84–92.

## Works by Other Authors

Bergmann, Merrie, James Moore, and Jack Nelson (2013). *The Logic Book*, McGraw-Hill, 6th edition.

Black, Max (1952). "The Identity of Indiscernibles." *Mind New Series*, Vol. 61, No. 242: 153–164.

Carey, Rosalind (2007). *Russell and Wittgenstein on the Nature of Judgment*. London and New York: Continuum International Publishing Group.

Donner, Wendy and Richard Fumerton (2009). *Mill*. Oxford: Wiley-Blackwell.

Dretske, Fred (1995). *Naturalizing the Mind*. Cambridge, Massachusetts: The MIT Press.

Eliot, T. S. (1914). *Philosophy 21: Advanced Logic*, Harvard University, 1914; Notes by T. S. Eliot. ed. Bernard Linsky (May 29, 2016).

Flowers III, F. A. and Ian Ground (eds.) (2016). *Portraits of Wittgenstein*. 2 vols., Second edition. London: Bloomsbury.

Frege, Gottlob (1984). *Die Grundlagen der Arithmetik*. Breslau Kober. *The Foundations of Arithmetic*, 2nd edition, tr. J. L. Austin. Oxford: Blackwell, 1980.

Frege, Gottlob (1903). *Grundgesetze der Arithemtic*, Volume 2, Jena: Verlag Herman Pohle.

Heck, Richard G., Jr. and Robert May (2010). "The Composition of Thoughts." *Nous* 45: 126–166.

Husserl, Edmund (1970). *Logical Investigations*, tr. J. N. Findlay. London: Routledge.

Kripke, Saul (1979). "Is there a Problem about Substitutional Quantification?" In *Truth and Meaning*, eds. G. Evans and J. McDowell. Oxford: Clarendon Press. 1976: 325–419.

Lenzen, Victor C (2016). *Theory of Knowledge: Philosophy 9c*, Harvard University 1914; Notes by Victor F. Lenzen. Edited by Bernard Linsky (September 8, 2016).

Mares, Edwin (2004). *Relevant Logic.* Cambridge: Cambridge University Press.

McGuinness, Brian (1996). Preface to the second edition of Wittgenstein 1971.

McGuinness, Brian (2002a). "Wittgenstein's 1916 'Abhandlung'." In *Wittgenstein and the Future of Philosophy: A Reassessment After 50 Years: Proceedings of the 24th International Wittgenstein-Symposium*, eds. Rudolf Haller and Klaus Puhl. Vienna: Hölder-Pichler-Tempsky. 2002: 272–282.

McGuinness, Brian (2002b). *Approaches to Wittgenstein: Collected Papers.* London: Routledge.

Moore, G. E. (1911–1912). "Notes on Russell's Lectures on the Philosophy of Arithmetic". Cambridge: Cambridge University Library.

Potter, Michael (2013). "Wittgenstein's pre-*Tractatus* manuscripts: a new appraisal." In Peter Sullivan and Michael Potter (eds.) *Wittgenstein's Tractatus: history and interpretation*, Oxford: Oxford University Press. 2013: 13–39.

Putnam, Hilary (1975). "The Meaning of Meaning." In *Mind, Language, and Reality: Philosophical Papers, Volume 2.* Cambridge: Cambridge University Press. 1979: 215–271.

Ramsey, Frank P (1920–1930). Online. Frank Plumpton Ramsey Papers. ULS Special Collections Department, University of Pittsburgh Library System. http://digital.library.pitt.edu/cgi-bin/f/findaid/findaid-idx?c=ascead;cc=ascead;q1=ramsey;rgn=main;view=text;didno=US-PPiU-asp198301

Van Heijenoort, Jean (1967a). *From Frege to Gödel.* Cambridge, Massachusetts: Harvard University Press.

Van Heijenoort, Jean (1967b). "Logic as Calculus and Logic as Language." *Synthese* 17: 324–330.

Wittgenstein, Ludwig (1929). "Some Remarks on Logical Form." *Aristotelean Society Supplementary Volume* 9.

Wittgenstein, Ludwig *TLP*$_2$ (1974a). *Tractatus Logico-Philosophicus*, Second edition. Tr. D. F. Pears and B. F. McGuinness. London: Routledge & Kegan Paul Ltd.

Wittgenstein, Ludwig (1974b). *Philosophical Grammar.* ed. R. Rhees, tr. A. Kenny. Berkeley, California: University of California Press.

Wittgenstein, Ludwig (1976). *Wittgenstein's Lectures on the Foundations of Mathematics: Cambridge, 1939.* ed. Cora Diamond. Ithaca, New York: Cornell University Press.

# Index[1]



[1] Note: Page numbers followed by 'n' refer to notes.





L. D. C. Elkind, G. Landini (eds.), *The Philosophy of Logical Atomism*, History of Analytic Philosophy,
https://doi.org/10.1007/978-3-319-94364-0

Kolibri 360 GmbH
Ferdinand-Jühlke-Straße 7
99095 Erfurt, Deutschland
produktsicherheit@kolibri360.de